CELLULAR

History of Computing
William Aspray and Thomas J. Misa, editors

A complete list of the titles in this series appears in the back of this book.

CELLULAR

AN ECONOMIC AND BUSINESS HISTORY OF THE INTERNATIONAL MOBILE-PHONE INDUSTRY

DANIEL D. GARCIA-SWARTZ
AND MARTIN CAMPBELL-KELLY

THE MIT PRESS CAMBRIDGE, MASSACHUSETTS LONDON, ENGLAND

The MIT Press would like to thank the anonymous peer reviewers who provided comments on drafts of this book. The generous work of academic experts is essential for establishing the authority and quality of our publications. We acknowledge with gratitude the contributions of these otherwise uncredited readers.

This book was set in Stone Serif and Avenir by Westchester Publishing Services.
Printed and bound in the United States of America.

Library of Congress Cataloging-in-Publication Data is available.

ISBN: 978-0-262-54392-7

10 9 8 7 6 5 4 3 2 1

CONTENTS

INTRODUCTION

This book presents a compact history of the international mobile-phone industry. Although it explores the technical dimensions of cellular systems, it emphasizes the economic and business dimensions of the industry's history.

Our book is structured in four major parts, each corresponding to a generation of cellular phones up to the present. These generations, in turn, are reasonably well matched to decades: the 1980s was the decade of the first cellular generation (1G), the 1990s the decade of the second (2G), the 2000s the decade of the third (3G), and the 2010s the decade of the fourth (4G). The caveat is that generations started at different points in time in different parts of the world. Two of the most widely diffused first-generation standards, for example, were the Advanced Mobile Phone Service (AMPS), created in the United States, and the Nordic Mobile Telephone (NMT) system, developed in Scandinavia. Both became operational in the early 1980s in their countries of origin. A few years later, they were exported to, and adopted by, countries around the world. AMPS became widely used all over Central and South America and the Caribbean, whereas NMT was heavily adopted in European countries outside Scandinavia. In practice, 1G cellular started later—and sometimes several years later—in standard-importing countries than it did in standard-creating countries.

Similarly, the second-generation Global System for Mobile Communications (GSM) standard was launched in several Western European countries in the early 1990s and later was adopted in Eastern Europe and other parts of the world. Thus, 2G cellular started later in Eastern Europe than it did in Western Europe. Further, some countries in Africa were still launching 1G systems in the early 1990s while some European countries were already operating on 2G standards.

Lags of this sort happened also in third-generation cellular, which started in many Organisation for Economic Co-operation and Development countries at the beginning of the 2000s. In China, however, and due mainly to industrial policy factors, 3G cellular was not introduced until the very late 2000s. Almost a full decade went by between the launch of 3G cellular systems in several Western European countries and 3G adoption in China.

With this caveat, each of the four parts in our book corresponds to a cellular generation and thus roughly to a decade. Each part, in turn, is organized into three chapters. The first chapter in each part presents an introduction to the relevant generation and its corresponding standard or standards. Thus, each part's first chapter identifies the novel elements associated with each generation (and its standards), including new technologies, new users, and new functions. In the second chapter of each part, we study in more detail the evolution of cellular in different countries, emphasizing the role that national governments played in shaping the cellular industry and fostering cellular growth. The final chapter of each part examines cellular product and service markets in the context of each standard, highlighting the markets for cellular infrastructure, cellular handsets, and cellular network services. In the last two parts of the book, which cover 3G and 4G, we also study the market for mobile semiconductors and the market for mobile operating systems.

THEMES OF THE BOOK

Although our book is organized chronologically, we focus on several important themes that have defined the cellular industry to thread our narrative. These themes include the importance of cellular standards, differences between closed and open standards, competition among standards and among firms within standards, the impact of technological change, and the role of national governments in shaping the evolution of the cellular industry. We also consider the changing roles that cellular phones have played in the everyday lives of people around the world and across time.

Although we usually associate the word *cellular* with the mobile devices we carry around every day, the cellular industry is a system-type industry. This means that different components of the cellular ecosystem have to interact smoothly with one another for us to be able to communicate with one another. It also means that companies operating in different markets have to cooperate with one another. From the early 1980s through the late 1990s, the cellular industry comprised three groups of companies. The first group supplied the cellular infrastructure, the second group the cellular devices, and the third group the cellular network services. Starting in the late 1990s, two additional groups of companies started playing an important role in the cellular industry: developers and sponsors of smartphone operating systems, and designers and makers of mobile semiconductors, especially baseband processors.

Because the interactions between base stations and cellular devices are especially crucial, organizations and companies have, from the beginning of cellular history, formed committees to define the rules governing such interactions (and other technical details). Those rules of interaction, usually called standards, are codified in detailed technical documents. Standards have contributed to defining the way the cellular industry has evolved because they have laid down the rules dictating how the various components of the cellular system interoperate. Without standards, the cellular industry would have been infeasible.[1]

A few standards in the early years of the industry were closed: they were developed by a single company, which owned all of the intellectual property—the technical specifications—associated with the standard. In such cases, only the standard owner was able to manufacture, or license other firms to manufacture, equipment adhering to the standard. In other cases, however, cellular standards were open. In such cases, several organizations interacted (often in committees) to create technical specifications, which were then made available to all comers, such that multiple companies were able to produce technology that adhered to the rules of interaction.

In the history of the cellular industry, standards have competed with one another and firms have competed with one another within specific standards. Between-standard competition happened when promoters of different standards competed with one another to gain adoption for their own standard in different countries and regions of the world. Within-standard competition happened when companies competed with one another by selling products, such as infrastructure technology and handsets, that adhered to a specific standard. By their very nature, closed standards tended to suppress within-standard competition, whereas open standards tended to promote it. Because they encouraged within-standard competition in the supply of cellular technology, open standards fostered wider adoption.

Open standards encouraged cellular adoption in two ways. First, the fact that firms competed with one another within specific standards led to lower prices for cellular technology, which in turn made it possible for wider segments of a country's population to access cellular products, handsets especially. Second, "network effects" kicked in. As more people acquired mobile phones and became cellular subscribers, cellular networks became more valuable for both existing and future subscribers. Large, and rapidly increasing, numbers of users attracted new users, and the size of cellular networks exploded. Both effects—lower technology prices and network effects—had an impact across countries as well. Closed standards were almost never adopted outside the country where they were created, but several open standards were widely adopted in countries other than the ones where they were developed. During the 1G era, AMPS and NMT were the main examples of open standards that were widely adopted outside the countries where they were created. During the 2G era, GSM was the leading example.

It has been argued that, with full interconnection among cellular networks and between each of the cellular networks and the fixedline telephone system, network effects should not matter. (Fixedline telephony is the traditional system of wired telephones, also known as the wireline system or the public switched telephone network.) Why would somebody benefit more from subscribing to a cellular system that has more subscribers if all systems are interconnected with one another and with the fixedline network? Historically, however, network effects have played a role in fostering the growth of cellular systems, among other things because carriers have implemented pricing structures in which calls that terminate on competing networks are more expensive than calls that terminate on each carrier's own network. These pricing structures have given rise to network effects based on differential pricing of services that, from the subscriber's perspective, are identical.[2]

Standards also exist in other high-tech industries, such as computers and semiconductors. Cellular standards, however, are different from computer standards in a fundamental way. Computer standards tend to be market based: a computer system, such as the IBM System 360, became an industry standard because it was successful in the marketplace.[3] Many cellular standards, by contrast, are committee based: they have been shaped by committees consisting of representatives of various organizations, and governments have frequently intervened to promote the adoption of a particular cellular standard.

This government involvement highlights another theme we explore in this book: the role national governments have played in shaping the way the cellular industry evolved in each country. First, they have allocated portions of the spectrum to cellular communications. The timing of this allocation and how generous it was shaped the evolution of cellular services in each country. Second, during the early years of the industry, some of the first cellular operators were often subsidiaries of a government-owned fixedline incumbent. Third, governments promoted the adoption of specific cellular standards—often just a single standard—in each country. Fourth, in some countries, governments encouraged the rise of "national champions"—large domestic corporations that became leading producers of cellular technology with government support. Fifth, governments licensed cellular operators and, by doing so, often shaped the structure of the network-services market in each country. Finally, government regulators frequently intervened in interconnection disputes between cellular operators and the incumbent fixedline telephone operator. In the early days of cellular, such disputes were frequently conflicts between private corporations (or consortia) and a state-owned entity. In short, national governments have played a much more influential role in the cellular industry than in other high-tech industries.

Another major theme in our book is the impact technological progress has had on the cellular industry. One aspect of this theme is the influence of changing semiconductor technology on the size, weight, and functionality of mobile devices. Semiconductor components became smaller over time, and this facilitated the transition from bulky, heavy, and rudimentary mobile phones to the smaller, lighter, and more sophisticated smartphones in use today. Another dimension of this theme, largely invisible to consumers, involves the changing approaches for allocating radio frequencies to multiple users so that many calls can be completed simultaneously. The cellular industry moved from frequency division multiple access, used in all 1G (analog) systems, to time division multiple access and code division multiple access, both used in the digital cellular systems adopted from the beginning of the 1990s. More recently, the 4G Long-Term Evolution (LTE) standard has relied on yet another approach, orthogonal frequency division multiple access.

A final theme we explore in our book is how people have used their mobile phones over time and across regions of the world. During the 1980s, in the 1G era of cellular, cellular devices facilitated voice communications when users were on the move. During the 1990s, in the 2G era, cellular subscribers started using short messaging service (texting), but voice communications continued to play a crucial role. Change accelerated during the 3G era of cellular, starting in the early 2000s, when texting became pervasive and mobile devices started supporting versions of the Internet and email. Change accelerated again in the late 2000s with the introduction of the Apple iPhone and the Android smartphones. These mature smartphones had many of the functions of a true computer. After Apple introduced the iPhone in 2007, and especially after it introduced the second version in 2008, mobile apps became the core feature of mobile phones. Apps captured the attention of users and shaped their choices in many areas of everyday life, from transportation to banking to entertainment, to mention just a few.

STRUCTURE OF THE BOOK

As explained earlier, our book is divided into four parts, each with three chapters. The first part covers roughly the 1980s and focuses on the first generation of cellular phones, also known as the 1G or analog era. Chapter 1 argues that cellular systems did not come into existence out of the blue: wireless telegraphs existed before there were wireless phones, and mobile phones existed before there were cellular phones. We track the evolution of mobile communications before the rise of cellular phones and explain the originality of the cellular concept. Chapter 2 focuses on the launch of the early cellular systems in Japan, Scandinavia, and the United States, as well as in other Western European countries that created their own standards. Among the countries that developed cellular standards, the United States and Britain were peculiar because they introduced competition in cellular services from the very beginning. Chapter 3 explores the early markets for cellular infrastructure, handsets, and network services.

The second part covers roughly the 1990s and focuses on the transition from first-generation to second-generation cellular, or from 1G to 2G. Second-generation cellular was the first generation of digital cellular systems. Chapter 4 highlights some of the new features of 2G cellular networks. In the 2G era, the existing standards for analog transmission were replaced with those for digital transmission. In addition, mobile phones were adopted widely by consumers (rather than just by business users and wealthy subscribers). Chapter 5 examines the transition from analog to digital cellular in some of the pioneering cellular countries, including the United States and Japan, as well as in several other large cellular markets in Europe. It also explores the rise of cellular systems in two countries that would eventually become massive cellular markets: China and India. In many of the countries studied in chapter 5, the evolution from 1G to 2G also involved the transition from monopoly to competition in cellular network services. Chapter 6 examines the dynamics of cellular product and service markets—infrastructure, handsets, and network services—during the 2G era.

The third part covers roughly the 2000s and focuses on the transition from second-generation to third-generation cellular networks, or from 2G to 3G. Chapter 7 explores how the 3G cellular standards were developed and the novel features they introduced. During the third cellular generation, feature phones and smartphones replaced voice-only phones, and data-related applications on mobile phones slowly became pervasive. Chapter 8 compares the evolution of cellular systems in several markets in the Asia-Pacific region, the Middle East, and Latin America. Chapter 9 analyzes changes in cellular product and service markets during the 3G era, tracking the growing importance of two ancillary markets: mobile semiconductors and mobile operating systems.

Lastly, the fourth part covers roughly the 2010s and examines the transition from third-generation to fourth-generation cellular systems, or from 3G to 4G. Chapter 10 tracks how the LTE standard was developed, and how smartphones came to play an increasingly important role in the cellular industry, especially after the introduction of Apple's iPhone in 2007 and of the first Android phones in late 2008. Chapter 11 analyzes cellular markets in Africa. Chapter 12 studies the dynamics of cellular product and service markets during the 4G era, highlighting the growing importance of Chinese companies in several of those markets.

In the concluding chapter of our book, we summarize the main themes developed in the book and reflect on 5G, the new generation of cellular systems. At the time of writing, the fifth cellular generation has become a topic of intense debate, not only because of its technical dimensions and the lifestyle changes it is expected to introduce but also for its national-security and geopolitical implications.

RELATED LITERATURE

Among the histories of mobiles phones, three excellent books stand out: Garry Garrard's *Cellular Communications* (1998), Guy Klemens's *The Cellphone* (2010), and Jon Agar's *Constant Touch* (2013).[4] Garrard presents a wealth of information on the evolution of cellular communications all over the world but covers only the first generation and a fraction of the second generation of cellular phones. Klemens devotes substantial portions of his book to the technical foundations of wireless and cellular communications. Agar tends to emphasize the cultural and social dimensions of mobile phones. In addition, Martin Cooper's *Cutting the Cord* (2020) contains valuable insights on the business and technological developments that led to the introduction of cellular systems in the United States.[5] We have also consulted dozens of peer-reviewed articles and books on the evolution of cellular networks in specific countries and regions of the world, as well as a variety of technical books on wireless and cellular communications. These sources are reflected in the endnotes for each chapter.

Two short and carefully written introductions to the history of cellular phones are John Meurling and Richard Jeans's *The Mobile Phone Book* (1994) and Karyn Poupée's *La téléphonie mobile* (2003).[6] The former covers the first generation and the beginning of the second generation of cellular systems, whereas the latter focuses on the second generation and the early years of the third generation.

Among the books that highlight the economic and business dimensions of cellular phones, four have been especially useful to us: Jeffrey Funk's *Global Competition between and within Standards* (2002), Harald Gruber's *The Economics of Mobile Telecommunications* (2005), Peter Curwen and Jason Whalley's *The Internationalisation of Mobile Telecommunications* (2008), and Curwen and Whalley's *Mobile Telecommunications in a High-Speed World* (2010).[7] We have also drawn from a collection of excellent book chapters, including Jerry Hausman's "Mobile Telephone" (2002), Jeffrey Church and Neil Gandal's "Platform Competition in Telecommunications" (2005), and Joshua Gans, Stephen King, and Julian Wright's "Wireless Communications" (2005).[8] Two books are full of insights on telecommunications economics more generally: Jean-Jacques Laffont and Jean Tirole's *Competition in Telecommunications* (1999) and Carlo Cambini, Piercarlo Ravazzi, and Tommaso Valletti's *Il mercato delle telecommunicazioni* (2003).[9]

In the process of writing this book we received superb research assistance from Nicolas Garcia-Vicente and Florencia Garcia-Vicente. Two anonymous reviewers made extremely helpful comments. The views expressed in this book are those of the authors and do not reflect or represent the views of Charles River Associates, Warwick University, or any of the institutions with which the authors are, or have been, affiliated.

I

THE FIRST GENERATION, THE 1980s

1

THE LONG ROAD TO THE FIRST CELLULAR SYSTEMS

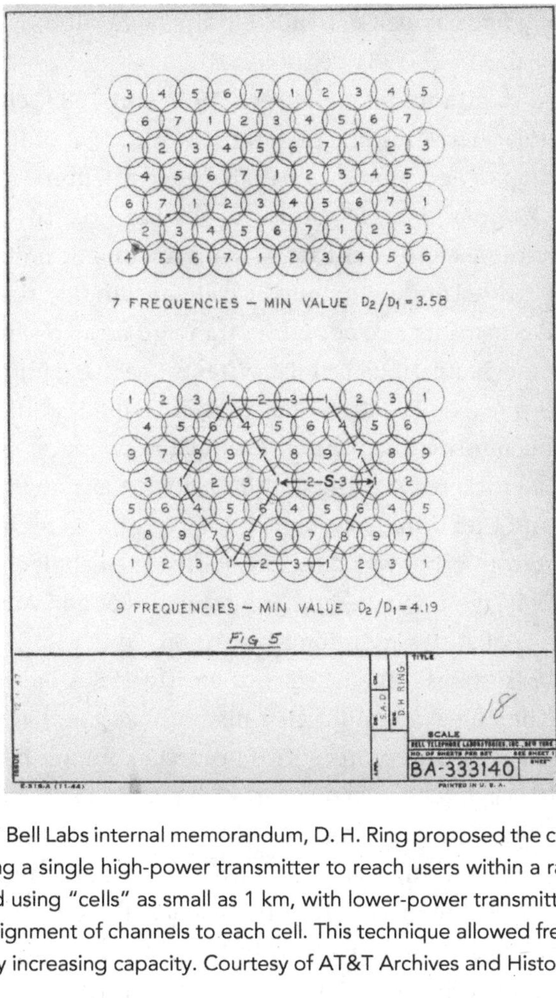

1.1 In a 1947 Bell Labs internal memorandum, D. H. Ring proposed the cellular concept. Instead of using a single high-power transmitter to reach users within a radius of 40 km, Ring proposed using "cells" as small as 1 km, with lower-power transmitters within each cell and an assignment of channels to each cell. This technique allowed frequencies to be reused, greatly increasing capacity. Courtesy of AT&T Archives and History Center.

In 1978, there was just one cellular-phone system in the world—a trial network designed to serve no more than 2,500 users in the Chicago area.[1] (There were many mobile-phone networks around the world, but only the test system in Chicago operated under cellular principles.) A decade later, by mid-1988, more than forty countries had installed at least one cellular-phone system, and these systems, combined, were serving no fewer than 2.5 million subscribers. By the end of 1988, there were more than 4 million cellular subscribers around the world.[2] The cellular revolution, which has transformed the way we live our lives, had started.

Table 1.1 presents a timeline of the worldwide development of cellular systems between 1979 and 1988. (Different sources have slightly different start dates for a few countries included in the table, although there is full agreement for the 1979–1984 period.)

A few facts stand out. First, between 1979 and 1983, eight countries launched cellular networks, the first in the world. Two of these countries were in Asia (Japan and Saudi Arabia); five were in Europe (Norway, Sweden, Denmark, Finland, and Spain); and the last was the United States. Second, six of these eight countries—Japan, the four Scandinavian countries, and the United States—adopted a cellular standard that they themselves created. The other two adopted a standard created somewhere else: Saudi Arabia and Spain set up cellular systems operating under the Nordic Mobile Telephone (NMT) 450 standard developed in Scandinavia.[3] Third, the number of countries that installed cellular networks grew slowly until 1984, but it skyrocketed starting in 1985. Finally, although many of the countries that set up cellular systems from 1985 on were in Europe, several countries from other world regions joined the cellular club as well—for example, Malaysia, Oman, and Tunisia in 1985, and Australia, Israel, South Africa, Thailand, and Indonesia in 1986.

The cellular systems established in the early 1980s did not come out of the blue. They have a multifaceted history that goes back to the early decades of the nineteenth century. This history is complex because it arises from the interweaving of several layers—the history of science, the evolution of technology, the search for the right business model, and the politics of regulation. The first three chapters of this book cover this history. This chapter focuses on the evolution of the science, the technology, and the business of wireless communications in the precellular world up to, and including, the first cellular systems. Chapter 2 concentrates on the launch

Table 1.1 Timeline of early cellular systems, 1979–1988

Year	Africa	Asia	Europe	The Americas	Oceania
1979		Japan			
1980					
1981		Saudi-Arabia	Norway, Sweden		
1982			Denmark, Finland, Spain		
1983				United States	
1984		Hong Kong, South Korea	Austria	Canada	
1985	Tunisia	Malaysia, Oman	Ireland, Italy, Luxembourg, Netherlands, Britain, West Germany, France		
1986	South Africa	Israel, Thailand, Indonesia	Iceland, Turkey	Virgin Islands	Australia
1987	Morocco	Bahrain, Singapore	Belgium, Switzerland	Cayman Islands, Dominican Republic, Bermuda	New Zealand
1988	Zaire	China, Macau	Cyprus	Venezuela	

Sources: Elaborated by the authors on the basis of information in US Department of Commerce, *A Competitive Assessment of the U.S. Cellular Radiotelephone Industry* (Washington, DC: US Department of Commerce, 1988), 76; J. Funk, *Global Competition between and within Standards: The Case of Mobile Phones* (Houndmills, UK: Palgrave, 2002), 43; and G. Garrard, *Cellular Communications: Worldwide Market Development* (Boston: Artech House, 1998).

of the first cellular systems, highlighting the role that national governments played in shaping the early cellular networks and markets. Chapter 3 covers the competitive interactions between the main actors in the early cellular markets, with markets defined by product (or service)—cellular network services, infrastructure equipment, and mobile devices—and geography.

RADIO TECHNOLOGY

Although this book is primarily a business and economic history of mobile telephony, it is useful to keep in mind a few aspects of radio technology and terminology that affected the development of the mobile-phone industry.

A topic that appears throughout the book is spectrum allocation.[4] *Spectrum* refers to the radio portion of the electromagnetic spectrum. The electromagnetic spectrum is a natural phenomenon: the highest frequencies of the spectrum are gamma rays and X-rays, while the lowest frequencies are those of radio waves. In between these extremes lie infrared, visible light, ultraviolet, and microwave radiation, among others. We are concerned only with radio waves.

The radio frequency spectrum is itself divided into "bands" such as VHF (very high frequency) and UHF (ultrahigh frequency). These terms are likely to be familiar to any user of a conventional broadcast radio or television receiver. These bands are also used by mobile phones. The radio spectrum is a natural resource for which there are competing demands from both broadcasters and mobile-phone service providers. In the United States, the radio spectrum is allocated by the Federal Communications Commission. Other countries have comparable organizations.

The earliest radio technology was the spark-gap transmitter invented by Heinrich Hertz (1857–1894) in the 1880s. (Electromagnetic frequencies are usually expressed in hertz in recognition of this scientist. *Hertz* is synonymous with cycles per second. Radio frequencies are usually stated in kilohertz or megahertz, abbreviated as kHz and MHz, respectively.) The transmitter used a rapidly repeating electrical spark to generate a radio signal. It was first exploited commercially by Guglielmo Marconi (1874–1937) in the early twentieth century. We briefly explore the Marconi system in the following section, both for its intrinsic interest and because the business issues he encountered and resolved foreshadowed some of the issues that would come up much later in the history of mobile phones.

The era of spark-gap transmission was short-lived for two reasons. First, it was a crude technology that consumed a large amount of the scarce radio spectrum. Second, it could only be used for telegraph messages using the dots and dashes of Morse code.

By the beginning of the twentieth century, so-called continuous wave transmission at a steady frequency had been achieved. This enabled many more transmitters to share the radio spectrum and to carry speech. In order to carry speech, the high-frequency radio signal, known as the carrier wave, had to be modulated by the much lower audio frequency. In the golden age of radio between the two world wars, the technique of amplitude modulation, customarily known as AM, was used for broadcast radio. AM was also used by the first mobile telephone systems in the 1930s and 1940s.

An alternative technique, frequency modulation (FM), was developed in the 1930s. Here, the frequency rather than the amplitude of the carrier wave was modulated (figure 1.2). FM had several advantages over AM: for example, it was much less prone to signal loss due to "fading," and noise and interference were much reduced. However, FM needed to use the VHF band, and radio receivers were more complex and costly to manufacture, so it was not until the 1950s that FM was widely available in domestic radio sets. As mobile telephony moved into the postwar era, it too adopted FM.

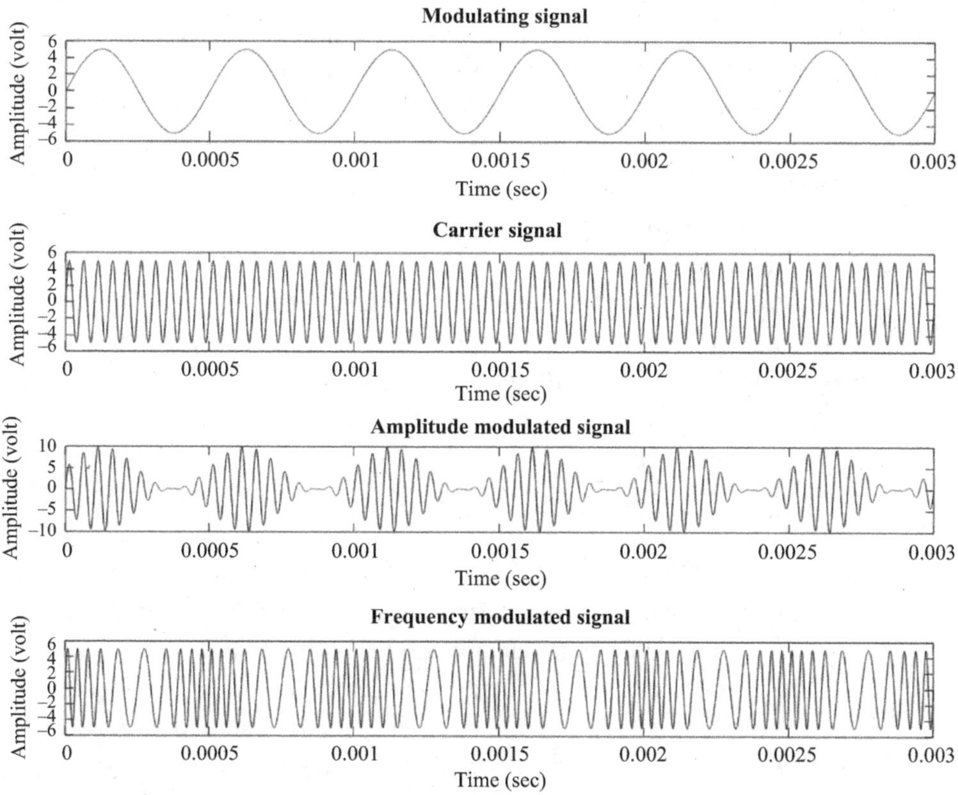

1.2 Amplitude and frequency modulation. *Source*: Constructed by the authors with Octave.

MARCONI AND THE BUSINESS OF WIRELESS TELEGRAPHY

Guglielmo Marconi was the son of a silk merchant from Bologna, Italy, and an Irish mother. The young Guglielmo developed an understanding of the foundations of electromagnetism from his studies under Augusto Righi (1850–1920). A professor at several Italian universities, Righi was a renowned Italian physicist and probably one of the few in Europe who, at the time, fully understood the significance of the theory of electromagnetism developed by James Maxwell (1831–1879) in Britain and of the experimental work of Hertz in Germany.[5] Interested in exploring further what he had learned at the Righi lectures, at a very early age Marconi started conducting his own experiments in his parents' country house outside Bologna. By 1894, he was convinced that wireless signaling across space was feasible.

After finding no interest in his wireless equipment and ideas on the part of the Italian Post Office, Marconi and his mother traveled to Britain in 1896 and managed to secure the support of William Preece, the chief engineer of the British Post Office. People who were keeping track of wireless developments in Europe understood that, except for some innovations in antenna design, the technology Marconi took from Italy to Britain was already available in Britain.

In 1896, Marconi filed his first patent, the first radio patent ever issued. It has been suggested that his core contributions to wireless in 1896 were not technological but of a different nature: first, unlike Hertz and other scientists, Marconi was confident that he could create a system for emitting and receiving "Hertzian" waves that would have military and commercial applications; and second, he was determined to do precisely that.[6]

In 1897, Marconi formed a private entity—the Wireless Telegraph and Signal Company Limited—to explore the commercial potential of his inventions, a decision that ended his collaboration with Preece and the British Post Office. At the beginning, Marconi and his partners viewed their business model as limited to selling wireless-telegraphy equipment. They soon realized, however, that such an approach had severe limitations: wireless was a new technology, and only organizations able to recruit competent personnel, such as the army and the navy, would buy Marconi's devices.

Thus, in 1899 Marconi's company made a fundamental business decision: it would stop selling wireless-telegraphy equipment and start providing communications services. The next year, a subsidiary was organized under the name of the Marconi International Marine Communications Company to implement the new business model. In the years that followed, a network of Marconi shore stations was built. Although these shore stations were owned by third parties, they were manned exclusively by Marconi personnel and they exclusively used Marconi equipment. Marconi's customers did not own the equipment; they leased it from the Marconi Company and thus gained access to a full-fledged communications system that relied on standardized apparatus and operating procedures.

The new approach had benefits for customers, since they were relieved from the uncertainties of owning and operating an unproven technology, and it was attractive for the Marconi Company as well, since it gave the firm control over the system in its entirety. Marconi took the idea of control one step further: the Marconi Company started refusing interconnection with all other wireless-telegraphy systems. Except in situations of distress calls, the Marconi operators were instructed not to accept messages coming from competitors operating stations not equipped with Marconi technology.

In the transition from the nineteenth to the twentieth century, Marconi did more than just experiment with new business models. In 1899, he transmitted signals across the English Channel, and two years later he did the same across the Atlantic. In between these two demonstrations, in the year 1900, he applied for an important patent—so important that it became the basis of much subsequent litigation.[7]

In 1901, Marconi signed a contract with Lloyd's, which sold maritime insurance and had a network of more than one thousand agents in all the major seaports. This was a crucial moment for the Marconi Company, because, by the terms of the agreement, Lloyd's agents were prevented from communicating with ships that used competing wireless-telegraphy equipment, such as the one manufactured by Slaby-Arco and De Forest. The Lloyd's contract, and the business derived from it, became the foundation for Marconi's domination of marine radio, especially in North Atlantic waters. By 1907, all major transatlantic liners carried radio installations, and all such installations relied on Marconi equipment (and Marconi operators). This was an early form of network effects in wireless telecommunications. Because the Lloyd's agents were only allowed to use Marconi equipment, ship owners tended to install Marconi devices on their vessels as well, and the rapidly growing stock of maritime users of Marconi equipment gave new users strong incentives to adopt it.[8]

Throughout his business career, Marconi remained focused on achieving wireless communications at increasingly longer distances. By the early twentieth century, he had conquered two important markets: wireless telegraphy among the military and in maritime civil communications. He then set his sight on a third market, the one served by transatlantic cable. This was a different market, and it catered mostly to the needs of financial and newspaper customers. By 1898 there were fourteen cables across the Atlantic, twelve of them in operation, and the companies that owned them were mostly profitable and had plenty of excess capacity. This was a market that Marconi was not equipped to conquer, at least not immediately. It took about thirty years for wireless to threaten the transatlantic cable business; and by the time this happened, wireless technologies were radically different from those Marconi had pioneered at the end of the nineteenth century.

FROM WIRELESS TELEGRAPHY TO WIRELESS TELEPHONY

Marconi's commercial success was based on wireless telegraphy rather than wireless telephony, and the wireless telegraphy that Marconi and his competitors practiced relied on a technology that was not appropriate for wireless telephony. The transition from wireless telegraphy to wireless telephony required a shift in technological paradigm, a shift that occurred in the first two decades of the twentieth century.

In the early days of wireless experiments and wireless signaling, the spark-gap transmitter was the only method for generating radio waves.[9] Scientists and entrepreneurs used an induction coil, or some other device, to place a high voltage across a spark gap. When a spark "jumped the gap," it generated an electromagnetic disturbance (or wave) that could be detected at a distance. A sequence of sparks created a chain of such disturbances, which could be interrupted to form the dots and dashes of the Morse code, thus conveying a message.

The waves generated by spark transmitters had a problem: they were not continuous. This fact had several implications, and the main one was that the signal emitted by a spark-gap transmitter was splashed across the radio spectrum. This made the task of locating the signal—while tuning out all other signals—considerably more difficult for the receiving apparatus. By the end of the nineteenth century, doubts were starting to arise whether this technology could serve as the foundation for a system of wireless communications.

In the first two decades of the twentieth century, devices were invented that generated continuous waves in the portion of the electromagnetic spectrum that is suitable for radio.[10] These devices made the transition from wireless telegraphy to wireless telephony possible.

PRECELLULAR WIRELESS TELEPHONY IN THE UNITED STATES

From around 1919, wireless telephony expanded at a steady pace in the United States. The early transmissions were maritime: in 1919 ship-to-shore radiotelephone service was launched for ships along the Eastern Seaboard of the United States, and by 1929 commercial radiotelephone service had begun for ships on the Atlantic Ocean.[11]

Police departments pioneered land-mobile radio communications.[12] In the early 1920s, the Detroit Police Department conducted the first experiments with mobile telephony on land vehicles. After almost seven years of failed attempts, in 1928 the first operational one-way mobile radio system went on the air in Detroit. The system was "one way" (or "simplex") because it operated in one direction only: police agents were able to receive calls in their cars, usually about crimes in progress, but were unable to respond to the calls they received. Soon the Cleveland Police Department set up a similar system, and in the early 1930s the Bayonne Police Department in New Jersey started operating a two-way system, the first one in the United States. The Bayonne system was "two way" (or "duplex") because it allowed agents in their cars not only to receive calls from headquarters but also to communicate back with headquarters and with other police cars.[13] By 1934, there were 194 municipal police radio systems and 58 state police systems in operation, and combined they served over five thousand police vehicles.

These early systems all used AM, but FM—invented by Edwin Armstrong (1890–1954)—quickly replaced AM as the foundation of mobile telephony. FM offered several advantages over AM. First, FM eliminated random noise ("static") from natural and human sources. In addition, it required considerably less power, which opened the door for better vehicular receivers. Lastly, FM exhibited the "capture effect": whereas in AM transmissions two competing signals are often superimposed over one another, an FM receiver tends to focus on the stronger of two competing signals while rejecting the other.[14]

In the late 1930s, the Connecticut state police inaugurated the first two-way FM mobile system, and by 1940 almost all police departments in the United States had adopted FM mobile telephony.[15] During the 1940s, individuals, companies, and public agencies other than police departments bought mobile units and land equipment and set up their own private FM mobile-phone networks. By the mid-1970s, there were about eight million mobile phones that relied on some form of private service in the United States, and most of them were not connected to the public switched telephone network. (The public switched telephone network, also referred to as a system of wireline, fixedline, or landline telephony, is the traditional network of wired phones.) As the number of users grew, the demands on the radio spectrum increased as well, and the Federal Communications Commission, which had been created in 1934, allocated increasing portions of spectrum to mobile telephony. Since radio demands for spectrum came from many different sectors of the economy and society, the allocative process became highly controversial and confrontational.[16]

A different type of mobile-phone service, one that enabled interconnection with the public switched network, started developing after World War II in the United States. For most of the history of telephony in the United States, the Bell System was the key player: it supplied local service in most, although not all, regions of the country and was also the monopolistic supplier of long-distance service. In 1946, the Bell System launched a program to supply "public correspondence systems"—a term describing a service provided by a common carrier that enabled communication among a variety of users and achieved economies of scale by combining different streams of traffic.

A public correspondence system would be different from a private mobile network in several ways: it would be set up and operated by a common carrier—a company that supplied communication services to the general public—rather than by a private individual, a private (non-telecommunications) firm, or a government agency, and it would allow mobile users to talk to fixedline phone users. The first of these systems, an "urban" network, was inaugurated in 1946 and served the city of Saint

Louis, Missouri. The next year, a system was set up to serve individuals traveling on the highway between Boston and New York City, and so it was called a "highway" system. Over the next several years, many urban and highway systems were installed in the United States. All of them relied on "push-to-talk" technology and were constrained by the limited number of channels available, so that demand for service consistently outstripped supply. Push-to-talk terminals required users to push a button to talk and to stop pushing the button to listen, so that no simultaneous communication was possible. All systems were operated in manual mode, such that each call to or from a mobile unit was handled by a special operator.[17]

In the mid-1960s, the Bell System took the next, and final, step in the evolution of precellular mobile-phone technology in America: in 1964, it launched the so-called MJ system to lower costs and improve efficiency. An improved version of this system, called the MK, was introduced in 1969 on a different frequency band. Both were part of the Improved Mobile Telephone System (IMTS), which was designed to roughly reproduce the features and convenience of traditional wired telephony.

IMTS featured automated "trunking," which dramatically increased total system capacity and improved spectrum efficiency while making life easier for mobile users. (Spectrum or spectral efficiency is a measure of traffic per unit of spectrum available.) In the earliest radio systems, both the transmitter and the receiver of radio waves were designed to operate on a specific frequency, with each radio channel dedicated to a specific user. In a trunked radio system, by contrast, all channels were made available to all users. Thus, to complete a call at a given point in time, it was a matter of finding which channels were available. For a given number of channels, a system could sustain considerably higher volumes of traffic if the channels were trunked.

The first mobile-phone systems—the ones set up in the 1940s—were manually trunked: each caller searched manually through the potentially available channels to determine which one was free at a particular point in time. IMTS, by contrast, offered automatic trunking: the mobile unit managed to find the available channels by tuning to a special tone, such that the user was able to initiate a call without delay. Apart from automatic trunking, IMTS offered direct dialing (so that users did not need to contact an operator to complete the call) and duplex communications (so that users could simultaneously send and receive radio signals). With direct dialing, full duplex, and automatic trunking, IMTS was the direct predecessor of modern cellular-phone systems, the culmination of three decades of development of FM mobile telephony.[18]

By the mid-1970s, before the advent of cellular systems, mobile communications in the United States encompassed a broad swath of services. First, there were many private wireless systems, usually not connected to the public switched telephone network. These systems, which were collectively known as Specialized Mobile Radio, were set up mainly to provide dispatch service, the sort of two-way communication frequently used by taxis, ambulances, and vehicles belonging to various commercial organizations. Second, there was a large array of paging systems that enabled one-way communications, with the message transmitted usually being a short signal to the recipient to call a specific number on a conventional wired phone. Third, there was the Citizens Band service, a broadcast system of limited range. And finally, there was mobile telephony *per se*—that is, wireless communications services supplied to the general public rather than just to users working for a government agency or private company. There were three types of providers of mobile telephony for the general public: the Bell System through its IMTS service; an array of independent wireline common carriers that had been competing with the Bell System in the (wired) domestic telephony market in certain regions of the country and had decided to confront the Bell System in wireless communications as well; and a variety of so-called radio common carriers—small, local operators that had never supplied wired-telephony services but had decided to enter wireless telephony.[19]

In the mid-1970s, there were several million private mobile telephones (not connected to the public switched telephone network) and several million pagers in operation. There were also about 143,000 mobile phones connected to the public switched network via either IMTS or some other system in the United States. Of the 143,000 users, IMTS accounted for about 31 percent, other independent wireline operators for about 13 percent, and the radio common carriers for the remaining 56 percent or so. At the time, there were about eighty-five million fixed-telephone subscriptions in the United States, which suggests that mobile telephony of the IMTS type was truly a minnow in the overall scheme of telecommunications in this country.[20]

PRECELLULAR WIRELESS TELEPHONY IN EUROPE AND JAPAN

Before the advent of cellular systems, people used mobile phones not only in the United States but also in several European countries and Japan. As happened in the United States, two types of precellular mobile telephony developed in Europe: a collection of private systems collectively known as Private Mobile Radio (PMR), most of them not connected to the public switched network, on the one hand, and various national mobile-telephony systems connected to the public network, on the other. PMR, the European equivalent to what in the United States was labeled Specialized Mobile Radio, started in Britain in the late 1940s, when a Cambridge taxi company installed equipment manufactured by the local W. G. Pye and Company Limited on its vehicles.[21] Soon thereafter, many European countries had their specialist manufacturers, particularly in the Nordic region.

Each PMR system was a closed communications network that allowed multiple users to stay in touch with a central controller (the dispatcher) or with one another (via the dispatcher). PMR systems were owned and operated by the organizations that used them—in the early days, ambulance services, taxi companies, and utilities. Until the 1960s, when solid-state electronics became widely available for mobile phones and other technologies, PMR equipment was built with vacuum tubes, and thus it was bulky, heavy, and quite expensive.

Although PMR systems were widespread in Europe, there were substantial cross-country differences in adoption depending on local factors, including the policies of the national postal, telegraph, and telephone (PTT) monopoly and the structure of the domestic economy. As of 1977, for example, there were 2.7 PMR subscriptions per 100 people in the United States, about 1.6 in Sweden, around 1.1 in Denmark, about 0.6 in Germany, and around 0.4 in Britain. (The United Nations defines *telephone density* as the number of telephone subscriptions per 100 people, and we will follow this approach in this book as well.) Sweden was the European country where PMR systems were most pervasive, but still trailed the United States by a wide margin.[22]

Mobile-phone systems connected to the public switched telephone network started developing in Europe in the mid-1950s, about a decade after they were installed in the United States, where the Bell System had launched the first such system in 1946. In 1955, Televerket, the Swedish PTT, was the first such entity to introduce a modern (noncellular) mobile-phone system.

Improved systems were set up in Sweden in the 1960s and 1970s, and by the early 1980s, the country had as many as twenty thousand users.[23]

In Britain, the General Post Office opened the first (noncellular) mobile-phone system in 1959 in South Lancashire, and a similar service was inaugurated in London in 1965. Over the years, various improved systems were installed, and the number of users reached about fourteen thousand in the mid-1980s. System growth was constrained by several factors: a limited number of frequencies was allocated to mobile phones; terminals had basic capabilities and relied on push-to-talk technology; and equipment and services were expensive.[24]

Mobile telephony was a prestige item for the wealthy: in a survey conducted in Britain in the mid-1980s, 65 percent of users drove (or were driven by private chauffeurs in) a Rolls-Royce, a Mercedes Benz, a BMW, or a Range Rover. Even though the service was rather primitive from today's perspective and prices were high, demand outstripped supply. Capacity constraints played a crucial role in preventing growth, since only a small number of channels had been allocated to mobile telephony and frequencies could not be reused at any distance less than one hundred miles from the transmitter. Cellular systems were created precisely to overcome this constraint.

In the 1960s and 1970s, many other European countries—including Norway, West Germany, Finland, Spain, Austria, Luxembourg, the Netherlands, Italy, and Switzerland—introduced (noncellular) mobile-phone systems such as the ones described for Sweden and Britain. Most of these systems shared some common features. First, they were all set up by the national PTTs, often in collaboration with their preferred national telecommunications-equipment suppliers. Second, they all worked at frequencies around 150–170 MHz and had only a small number of channels allocated for service. Third, many of them required operators to connect phone calls to or from mobile devices, and even the "automatic" ones required the caller to know roughly where the mobile user was located. Fourth, most of them relied on push-to-talk technology rather than full duplex. And finally, all of them were expensive.

Outside Europe and the United States, there were significant mobile-phone developments in Japan before the introduction of cellular phones. The first maritime mobile-phone system was inaugurated in that country in 1923. The transition of mobile phones to land vehicles took about a quarter of a century: the first mobile-phone system on land, a police radio

1.3 A British Post Office wireless telephone in use in 1960. Courtesy BT Heritage & Archives.

system, was launched in 1948. In the next several years, mobile systems for fire and flood protection became pervasive, and later a host of businesses, including taxi and construction companies, adopted them.[25] A paging system was set up in 1968, and by 1977 it had 640,000 subscribers.

Nippon Telegraph and Telephone (NTT), a public corporation that had a monopoly on the supply of telephone services, formulated plans for a car telephone in the early 1960s, and a system like Bell's IMTS was introduced in 1967. From the late 1960s, NTT started considering the development of a cellular system for Japan. NTT expected that the system would be roughly identical to the one proposed by AT&T to the Federal Communications Commission in the United States in 1971, and that it would eventually have about a million users.

THE CELLULAR MODEL

Although the features available on precellular mobile systems were limited, all of them faced the same problem: demand consistently exceeded capacity. This happened partly because not enough spectrum had been allocated to mobile systems, but the problem was much deeper than that—precellular systems used whatever spectrum was available in a highly inefficient manner.

Since at least Marconi, wireless communications had followed the broadcasting model: in order to serve a given area, system engineers placed a high-power transmitter (an antenna) in a carefully chosen location, usually a relatively elevated portion of terrain, and sent a signal out as far as it was feasible with the available technology, usually forty or fifty miles away.[26] The upside was that fair coverage could be provided over a reasonably large area, but the downside was that the system could be expanded only if additional channels were made available, since the existing ones became quickly occupied by a small number of calls. During the 1970s, the Bell mobile system was able to support only about a dozen simultaneous conversations in New York City, a metropolis of almost twenty million people that covered over a thousand square miles.[27]

In the second half of the 1940s, as the first true mobile systems were set up in the United States, planners working for the Bell System started to consider alternative wireless models that would enable the company to overcome the constraints created by scarce spectrum. The "cellular" idea, developed by Douglas Ring in a 1947 Bell Laboratories internal memorandum, was one of those models.[28] The cellular concept did not require the Bell System (or any other) to introduce a brand-new communications technology: mobile telephony would still be based on FM transmission. The idea entailed not a new piece of hardware but rather a new architecture, a new approach to using the existing technologies. Instead of covering a broad area with a high-power transmitter, the cellular framework proposed providing service over the same area with a collection of transmitters of moderate power, each designed to cover a small portion of the original region. Put differently, the large area would be broken up into a collection of "cells," each served by its own antenna.

Such an arrangement had the potential to become considerably more costly than the original one because, for any given region, the cellular model necessitated many transmitters—also referred to as "cell sites" and "base stations"—whereas the broadcasting approach required only one. A focus on cost, however, only captured half of the picture, since the cellular model offered the potential for "frequency reuse," which was infeasible in the broadcasting model. The upside of the cellular idea was that the same channels or frequencies could be used more than once within the same area—in different cells, that is, and only if the cells were not too close to one another.

In the early stages of cellular design, Bell's planners conceived of cells as circles. This was quite reasonable, since, in a system based on antennas transmitting in all directions, a cell was bounded by a contour of constant signal level. Circles are impractical for cellular design, however, since they cover a given area only imperfectly—a collection of circles has the potential to create some areas that are covered by more than one and some that are covered by none at all—and hexagons replaced circles in the minds and drawings of cellular designers. A metropolitan area could be subdivided into, say, 99 hexagons, each served by an antenna of moderate power. (The number 99 is arbitrary.) If 12 channels were allocated to the mobile system in the city, and if all 12 channels could have been reused in every cell or hexagon, then going from a broadcasting architecture to a cellular architecture would have increased the theoretical capacity from 12 to 1,188 (that is, 12×99) circuits. In practice, not every frequency could be reused in every cell—mobiles operating on the same channel in adjacent cells generated interference and degraded the system's quality of service. Frequencies could, however, be reused in nonadjacent cells. If, say, the cells were organized in clusters of three—so that there were "A," "B," and "C" cells—and each cell type was allocated 4 channels or circuits, then there would be 33 clusters, each with 12 frequencies, for a total of 396 channels in the area.[29] As long as the cellular architecture facilitated an increase in capacity (from 12 to 396 channels in the example) that more than compensated for the enhanced cost, the system would make economic sense.

Figure 1.4 shows a hypothetical cellular architecture for a mobile network with cells of three types (A, B, and C). The A cells have been allocated channels 1 through 4, the B cells channels 5 through 8, and the C cells channels 9 through 12. Cells are organized in clusters of three, and no two cells of the same type are adjacent to one another—which is necessary to avoid interference.[30]

Beyond its advantage of increased capacity and disadvantage of enhanced cost (in comparison with the broadcasting approach), the cellular model had one additional virtue and one additional problem. The virtue was

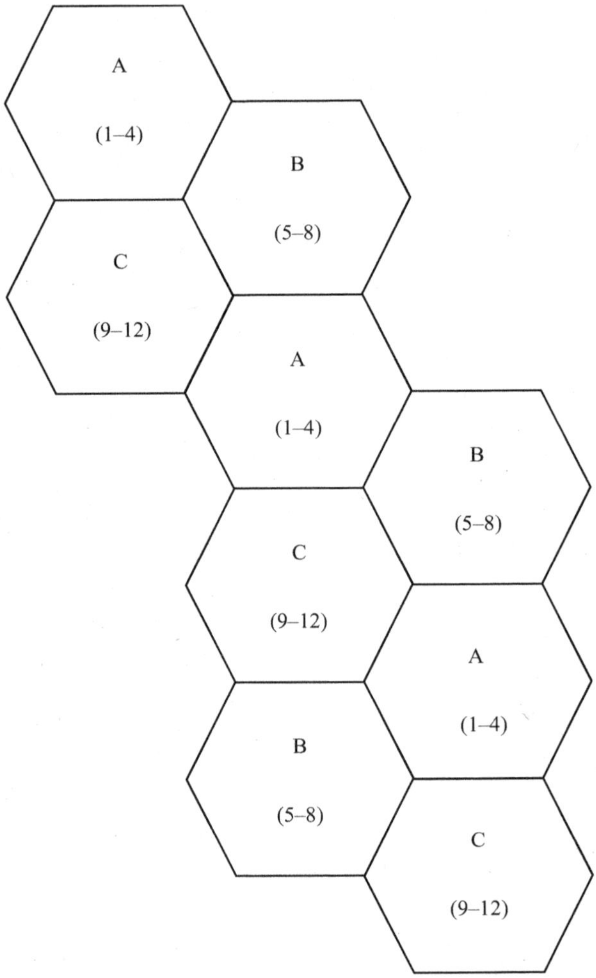

1.4 The (hypothetical) cellular architecture for a mobile network, with clusters of three cells each and channel numbers between parentheses within each cell. *Source*: Constructed by the authors.

that the system could grow its capacity over time depending on need for a given allocation of channels. As the use of mobile phones grew, and the existing channels and architecture became insufficient to maintain quality of service, cells could always be split to generate new, smaller cells. Each of the original cells in a system could be broken up into two or more, which would create new opportunities for frequency reuse in nonadjacent cells. Interference could be avoided—or kept at reasonably low levels—by managing the strength of the signal transmitted by each antenna and by carefully redesigning the frequency-reuse patterns considering the larger number of (smaller) cells. Further, not all the original cells needed to be subdivided—only the ones where mobile traffic was too intense for the original configuration. In theory, the model allowed for almost unconstrained expansion within the limits of the existing frequency allocation.

There was an additional problem (beyond the increased cost arising from multiple cell sites), and it was that mobile phones tend to be, by definition, mobile: users were quite likely to transition from one cell to another during a given call, and the frequency of such transitions would naturally increase as traffic volumes grew and cell splitting became pervasive. This created the need for a central control system to keep track of users and "hand off" or "hand over" a call from one cell (and transmitter) to another. This control system had to continuously measure the signal strength coming from the individual cell sites and use those measurements in real time to "sense" when a user was in the process of moving from one cell to another. At that point, it had to be able to switch the call from the first hexagon to the second without interrupting the call. In addition, this unit connected cellular users with the public switched telephone network, which was crucial in the early years of cellular when most calls originating or terminating in mobile phones involved a wireline user.

The cellular model, in short, was theoretically simple and relied on a small set of basic tenets: small hexagons or cells served by transmitters of moderate power, frequency reuse in nonadjacent cells, the potential for almost unlimited expansion via cell splitting within the framework of an existing spectrum allocation, and a powerful central switching system that handed off calls from one transmitter to another when necessary and connected mobile users to the public switched phone network.

THE EARLY CELLULAR SYSTEMS

It has been said that "all models are wrong, but some are useful," and the dictum applies well to the cellular model.[31] In practice, the real-world cellular systems set up in the late 1970s and early 1980s in the United States, Scandinavia, and Japan did not work as smoothly as the idealized geometry of hexagons predicted they would.[32]

There were, among other things, coverage problems. The task of cellular engineers was considerably more challenging than that faced by those in charge of precellular mobile-phone systems.[33] In an advanced precellular system such as IMTS, established by the Bell System in the 1960s, the goal was to cover as large an area as possible with a single transmitter, and thus the engineer's task was to find the highest elevation in the region, place the transmitter there, and send out the signal at maximum power. In cellular systems, by contrast, there were many cells and transmitters, and in each cell an optimal location had to be found for the antenna serving the theoretical hexagon. Although such optimality could be achieved on paper (or on a computer screen), in practice many factors tended to attenuate wireless signals—including tall buildings, the way the streets were oriented, large trees, and a particularly complex local geography.

Nonetheless, the early cellular systems substantially increased the efficiency with which the existing spectrum allocation was used.[34] Because they were scalable by cell splitting, cellular networks were able to accommodate rising mobile traffic during the 1980s. Table 1.2 shows that, by the late 1980s, the number of cellular users in several European countries was an order of magnitude larger than the maximum number of mobile users achieved by the precellular systems in those countries.

Sweden, for example, where the number of mobile users in the precellular world had reached a maximum of 20,000, had about 349,000 cellular users in 1989, and the Netherlands, which had reached a maximum number of precellular mobile users of 5,000, had about 56,000 cellular subscribers only four years after cellular launch.

THE EARLY CELLULAR SYSTEMS: TAKING STOCK

In the late 1980s, cellular phones were more a promise than a reality as a communications medium. To convey the extent to which this was true, table 1.3 presents the evolution of the number of cellular subscribers per 100 people (that is, cellular density) in the pioneering countries.

The message of the table is twofold. First, cellular density was considerably higher in the Scandinavian countries than in the United States and Japan. In 1989, for example, cellular density in the United States was under 1.5, whereas in Scandinavia it varied between 2.4 and 4.1 depending on

Table 1.2 A comparison of precellular and cellular mobile-phone services in Europe

Country	Introduction of precellular mobile service	Maximum number of users on precellular mobile networks	Introduction of cellular mobile service	Cellular subscribers in 1989
Sweden	1955	20,000	1981	349,000
Britain	1959	4,000	1985	975,000
Norway	1966	n/a	1981	167,651
West Germany	1968	30,000	1985	163,619
Finland	1970	35,000	1982	190,031
Spain	1972	1,000	1982	29,783
Austria	1972	10,000	1984	50,721
Luxembourg	1972	n/a	1985	665
Netherlands	1972	5,000	1985	56,000
Italy	1973	6,000	1985	66,070
Switzerland	1978	n/a	1987	72,735

Sources: G. Garrard, *Cellular Communications: Worldwide Market Development* (Boston: Artech House, 1998), 21; J. Funk, *Global Competition between and within Standards: The Case of Mobile Phones* (Houndmills, UK: Palgrave, 2002), 43; International Telecommunications Union (ITU) database.

country. Second, and more importantly, despite the rapid growth they experienced during the 1980s, by the end of the decade cellular-phone systems were of insignificant size in the overall scheme of telecommunications in America, Japan, and Scandinavia. At a time when at least 49 out of every 100 people in the United States and Scandinavia had fixedline phones, no more than 4 out of every 100 had cellular phones.[35] The same was true in all other countries where cellular systems had been set up during the 1980s.

The cellular idea was developed, at least in primitive form, in the mid- to late 1940s, and it took about three decades for the idea to become a reality in the United States and elsewhere. Technological change was required to make the cellular concept operational in the real world. Several technologies essential for cellular communications became available between

Table 1.3 Cellular subscriptions per 100 people in pioneer countries, 1980–1989

Year	United States	Denmark	Finland	Norway	Sweden	Japan
1980	n/a	n/a	0.49	n/a	n/a	n/a
1981	n/a	n/a	0.59	0.04	0.24	0.01
1982	n/a	0.14	0.70	0.27	0.33	0.02
1983	n/a	0.31	0.87	0.57	0.44	0.02
1984	0.04	0.60	1.07	0.94	0.69	0.03
1985	0.14	0.90	1.38	1.52	0.87	0.05
1986	0.28	1.13	1.73	2.09	1.34	0.08
1987	0.50	1.51	2.14	2.87	2.06	0.12
1988	0.83	1.98	2.79	3.62	2.87	0.20
1989	1.39	2.41	3.83	3.97	4.10	0.40

Source: ITU database.

the late 1940s, when the cellular idea was born, and the early 1980s, when the first cellular systems were launched.[36] For example, semiconductor technology evolved rapidly starting in the late 1940s: the transistor was invented in 1948, the integrated circuit in 1959, and the microprocessor in 1971. Improvements in semiconductor technology affected the evolution of mobile communications in two ways: they played an essential role in the progressive miniaturization of mobile phones, and they enabled digital switching of phone calls, which was essential in cellular systems, especially in connection with the call hand-off function.[37]

2

THE FIRST CELLULAR SYSTEMS: JAPAN, EUROPE, AND THE UNITED STATES

2.1 Picture of a BT Cellnet "Jade" mobile phone in use in 1985. Courtesy BT Heritage & Archives.

National governments have played a more influential role in shaping the evolution of cellular-phone markets than they have in other high-technology industries. This is so because governments have traditionally fulfilled several responsibilities in connection with cellular phones. First, when radio signals travel through space they use radio spectrum, and governments have allocated portions of that spectrum to various uses, including mobile-phone service. Second, having allocated spectrum bands for use in cellular systems, governments have relied on beauty contests, lotteries, and auctions to award licenses to would-be cellular operators. Third, they have selected specific cellular standards and have influenced the process by which such standards are developed. Fourth, in the early days of cellular, governments frequently extended their fixedline phone monopoly into the cellular market through a cellular subsidiary of the fixedline incumbent. Finally, they have often intervened in disputes between the cellular operators and the operators of the fixedline telephone network. There is nothing akin to spectrum allocation and to government licensing of entrants in, say, the computer industry. Moreover, whereas in the computer industry most standards are defined by market mechanisms, many cellular standards have been developed through a long process of interaction among a variety of stakeholders, often in the context of standard-setting organizations sponsored by national governments.

This chapter examines the launch of the first cellular systems, highlighting the role that national governments played in the process. We begin with a historical account of the development of the Bell System in the United States; the consolidation of the national postal, telegraph, and telephone (PTT) administrations in Europe; and the formation of the Japanese telecommunications system. These are the three areas of the world where cellular-phone systems became operational for the first time.

The core of the chapter is an analysis of the many ways national governments shaped cellular standards and market structures in the 1980s in the countries that created their own standards. All the systems discussed in this chapter were launched during the first era of cellular history, which started in the late 1970s and ended in the early 1990s. For this reason, all of them are known as first-generation (1G) systems. They are also known as "analog" cellular networks, to distinguish them from the digital networks that were set up starting in the early 1990s. In the analog era of cellular, voice was transmitted as a continuous wave varying in frequency (pitch) and amplitude (volume or strength). In the digital era of cellular, which started in the early 1990s, the analog signal (voice) was digitized—it was represented as a binary code and was transmitted as bursts of zeros and ones.[1]

Within the first era of cellular history, there were two groups of standard creators. The first cellular systems were launched in Japan, Scandinavia, and the United States, in that order, starting in 1979. Each of these regions created its own cellular standard, and we call them the "first wave" of cellular-standard creators. There was another group of countries—Britain, France, Germany, and Italy—that also developed their own standards and launched their cellular systems in the mid-1980s. We call them the "second wave" of standard creators.

There was a third group of countries—in Europe and elsewhere—that introduced cellular systems during the 1980s. These countries did not create their own standards but rather imported ones developed mostly by the first wave of standard creators. We introduce this third group of countries—the standard importers—at the end of the chapter but do not study their cellular systems and markets in detail.

PRECELLULAR TELECOMMUNICATIONS SYSTEMS IN THE UNITED STATES, EUROPE, AND JAPAN

Cellular phones did not come into existence in a vacuum. They were preceded by fixedline telephony systems set up in the late nineteenth century. (Each of the wired-telephone systems established in the second half of the nineteenth century is usually referred to as a landline, wireline, or fixedline system, and also as the public switched telephone network in each country.) In most cases, the companies that introduced the first cellular phones were the same firms that had acted as (often monopolistic) providers of fixedline telephony for decades. Thus, an overview of the development of fixedline systems in the United States, Europe, and Japan is necessary to understand the rise of the first cellular networks.

THE UNITED STATES

On February 17, 1876, Alexander Graham Bell, a teacher of speech for the deaf, filed a patent for a telephone. A few hours later, Elisha Gray, an electrician, submitted a preliminary application (or caveat) for a telephone patent. Gray happened to work for Western Electric, the manufacturing arm of Western Union, which at the time was the largest carrier of telegraph messages in the United States.[2]

The stage was set for a confrontation between Bell and Western Union over the commercial exploitation of wired telephony. Bell, who had arrived first to the patent office, owned the intellectual property rights over the device, but Western Union already had a network infrastructure deployed across the country. Bell and a group of financial supporters founded the Bell Company as a manufacturing organization that leased equipment to local exchanges operated by private companies, which in turn assigned a portion of their stock to the Bell Company for the right to use its innovations. The first Bell Company franchisee was New England Telephone, launched in 1878. Between 1876 and 1879, the Bell Company and Western Union rushed to set up telephone exchanges all over the United States.

In 1879 the Bell Company and Western Union reached an agreement, according to which they pooled all of their telephone patents under the Bell Company's control. In practice, the deal allocated markets: Western Union sold its telephone network to the Bell Company, which in turn committed to stay out of the wired-telegraphy business. Western Union agreed to finance 20 percent of the Bell Company's research and development in exchange for 20 percent of Bell's rental income on telephones during the seventeen-year period of the agreement. Because the Bell patents were upheld in court as controlling all voice communications over wire (and not just the original device invented in 1876), the deal granted the Bell Company a monopoly over telephone service until the expiration of the patents in 1893. The agreement was advantageous for Western Union as well, since it consolidated its dominance in the telegraphy market, the long-distance complement to local telephony.

From 1879 to 1893, the Bell Company took steps to solidify its position in the telephony market: it obtained additional patents to cover all aspects of telephone equipment, equipped many desirable locations with telephone exchanges, and vertically integrated equipment manufacturing and telephone operation. In 1880 it was reorganized as the American Bell Company and entered into permanent agreements with its associated local companies. At the same time, the International Bell Telephone

Company was created in Brussels, Belgium, to sell telephones and switchboards in continental Europe. Later, the International Bell Telephone Company became a telephone-service provider, with operations in several European countries.

In 1882 Bell acquired Western Electric, the manufacturing subsidiary of Western Union, and signed an exclusive-dealing arrangement according to which American Bell would only purchase equipment from Western Electric, which in turn would only sell it to companies licensed by American Bell. In 1885 American Telephone and Telegraph (AT&T) was formed as a subsidiary to develop long-distance service across the United States. In short, the period witnessed the rise of a business organization that combined corporate management at the national level (provided by American Bell), communications equipment (supplied by Western Electric), local and regional service within specified territories (supplied by the local operating companies licensed by Bell), and long-distance service (provided by AT&T).

At the end of the nineteenth century, the Bell group of companies, usually referred to as the Bell System, underwent several transformations. In a corporate reorganization, AT&T bought the assets of its parent, American Bell, and thus became the parent company of the Bell System. From then on, the AT&T name became synonymous with telecommunications in the United States. More importantly, once the Bell patents expired, several independent firms and cooperatives entered the market to provide local telephony, first in small towns and later in larger cities. A few independent providers entered the long-distance market as well. Entrants weakened AT&T's market position to such an extent that, by 1900, 38 percent of phones installed in the United States were controlled by independent operators.

The entry of independent operators into the market for local telephony raised the interconnection issue—a problem that would come up often, decades later, in the history of cellular networks. The Bell System, which had telephone exchanges in the major cities, refused interconnection between its telephones and those of the newly arrived competitors. Because the value of having a telephone rose with the size of the network, preventing interconnection handicapped new entrants and gave a competitive advantage to the incumbent. New entrants were forced to resort to one of two strategies: either they focused on geographic areas not yet reached by the Bell network, or they focused on social groups not served by Bell in regions where Bell already had an exchange. Either way, competition accelerated the growth of fixedline density in the United States.[3] (The United Nations has identified a set of "core information and communications technologies [ICT] indicators," including fixed-telephone or fixedline density, the number of fixed-telephone lines per one hundred inhabitants, and mobile density, the number of mobile subscribers per one hundred inhabitants. In this book, we use both terms, *fixedline density* and *mobile density*. We also use *cellular density* to refer to the number of cellular subscribers per one hundred inhabitants.)[4]

Starting in 1907, Theodore Vail, newly elected president of AT&T, implemented a multifaceted strategy to restore the company's dominance in fixedline telephony. One strategy was a merger with Western Union and the acquisition of independent telephone operators. In 1909, AT&T purchased 30 percent of the Western Union stock and started buying competing phone operators. Another strategy was a comprehensive process of defensive patent acquisition, especially in the nascent field of wireless communications: in 1913, for example, AT&T acquired all of the de Forest triode patents and continued developing the technology for use in telephone repeaters and radio. Alarmed by AT&T's acquisition spree, the government intervened, and in 1913 the company and the attorney general signed the Kingsbury Commitment, by which AT&T agreed to sell its Western Union stock, to allow independent competitors to interconnect with the AT&T network, and to stop buying competitors. In practice, AT&T continued acquiring independent operators after 1913, but at a slower rate.

By the early 1930s, the Vail strategy had restored AT&T to its position as the dominant firm in the American telecommunications industry. The company owned 80 percent of phones in the United States and the only long-distance network. The remaining 20 percent was owned by many small companies that depended on AT&T for long-distance service and for interconnection with other operators. In 1934 Congress passed the Communications Act, which created the Federal Communications Commission (FCC) and inaugurated an era of federal regulation of telecommunications, including telephony, radio transmission, and broadcasting. The FCC was responsible for allocating radio spectrum and for granting licenses to spectrum users. Between 1934 and 1956, American telephony was an era of regulated monopoly, with AT&T at the helm.

After World War II, the development of microwave transmission started to pose a significant threat to AT&T's control of long-distance communications. In addition, in 1949 the Department of Justice filed an antitrust suit against AT&T and Western Electric, highlighting the need to separate the provision of telephone services, which was regulated, from the supply of telephone equipment, which was not. The government demanded that AT&T's ownership of Western Electric come to an end, that Western Electric be broken up into three new companies, and that all restrictive agreements between the Bell operating companies, AT&T, and Western Electric be terminated.

In 1956 the government and AT&T signed a consent decree: this settlement agreement allowed the existing arrangements among the various Bell System companies to continue, but also restricted the scope of the system's activities and demanded that AT&T license the system's patents under generous terms. The decree also had practical implications for the development of mobile telephony.[5] It required that the Bell System abandon the mobile-radio manufacturing business, which opened the door for Motorola to become the leading US supplier. Additionally, the decree strengthened the position of the radio common carriers, independent operators that offered (noncellular) mobile-phone service to the general public.

By the time cellular communications came to the fore in the United States in the early 1970s, AT&T was a vertically integrated supplier of telecommunications services and equipment. It had a monopoly on long-distance service and was also the single provider of local telephony in most of the United States. It relied on equipment designed by Bell Labs and manufactured by Western Electric.

At around the same time, the Department of Justice became concerned that AT&T was once again abusing its market position and, in 1974, filed an antitrust lawsuit against the company. The complaint argued that AT&T used its (legitimate) market power in local telephony, which was viewed as a natural monopoly, to (illegally) restrain competition in other markets that were potentially competitive, such as long-distance service. The parties reached a settlement in 1982, which led to the divestiture of AT&T's local operating companies in 1984. The "new" AT&T retained only the potentially competitive businesses, including long-distance service, whereas the local operating companies were organized into seven Bell regional holding companies, whose exclusive focus was to deliver local-telephone services. These seven companies, which were barred from manufacturing telephone equipment and from offering long-distance service, were the ones that introduced cellular service in the United States in the early 1980s.

EUROPE

Between the late nineteenth century and the late twentieth century, fixed-line telephony in Europe was a service operated by a monopolist, and the operating entity was a state administration for post, telegraph, and telephone (PTT). The arrangement was known as the PTT System.[6]

The European PTT System came into existence through a process that lasted centuries and involved confrontations between the state and private providers of postal, telegraph, and telephone services. The history of the German PTT, usually considered the archetypical state system, provides insights into the nature of these conflicts. It had its remote origins in 1614, when Prussia established a state-run postal monopoly with the declared goal of generating revenue for the state.

The Prussian postal system had its competitors: the main one was the Taxis System, a private postal network that operated mainly in western and central Europe starting in the late fifteenth century. Between the seventeenth and nineteenth centuries, the Prussian state had frequent confrontations with the Taxis over the rights to supply postal services. The end of the Taxis came in 1866, when Prussia nationalized the system.

In the years that followed, Heinrich von Stephan, who began his career as a local postal clerk, rose to become postmaster general and integrated the German postal systems. In 1876 the newly established German state appointed von Stephan to also head the telegraph office, and he quickly merged the two services despite the objections of the technical personnel of the telegraph system.

The next year, 1877, von Stephan introduced the telephone in Germany. Whereas in the United States the telephone was installed in private businesses and homes, the German postal and telegraph authority viewed the device as an extension of state telegraphy—it was to be used only in rural post offices. The American Bell Telephone Company started setting up a system to provide private telephone service in the state of Württemberg while waiting for a license but was soon stopped by the police. This was the only example of private provision of phone service in Germany until the 1990s. By the end of the nineteenth century, the German PTT was consolidated as the monopolistic supplier of postal, telegraph, and telephone services in the country.

Most countries in Europe followed the German pattern: first, the state monopolized postal services, and later, when the telegraph and the telephone were invented, the state integrated them into its monopoly. There were, however, cross-country differences in the role that private organizations played in the provision of telephone services. In the early years, private companies were established in most European countries to supply fixedline phone services—the exceptions were Germany and Switzerland. In the late nineteenth century, however, national governments intervened and monopolized the supply of such services, as they had done with postal and telegraph services years earlier. Telephone systems were fully nationalized in 1888 in Hungary, in 1889 in France, in 1895 in Austria, in 1896 in Belgium, in 1907 in Italy, and in 1911 in Britain. By the beginning of World War I in 1914, a PTT administration had taken shape in most European countries.

In some countries, notably in Scandinavia, private participation in the supply of telephone services persisted well into the twentieth century, and in a few it never completely disappeared. In Norway and Denmark, for example, an array of small joint-stock companies, mutual associations, and cooperative societies played an important role in the evolution of the national telephone network.

In Sweden, the General Telephone Company, founded by local entrepreneurs, played a crucial role in establishing fixedline phone exchanges, in competition first with the International Bell Telephone Company and later with the Swedish government. The government finally took control of the General Telephone Company in 1918 and integrated it into the state-run phone network in 1923. From then on, the Swedish PTT, later called Televerket, became the single actor in Swedish telecommunications, although it never held a legal telephone monopoly.

Finland did not follow the traditional PTT path either. After 1917 a telephone system emerged that combined a role for small companies—private enterprises, subscriber cooperatives, and municipal undertakings—at the local level with a role for the Finnish government at the long-distance level.

In most European countries, the PTT System included not only the government entity that functioned as the supplier of services but also one or more private companies that functioned as equipment suppliers. This role was played, for example, by Siemens in Germany and Ericsson in Sweden. After World War II, these domestic equipment firms had access to large markets that were frequently protected from foreign competition by buy-domestic policies.[7]

JAPAN

The Japanese telecommunications system followed a path similar to that of the traditional (non-Scandinavian) European PTTs. Although a number of private companies showed interest in entering the telephone business in the late nineteenth century, in 1889 the Japanese government decided that the Communications Ministry, which had a monopoly over telegraph services, would be the sole provider of telephone services as well.[8]

World War II decimated the telephone network. The US occupation forces, which were in Japan from 1945 through 1952, ordered the reorganization of the telecommunications system, and the Communications Ministry was split into two, one for posts and the other for telecommunications. Toward the end of the occupation, the prime minister and others pushed for the creation of a private telecommunications carrier, whereas the minister of telecommunications opposed the plan, arguing that all telecommunications services should be provided by the ministry. The parties reached a compromise in 1952. The deal had three components: first, the Ministry of Telecommunications became Nippon Telegraph and Telephone (NTT), a public corporation with a monopoly over domestic telecommunications and the supply of telephone equipment; second, the Ministry of Posts became the Ministry of Posts and Telecommunications and was charged with supervising NTT; and finally, another government-regulated corporation monopolized the supply of international telecommunications services.

Like many European PTTs, NTT developed cooperative relationships with a family of domestic suppliers, which included Nippon Electric Company, Hitachi, Fujitsu, and Oki Electric. NTT's Electrical Communications Laboratory worked with these companies to design, manufacture, and test equipment. In addition, NTT fostered cooperation in research and competition in manufacturing among the companies.

Table 2.1 presents statistics on fixedline density for the 1960–1980 period in all the countries that created a cellular standard in the late 1970s and the 1980s.

A few facts stand out. By the time the first cellular systems were established, around 1980, countries belonging to the first wave of cellular-standard creators (the United States, the Scandinavian countries, and Japan) had higher fixedline density levels than those belonging to the second wave. In addition, among the cellular pioneers, Scandinavia was the region with the highest fixedline density. The standard creators of the first wave were among the twenty countries with the highest fixedline density in the world at the time—and Sweden was number one.

The rest of the chapter focuses on the introduction of the early cellular systems. We first cover the United States, Scandinavia, and Japan—the pioneers that created the first cellular standards and launched the first cellular networks. Then we study the second wave of standard creators:

Table 2.1 Fixedline density in cellular-standard creators, 1960–1980

	1960	1965	1970	1975	1980
United States	26.70	29.37	32.96	36.75	41.02
Scandinavia	18.96	24.77	30.10	37.24	44.77
Japan	3.88	7.53	15.82	29.22	34.45
Britain	9.62	12.02	16.56	23.53	32.26
Germany	4.42	6.38	10.99	16.31	26.10
France	4.71	6.15	8.22	13.17	28.85
Italy	6.13	8.79	12.12	17.53	23.15

Source: Calculations by authors from International Telecommunications Union (ITU) database.

Britain, Germany, France, and Italy. Toward the end of the chapter, we briefly discuss the standard importers in Europe and the rest of the world.

REGIONAL CELLULAR DUOPOLIES UNDER A NATIONAL STANDARD: THE UNITED STATES

Between the late 1940s and the late 1960s, the FCC's spectrum-allocation decisions were consistently against the mobile-phone interests and in favor of the television broadcasters.[9] In the course of those two decades, however, mobile-phone technology evolved considerably and demand for mobile telephony increased. In 1968 the FCC opened a rule-making procedure that became known as the Cellular Docket (or Docket 18262), and in 1970 the FCC allocated a band of spectrum to mobile-phone services.

The 1970 allocation did not lead to the immediate launch of a cellular system in the United States.[10] AT&T made a formal cellular-system proposal to the FCC in 1971, and the agency made an allocation of spectrum the following year. Another battle, however, dominated the 1970s: the discussion whether cellular service would be a monopoly. AT&T assumed that the new cellular service would be supplied by AT&T only, and the FCC shared this assumption in the early 1970s. During that decade, however, the radio common carriers—independent providers of mobile telephony to the general public—contested this assumption and, in addition, they received the support of the Department of Justice, an agency that viewed AT&T's role in the American telecommunications system with increasing suspicion. The FCC's view on cellular evolved considerably during the decade, leading to the regional-duopoly approach that was finally implemented.

In 1975 AT&T was granted a license to run a trial cellular system in Chicago. The network, which operated under the Advanced Mobile Phone Service (AMPS) standard and was launched in December 1978, reached its maximum capacity of two thousand users in about three years.[11] Additional test systems were set up by Motorola in the Baltimore–Washington, DC, area and by Millicom in the Raleigh-Durham region. (Motorola was an electronics firm founded by the Galvin brothers in the late 1920s that accumulated substantial expertise in wireless communications, and Millicom was a telecommunications company set up by the Swedish entrepreneur Jan Stenbeck.) Once the trial systems showed that cellular networks were technologically feasible and that there was a demand for cellular service, the FCC decided to grant regional licenses for the operation of cellular systems. The geographic scope of each license would be limited to one of the so-called metropolitan statistical areas (MSAs), regions made up of one or more adjacent counties with at least one urban core area of fifty thousand people or more.

By the early 1980s the FCC had decided that it would grant two licenses in each of the 306 MSAs: one of the licenses would go to the existing fixed-line operator, on the condition that it set up an independent subsidiary to run the cellular network, and the other would be reserved for a nonwireline competitor, a "pure" cellular operator.[12] The main restriction the FCC imposed was that all operators in all regions would have to offer services compliant with the AMPS standard developed by AT&T. Licenses would be awarded through comparative hearings, popularly known as beauty contests, with the criteria for winning vaguely defined, if at all.

Beauty contests were popular in the early days of cellular history. Each candidate submitted a detailed document outlining the characteristics of the service it promised to provide in the relevant spectrum band. Usually, one of the key features of the proposal was the extent of geographic coverage the candidate committed to achieve in a prespecified time period. Unlike what happens in auctions, the winner or winners of a comparative hearing usually received the spectrum band free of cost (other than the cost of preparing the application). By their very nature, beauty contests involved a subjective assessment of the quality of the applicants and of the service they promised to supply, which opened the door to legal challenges by the losers.

The ninety most populous MSAs were allocated in three rounds between June 1982 and March 1983. The number of applications grew considerably, from 194 in the first round to 567 in the third round. This growth was driven mainly by the rise in the number of nonwireline applicants: there were fewer than five such applicants per license in the first round and more than sixteen per license in the third round.

Overwhelmed by the number and volume of the applications it received, the FCC changed the procedure for future awards in October 1983: going forward, licenses would be granted by lottery rather than by beauty contests. The agency expected that the number and volume of applications would decline, and that the agency's decisions would not be subject to legal challenge by companies that failed to obtain a license. The introduction of the lottery, however, had an unexpected consequence: individuals and companies that would not have applied under a competitive hearing decided to apply under the lottery. Because winners could always resell their licenses in the secondary market that was developing quickly, the lottery created participation incentives for companies that never intended to operate a cellular system. Both the competitive hearings used earlier and the lottery that replaced them involved a windfall for the winners, since, leaving aside the cost of preparing the application, they received a valuable band of radio spectrum for free.

Between 1984 and 1986, the FCC received more than ninety-seven thousand applications for MSAs numbered ninety-one and higher. The FCC also relied on a modified lottery to award licenses in 428 rural statistical areas (RSAs), regions that included all counties that did not belong to an MSA. For the RSAs, the FCC received no fewer than three hundred thousand applications. It took the agency about four years to award cellular licenses in the 306 MSAs, and an additional two years to do the same in 428 RSAs.

In 1984 AT&T had been forced to divest local-telephone service: going forward, the company would focus on long-distance communications only, and seven Regional Bell Operating Companies were formed to supply local fixedline telephone services, each in a regional market.[13] As a direct consequence of the divestiture, AT&T was not able to enter the market for cellular services in the early 1980s: each of the operating companies set up an independent subsidiary in its own region to develop its own cellular network in competition with a nonwireline operator. In most of the top ninety MSAs, cellular service was launched in 1984 and 1985.[14] By 1991, cellular density in the United States was about 2.95, higher than in Japan but lower than in the Scandinavian countries.

NATIONAL CELLULAR MONOPOLIES UNDER A SHARED STANDARD: THE SCANDINAVIAN COUNTRIES

By the 1950s, the Swedish telephone system was run by Televerket, the national PTT.[15] Although Televerket did not have a legal monopoly on

2.2 A Motorola Dynatac 8000X ("Brick") cell phone. Released in 1983, it was priced at USD 3995 in the United States. Courtesy of AT&T Archives and History Center.

the provision of fixedline telephone services, in practice it operated as one. It had provided phone services of good quality at reasonable prices for decades. In fact, fixedline density in Sweden had been, since the beginning of the twentieth century, among the highest in the world.

Televerket started elaborating plans for mobile-telephony trials in the 1940s. The first (noncellular) system was installed in Stockholm in 1950 and remained in operation until the late 1960s. In 1961 Televerket started trials for a second (noncellular) system, formally inaugurated in 1965.

In 1964 a study group was formed at Televerket to examine the mobile-phone landscape. Three years later, the group produced the "Land Mobile Radio Communication" report, which surveyed all mobile-phone systems in operation around the world, suggested that in the future a frequency-reuse method would have to be developed, and proposed a cellular approach to improve the efficiency of spectrum usage (although the term *cellular* was not used). The report also recommended the development of a new mobile-phone network with national coverage. Laboratory work to develop such a system started at Televerket in 1968.

At the Nordic Telecommunications Conference in 1969, Televerket staff reported on the mobile-telephony research that had been conducted in Sweden. Mobile communications were immediately identified as an area for Nordic cooperation, and a joint working group, which came to be known as the Nordic Mobile Telephone (NMT) Group, was formed. In its first report, published in 1970, the NMT Group recommended that a pan-Nordic automatic mobile-phone system be developed, a process that would take about a decade, and that manual systems be implemented in the interim. (Automatic systems were those that did not require operator assistance.) Noncellular, operator-assisted mobile systems were soon set up in the Scandinavian countries. In Sweden, for example, a system—the third mobile, noncellular network in the country—was in operation between 1971 and 1987 and reached a maximum of twenty thousand users.

Between 1971 and 1975, the NMT Group worked on defining the features of a future pan-Nordic cellular network. The 1975 report from the NMT Group to the Nordic Telecommunications Conference included specifications of all the technical elements of the new system, with procedures for call hand-off and roaming. The main goal would be coverage rather than capacity, and consequently, frequency bands around 450

MHz were selected for implementation: it was a frequency range that would deliver long wavelengths and large cells, but limited capacity to serve areas with high population density. (By contrast, the American AMPS standard operated in frequency bands between 800 MHz and 900 MHz.) Commercial NMT 450 services started in late 1981 in Sweden and Norway, and in early 1982 in Denmark and Finland. NMT 450 was the first cellular system in operation in Scandinavia and in Europe more generally, and one of the first in the world.

The network grew rapidly after 1983, when people became fully aware that widespread coverage would be available, and that roaming across all Scandinavian countries would be feasible. In areas with high population density, such as Stockholm, NMT 450 reached its capacity limits quickly. In the mid-1980s, a new version of the NMT system (NMT 900) was inaugurated. Because it operated on a higher frequency band (900 MHz versus 450 MHz), it had shorter wavelengths and smaller cells, and was thus more appropriate for high-density areas.

Sweden was peculiar among the Scandinavian countries because it awarded a second license to Comvik, a private firm that, for several years, was the only private cellular operator in Europe. For several reasons, however, Comvik never acquired enough subscribers to challenge Televerket's leadership in the Swedish cellular market. Rather than adhering to the Scandinavian cellular standard, Comvik chose a proprietary technology for its network, which forced subscribers to use terminals that were incompatible with the prevailing standard and thus quite expensive. In addition, Televerket was both Comvik's competitor and regulator, which handicapped the private operator in two ways. First, Televerket granted Comvik a limited allocation of cellular frequencies, which restricted Comvik's opportunities for growth. Second, whereas Televerket's cellular users paid only for calls they made, Comvik's users paid for both calls they made and calls they received. This gave the company's customers incentives to keep their phones turned off, which limited usage.

The NMT system grew at a brisk pace, especially in the second half of the 1980s. By 1991, Scandinavia had reached a high level of cellular density—between 3.41 (in Denmark) and 6.59 (in Sweden), much higher than the levels achieved in the United States and Japan at the time.

A (TEMPORARY) NATIONAL CELLULAR MONOPOLY UNDER A NATIONAL STANDARD: JAPAN

The Ministry of Posts and Telecommunications, which oversaw the communications industry in Japan in the late 1970s, relied on a wholly government-owned public corporation, NTT, to introduce cellular service in the country in 1979. The system, which came to be known as NTT's high-capacity system (HCS) and operated in frequency bands around 900 MHz, was the first cellular network in operation anywhere in the world.[16] (The Chicago system set up in 1978 was a trial network.) Because NTT made little effort to commercialize the system, the early launch may have been driven more by the desire to do research on practical cellular systems than by an interest to profit from them. Service tariffs were high, as were terminal prices, all of which contributed to the slow growth of cellular density, especially in the first half of the 1980s.

The situation changed somewhat in 1986, when the ministry licensed two new service providers to compete with NTT: Nippon Ido Tushin (IDO), backed by Toyota and Nissan, and Daini Denden Inc. (DDI). Unlike NTT, however, neither company received a nationwide cellular license. IDO was awarded an operating license only in the Tokyo-Nagoya region, and DDI received a license for the rest of the country. Both were at a disadvantage relative to NTT. IDO decided to use the NTT standard for its network, whereas DDI chose to use the British Total Access Communications System (TACS) standard, modified to suit the Japanese spectrum allocation, thus creating a second standard, JTACS, in a portion of the country.

Competition put pressure on monthly tariffs, and cellular density grew at a faster pace. The number of cellular subscribers doubled in 1989 and almost doubled again in 1990. Nonetheless, Japan was not able to catch up to the Scandinavian countries: by 1991 cellular density in Japan was 1.12.

Figure 2.3 presents the evolution of cellular density in all the cellular pioneers (the United States, the Scandinavian countries, and Japan) through 1991. The figure shows that, by 1991, Scandinavia had the highest density, followed by the United States and then Japan.

In the mid-1980s, there was a second wave of standard creators that consisted of Britain, France, Germany, and Italy, and we focus on them next.

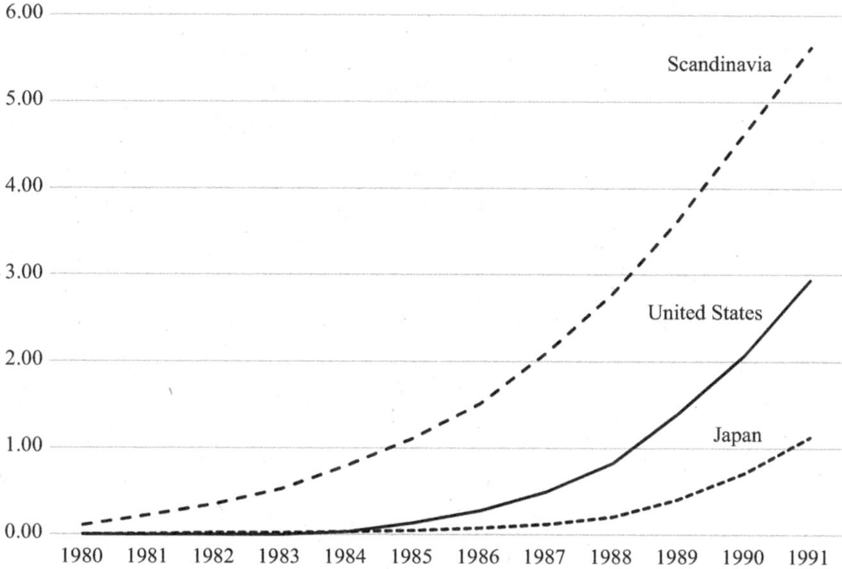

2.3 Cellular density in Scandinavia, the United States, and Japan, 1980–1991. *Source*: ITU database.

A NATIONAL CELLULAR DUOPOLY UNDER A NATIONAL STANDARD: BRITAIN

The evolution of the British cellular system in the 1980s was unusual in Europe because the British government introduced cellular competition from the moment cellular systems were launched. In most European countries at the time, the PTT was the monopolistic provider of cellular services.

In Britain, competitive cellular networks were launched in the broader context of liberalization of the country's telecommunications services. Under Prime Minister Margaret Thatcher, who dominated British politics in the 1980s, the right wing of the Conservative Party set economic policy.[17] The Thatcher policy for telecommunications had three goals: divesting telecommunications from the Post Office, introducing competition in the supply of terminal equipment and in the provision of network services, and privatizing the public provider of network services.

The government started implementing its transformational policies by means of the British Telecommunications Act of 1981. The act separated telecommunications from postal services and created a new state-owned corporation, British Telecom (BT), responsible for telecommunications. In addition to ending the terminal-equipment monopoly, the act permitted the secretary of state for the Department of Trade and Industry or BT itself to grant licenses for the operation of competitive fixedline networks. A consortium that included Cable & Wireless, Barclay's Merchant Bank, and British Petroleum established Mercury Communications, the first alternative carrier for long-distance services in Britain.

The government decided to sell a majority of BT's shares, and a second telecommunications bill was introduced in 1982 to carry out the privatization process. In 1985, 51 percent of BT's shares were offered to the public in what was, at the time, the largest stock sale in British history. At the same time, the government established the Office of Telecommunications, usually referred to as Oftel, a regulatory body in charge of supervising existing licensees and making recommendations on applications for new licenses. The Department of Trade and Industry, however, retained considerable authority in telecommunications matters, including issues related to spectrum allocation.

The government required that competitive cellular networks be introduced in the 1980s.[18] In 1982 the Department of Trade and Industry announced that it would award two cellular licenses. One of them would go to BT, on the condition that it collaborate with another company to set up an independent cellular subsidiary. BT partnered with Securicor, a leading private security company, to establish Telecom Securicor Cellular Radio, in which BT had a 51 percent stake (later increased to 60 percent). The second license would be open to all applicants. No standard was defined, but it was decided that cellular service would operate on the 900 MHz spectrum band.

Several of the principal British electronics firms—including GEC, Ferranti, and Plessey—applied for the second license, as did Racal, an innovative company focused on military communications, and Aircall, the only firm that had practical experience in the supply of (noncellular) telecommunications services. The government used a beauty contest to select the winner: the second license went to a consortium led by Racal, which had partnered with Millicom and Hambros Advanced Technology Trust.

It was decided that, rather than designing a completely new standard, an existing standard would be modified for the British market. The

American AMPS was chosen as the foundation for the British standard, primarily because AMPS equipment was available from a variety of suppliers and the standard operated at a frequency that was close to the 900 MHz band. The modified version of AMPS that was adopted in Britain was named Total Access Communications System (TACS): the crucial modifications made to AMPS were in the frequencies to be used (slightly higher in Britain than in the United States), in the number of channels that would be available, and in the spacing between channels.

The firms receiving cellular licenses—Telecom Securicor Cellular Radio, which traded under the Cellnet name, and Racal-Millicom, which traded as Vodafone—would have to meet three conditions. First, service would have to be provided to 90 percent of the British population by the end of 1989; second, operators would not be allowed to manufacture equipment or to provide value-added services; and third, services on the networks would be sold only through service providers (or airtime resellers). Cellnet, which chose Motorola as its equipment provider, conducted a limited test of its cellular network in mid-1984, whereas Vodafone, which went with Ericsson, started supplying services in early 1985.

Intense competition between Cellnet and Vodafone led the operators to invest heavily in marketing and promotion, which increased market awareness of the nature and scope of cellular services. Additionally, because the licensees were prevented by the terms of the license from selling services directly to subscribers, they were forced to rely on the service providers, which became a powerful marketing and sales force. Further, since there were plenty of manufacturers with experience in the production of terminals for the AMPS standard, and the AMPS terminals worked in Britain, the TACS terminal market quickly became competitive. By 1988 there were more terminal models approved for the British market than for any other cellular market in Europe, which led to rapid declines in terminal prices.

The pace of cellular growth in Britain was brisk. By 1991 there were 1.2 million cellular subscribers in the country, the equivalent of about 2.2 percent of the population. This was a remarkable achievement for a cellular system that had been launched relatively late in the decade, and it compared favorably to the cellular density achieved by most other European countries that inaugurated their systems in the mid-1980s, including Germany, Italy, and France.

NATIONAL CELLULAR MONOPOLIES UNDER NATIONAL STANDARDS: GERMANY, ITALY, AND FRANCE

The remaining members of the second wave of standard creators were France, Germany, and Italy, and they shared several features. First, in each country the domestic PTT was a monopolistic supplier of cellular network services.[19] Second, each of these countries developed its own cellular standard, incompatible with all systems in operation outside the country. Third, in each case the standard was developed in cooperation with a powerful local electronics supplier. And finally, even though in the early years terminal prices and service tariffs were high, the cellular capacity deployed at the beginning was not enough to satisfy demand.

Germany, Italy, and France are examples of countries where "industrial policy" reasons dominated over other considerations. Governments attempted to facilitate the growth of a domestic cellular-products industry in each country, and they viewed closed standards as the right approach for achieving this goal.

GERMANY

Siemens, the main provider of telecommunications equipment to the Deutsche Bundespost, started testing cellular systems in 1978.[20] By 1984 it had a complete set of cellular network products available, and the Bundespost, the domestic PTT, decided to launch a network based on the Siemens design. The system, known as Netz-C, was inaugurated in 1985 and relied on Siemens's C450 standard.

In the early years, Siemens dominated the terminal market with a share of around 45 percent, and terminals were expensive. Moreover, the PTT set cellular tariffs at high levels, only exceeded in France, Luxembourg, and Spain, which stunted cellular growth. Nonetheless, by 1988 there were capacity problems, and the PTT was forced to redesign the system—first by adding new channels in existing cells, and later by increasing the number of cells by cell splitting. In relative terms, cellular growth in Germany was slow, even after the system redesign: by 1991 there were about 532,000 subscribers, which represented only 0.67 percent of the population.

ITALY

In Italy, the Società Italiana per L'Esercizio delle Telecomunicazioni (SIP), the domestic PTT, and Italtel, a large domestic telecommunications supplier, collaborated to introduce a cellular system in 1985.[21] SIP was the state-owned company that operated local and long-distance fixedline telephone services, and had earlier operated a noncellular mobile-phone system in the country. When the Italian cellular system was inaugurated under the Radio Telephone Mobile System (RTMS) name, Italtel supplied all the infrastructure and many of the terminals.

For several reasons, RTMS was handicapped from birth. In the early years, only five digits were used to address cellular phones. The system retained the concept of calling areas, so that, for a call to be completed, the caller had to know where the cellular user receiving the call was located. Further, terminals were manufactured only by Italian companies, and SIP was the exclusive supplier through which terminals were rented to subscribers. Despite all these limitations, in the large Italian cities the system reached capacity in just a few years, and cells were redesigned to accommodate more users.

The Italian market underwent a transformation in 1990, when the PTT decided to install a TACS network created by a consortium led by Ericsson. The TACS system introduced in Italy benefited from the experience accumulated in prior years in other countries: it was reliable, and there were plenty of international manufacturers producing terminals for it.

After the introduction of the TACS network, terminal prices collapsed, and tariffs declined. The number of cellular subscribers jumped from about 66,000 in 1989 to about 568,000 in 1991—that is, from 0.12 percent of the population to roughly 1 percent.

FRANCE

Telecommunications services in France were operated by the Direction Générale de Télécommunications (DGT), the domestic PTT, which had launched its first mobile-telephone service in the 1970s, and a second, Radiocom 200, in 1982, both noncellular systems. In 1981 DGT started planning for a third network, which eventually came to be known as Radiocom 2000, the first to operate on a cellular configuration.[22]

The development contract for Radiocom 2000 was awarded to Matra, a leading French aerospace and military electronics company that had been responsible for developing the Radiocom 200 system. Radiocom 2000, which started operations in 1985, was a peculiar mobile network that mixed features of a cellular network with those of a private mobile-phone network. For example, hand-over of calls from cell to cell was considered in the system's technical specifications but was not implemented for many years. In addition, the operating frequencies were so low that the kind of frequency reuse that was typical of cellular networks at the time was infeasible in the French system.

Radiocom 2000 was made up of two systems: a dispatch radio service that offered optional communication with the fixed-telephone network, and an automatic mobile telecommunications network without call hand-over from cell to cell. The tariff structure was especially complicated: prices depended on the type of service and on whether the user wanted national or regional coverage. At the time of launch, Matra was the exclusive supplier of terminals for the system, although later other vendors entered the market.[23]

Because congestion quickly became an issue on the Radiocom 2000 system, in mid-1987 the French telecommunications regulator invited bids for the operation of a second network. The license was awarded to a consortium that launched the Société Française du Radiotelephone (SFR) system in 1989. For this network, the consortium relied on a modified version of NMT 450, with the modifications significant enough that the Scandinavian terminals became unusable in France.

Contrary to expectations, the SFR tariffs were not much lower than those that prevailed on the Radiocom 2000 network, and this was due to an issue that would reappear often in the history of cellular networks: the interconnection charges imposed by the owner of the fixedline network. In the French case, DGT—renamed France Télécom in 1988—charged high fees to the SFR consortium for interconnection and leased lines, which forced SFR to charge high cellular tariffs to its customers.

Despite these limitations, the introduction of a second cellular network in France accelerated subscriber growth. In 1988, before the SFR launch, there were about 98,000 cellular subscribers in the country, which represented 0.17 percent of the population. By 1991, two years after the SFR

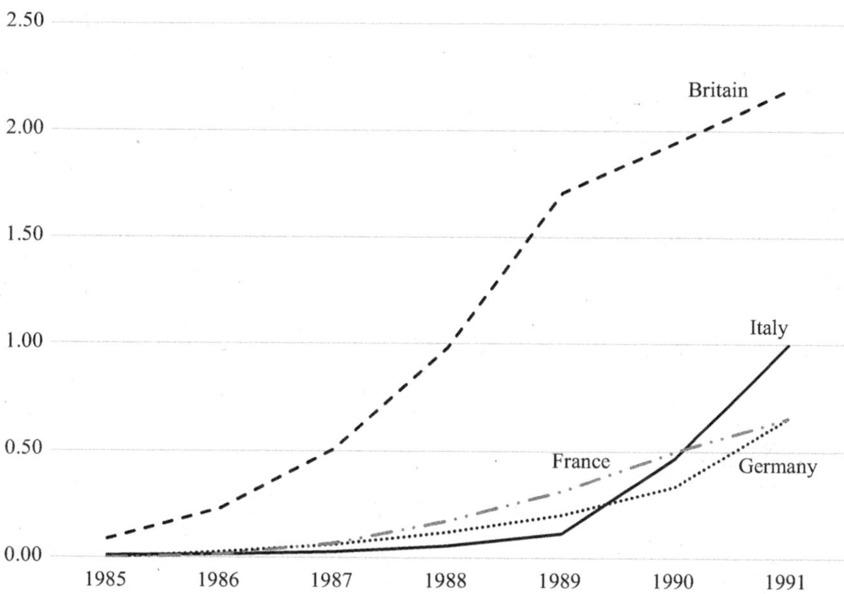

2.4 Cellular density in Britain, Germany, Italy, and France, 1985–1991. *Source*: ITU database.

inauguration, the number of subscribers had grown to 375,000, or about 0.64 percent of the population.

Figure 2.4 compares cellular density among the members of the second wave of standard creators.

Two facts stand out in the figure. First, Britain—the only country that introduced competition from the beginning—clearly outperformed all other second-wave standard creators. Second, the introduction of a second network in France and Italy in the late 1980s accelerated the growth of cellular density in both countries.

CELLULAR-STANDARD IMPORTERS AROUND THE WORLD

As pointed out earlier, during the 1G era, there were many more standard importers than creators. Table 2.2 presents standard creators and importers around the world during the 1980s.

The table shows that, during the 1980s, there were standard importers all over the world, but they were unevenly distributed. There were

Table 2.2 Standard creators and importers, 1981–1989

Standard	**1981**	**1982**	**1983**	**1984**	**1985**	**1986**	**1987**	**1988**	**1989**
NMT	*Sweden* *Norway* Saudi Arabia	*Denmark* *Finland* Spain		Austria	Luxembourg Netherlands Oman Tunisia Malaysia	Iceland Turkey Thailand Indonesia	Belgium Morocco Switzerland	Cyprus	France Algeria Cambodia Faroe Islands
AMPS			*United States*	Canada South Korea		Australia Virgin Islands Israel	Cayman Bermuda Dominican Republic Singapore New Zealand Thailand Bahrain	Venezuela Zaire	Chile Curacao Argentina Antigua Saint Kitts Mexico Dutch Antilles Brunei Taiwan
TACS				Hong Kong	Ireland *Britain*			Macao China	Sri Lanka Japan Malaysia Taiwan Mauritania United Arab Emirates
Other	*Japan (1979)*				*Germany* *France* *Italy*	South Africa			Portugal

*Standard creators are in italics

Source: J. Funk, *Global Competition Between and Within Standards: The Case of Mobile Phones* (Houndmills, UK: Palgrave, 2002), 43

large nuclei of standard importers in Western Europe, the Middle East, the Asia-Pacific region, and the Caribbean. By contrast, very few countries launched cellular systems in Africa, Latin America, and the Asian continent outside the Asia-Pacific region.

The table reveals certain patterns about the standards that were imported in different regions of the world. First, the two most frequently imported standards were the Scandinavian NMT and the American AMPS, with the British TACS a distant third. Second, the Japanese, French, and Italian standards were not adopted outside their home countries, and the German standard was only imported by South Africa and Portugal. Third, all Western European countries that imported a standard adopted one created in Europe—in most cases, the Scandinavian NMT. Fourth, all Caribbean and Latin American countries that imported a standard adopted the American AMPS standard, and the same was true about countries that had historical connections with the United States, such as Canada, South Korea, and Israel.

Other regions of the world, such as Africa, the Middle East, and Asia (including the Asia-Pacific countries) imported a variety of standards. For example, some countries in the Asia-Pacific region (Indonesia, Malaysia, and Thailand) adopted the Scandinavian NMT, others (Singapore and Taiwan) imported the American AMPS, and at least one (Hong Kong) adopted the British TACS.[24]

The standard importers of the 1980s shared some economic and demographic features. Most of them were among the wealthiest countries on the continent where they are located: this applies to the few Latin American countries that introduced cellular phones in the 1980s, such as Argentina, Mexico, and Venezuela, and to the few African countries that did, such as South Africa, Morocco, Algeria, and Tunisia.

Further, many standard importers—including most of the Western European and Middle Eastern countries, and some of the Caribbean nations—were among the richest countries in the world at the time, and still are. In addition, many standard importers in the Asia-Pacific region—and especially Hong Kong, Singapore, Taiwan, and South Korea, the so-called Asian Tigers—were among the fastest-growing economies in the world. Finally, the Caribbean nations that imported a cellular standard in the 1980s are small countries, and thus relatively easy for cellular operators to cover.

The table reveals that not all standards were created equal.[25] Four countries—Germany, France, Italy, and Japan—viewed cellular-standard creation as an opportunity to do "industrial policy." National governments in these countries entrusted the domestic electronics "national champion"—Siemens in Germany and Matra in France, for example—with the task of developing the standard as a proprietary solution, which would potentially generate an increase in exports, or so they hoped. The national governments of the Scandinavian countries, the United States, and Britain, on the other hand, conceived of their own cellular standards as "open"—that is, as ones that would be freely available to any equipment manufacturer willing to create products that adhered to the standard.

Differences between closed and open cellular standards translated themselves into cross-country differences in the adoption of both cellular phones and cellular standards. Cellular phones were adopted at a faster pace in countries with open cellular standards than in those with proprietary standards. Among the first-wave standard creators, Scandinavia and

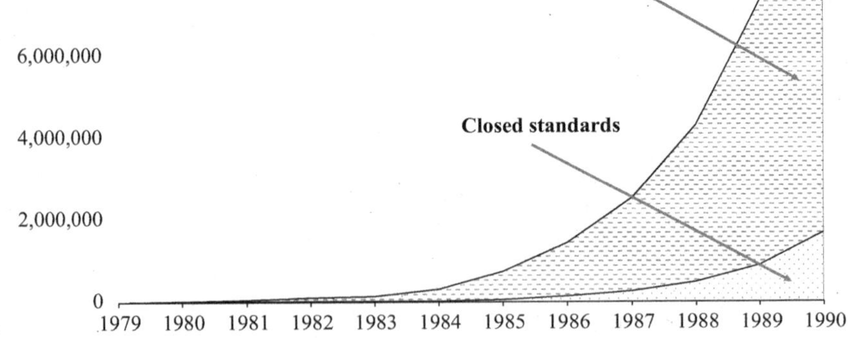

2.5 Adoption of closed and open cellular standards, number of subscribers, 1979–1990. *Source*: Calculations by the authors from information in the ITU database.

the United States outperformed Japan. Among countries in the second wave, Britain outperformed all others.[26] This was so because open standards facilitated competition among equipment manufacturers, which led to declining prices for terminals and infrastructure, which in turn led to faster adoption. For the same reason, open cellular standards such as NMT 450 and AMPS were heavily adopted outside the countries where they were created, whereas closed standards were not.

Figure 2.5 shows the extent to which open and closed cellular standards were adopted across the world (in number of subscribers). Open standards include AMPS, NMT, and TACS; closed standards include the proprietary standards developed in Japan, Germany, France, and Italy.

The figure shows the dominance of open standards. By 1990 about 85 percent of world cellular subscribers were using mobile phones that adhered to open standards.

3

COMPETING IN THE EARLY CELLULAR MARKETS

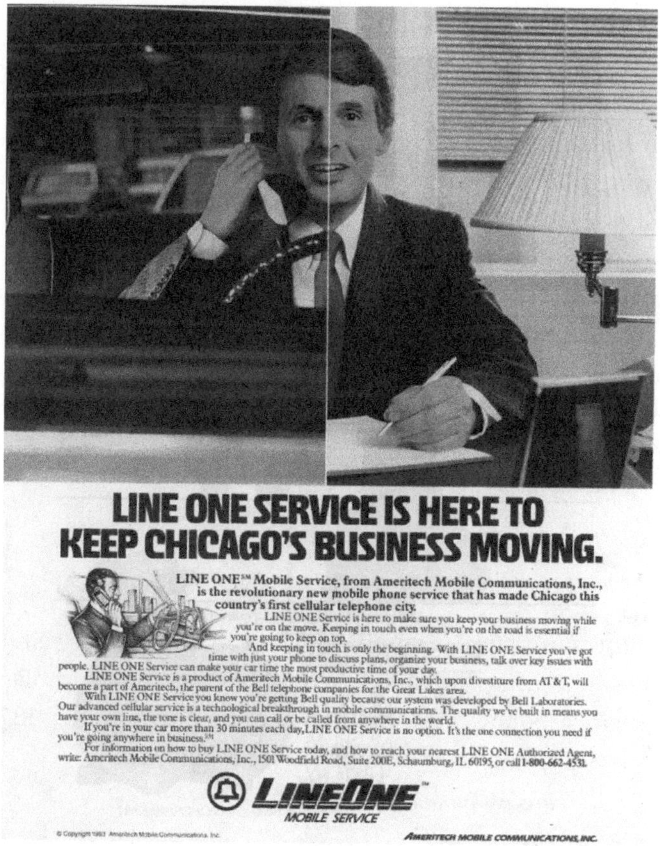

3.1 A 1983 advertisement promoting Ameritech mobile communications, the first company in the United States to provide cellular mobile phone service to the public. Courtesy of AT&T Archives and History Center.

In the early history of the cellular industry, organizations in three different markets—network services, terminals, and network infrastructure—interacted to deliver cellular services. Firms cooperated and coordinated with one another across markets while at the same time fending off competitors within each market.

This chapter focuses on competition in the early cellular markets. We introduce the major players and study the nature and dynamics of competition in each market. We examine how competition shaped the prices and quality of products and services.

Three companies became key equipment suppliers during the first era of cellular evolution (and beyond): Ericsson, Nokia, and Motorola. We introduce them first, highlighting their similarities and differences. Then we examine how companies competed in the market for network infrastructure and in the market for terminals. Although individual companies may have had advantages in their own domestic markets, including those arising from government actions, the markets for cellular terminals and infrastructure were global.

In the second half of the chapter, we focus on competition in cellular-network services in the two largest markets where such competition existed during the 1980s—Britain and the United States. Here we examine specific national markets, such as Britain, and regional markets, such as those that existed in the United States in the 1980s. Toward the end of the chapter, we address the question whether competition had an impact on service prices and cellular-phone adoption.

THE BIG THREE: ERICSSON, NOKIA, AND MOTOROLA

Three companies played a crucial role in the markets for cellular handsets and cellular infrastructure: Ericsson, Nokia, and Motorola. Here we explore their corporate history in the precellular period, highlighting the strategic decisions that led them into cellular markets.

ERICSSON

Lars Magnus Ericsson (1846–1926) founded his eponymous company in Stockholm in 1876.[1] At first the firm was an electromechanical workshop that specialized in repairing electrical equipment, but it soon transitioned to designing its own telegraphy equipment and later to manufacturing telephony equipment.

From the late 1870s, L. M. Ericsson and Company grew both in its home market and internationally. In Sweden, Ericsson supplied switching equipment to the local postal, telegraph, and telephone (PTT) administration and to private competitors. In 1881 the company started its international expansion with exports to Russia and Norway. By 1894 the firm was exporting to China, and by 1902 it had an office in the city of New York and a factory in Buffalo, in upstate New York. This process continued throughout the twentieth century, with interruptions during the world wars. A key driver of the firm's international forays was Ericsson's Telephone Exchange Division, also known as X, which helped the company become one of the core players in the worldwide telephone-switching market.

Telephone-switching technology evolved considerably in the course of the twentieth century.[2] By the beginning of World War I in 1914, most switching was done manually via plug-and-jack switchboards. Although the American entrepreneur Almon Strowger had invented an automatic switching device in the late nineteenth century, by 1915 only about 4 percent of calls in the United States involved automatic switching.

The movement from manual to automatic switching did not take off until after World War I, and it entailed more than just the adoption of the Strowger device. The Strowger switch, creative as it was, had severe limitations. In it, the switches were directly controlled by the users. When a user dialed, say, the number two, two pulses were transmitted to the switch, a part of which moved two positions either horizontally or vertically. The switch only operated as fast as the digits were dialed, and it remained occupied by that user for the duration of the call.

The crossbar switching system, installed for the first time in the late 1930s, overcame the limitations of the Strowger switch. The crossbar switch was a "central control" made up of relays (that is, mechanical switches controlled electrically). When the user dialed a number to start a call, the crossbar switch received the number, stored it, searched for a line not in use at the time, and activated several switches to make the connection. The crossbar switch separated control of the switching process from the switches themselves.

Starting in the late 1940s, progress in semiconductor and computing technology facilitated the transition to a new phone-switching technology. In the late 1960s both Ericsson and Televerket, the Swedish PTT, developed versions of the stored program control (SPC) switch, a technology in which a digital computer controlled the telephone exchange. In 1968 Ericsson installed its first SPC system in Tumba, not far from Stockholm.

Ericsson and the PTT, however, were concerned about the high research and development (R&D) costs of SPC projects and decided to pool their resources. In 1970 Ericsson and TELI, Televerket's wholly owned manufacturing arm, formed the ELLEMTEL joint venture to carry out R&D in switching technologies, especially digital switching.[3] One of ELLEMTEL's first projects, named AX, was a proposal for an SPC local exchange system that would provide cost benefits over its predecessors. By 1975 the joint venture had developed what would become Ericsson's flagship switching product, the AXE system, which consolidated the company's position as a leading competitor in the global fixedline switching market. The AXE system, with some modifications, would also provide the foundation for Ericsson's entry into the market for cellular switching systems in the 1980s.

Expertise in switching technology was just one of the avenues that led Ericsson into mobile phones. Very early on, in 1919, the company cofounded a small firm called Svenska Radio Aktiebolaget (SRA), which provided Ericsson with a second entry path into mobile phones. SRA was set up by Ericsson and two other Swedish companies.[4] Soon, the British Marconi company became a part owner as well, which allowed SRA to use some of the Marconi patents in the Swedish market. Initially, the firm made radio transmitters and radio receivers for home use, and later it explored other applications of radio technology, both civil and military.

In the 1940s SRA entered the business of land mobile radio, supplying radio equipment to trucking companies, police forces, and the military. In the early 1960s SRA incorporated paging systems into its portfolio and started selling some of its products outside Sweden. Although the company grew consistently throughout the 1960s and 1970s, by the early 1980s, when a cellular network was launched in Sweden, SRA was still a small part of Ericsson's overall business, which focused on selling transmission and switching equipment to telecommunications operators around the world. In 1981, as the Nordic Mobile Telephone (NMT) standard was about to take off in Scandinavia, Ericsson gave SRA full responsibility for the development of cellular-phone capabilities. The next year, SRA became Ericsson Radio Systems.

By the beginning of the 1980s, Ericsson had acquired expertise in two important areas of telephone technology: switching and precellular mobile radio. Both would play a crucial role in paving Ericsson's way into cellular-phone markets.

NOKIA

Although Nokia and Ericsson both eventually became leading Scandinavian multinational corporations in telecommunications, their histories are different. Ericsson entered the telecommunications business, and became a powerful multinational, much earlier than Nokia. By the early 1980s, when cellular phones arrived on the scene, Ericsson already had a long history of dealing with the national PTTs and continued having them as its core customers during the early cellular era. Nokia, which lacked such a history, focused on non-PTT customers. Although both had expertise in the design and manufacturing of infrastructure and terminals, Nokia concentrated on terminals (and used them as an opportunity to sell infrastructure), whereas Ericsson focused on infrastructure (and used it as a channel to sell terminals).[5]

Nokia was founded by Fredrik Idestam (1838–1916), an entrepreneur, and Leo Mechelin (1839–1914), a politician. In 1863 Idestam traveled to Germany to observe how pulp was manufactured and, after returning to Finland, built his own mill in Tampere, not far from Helsinki. The history of Nokia, from its founding in 1865 through the end of the first cellular era in the early 1990s, has three well-defined periods. From 1865 through 1910, the company focused on paper and pulp manufacturing. Between 1910 and 1967, Nokia was made up of three different companies: the forestry business, the Finnish Rubber Works, and the Finnish Cable Works. The third period started with the merger of the three companies into one in 1967.

Nokia's forays into electronics started slowly in the mid-twentieth century.[6] Toward the end of the 1950s, the Finnish Cable Works established an electronics department that originally marketed computers from foreign manufacturers. The company also set up a data-processing center that provided services to some of the leading Finnish companies. Additionally, the electronics department of the Finnish Cable Works began manufacturing electronic equipment, including instruments for nuclear power research and various devices for military purposes. In 1971 Nokia launched its own computer R&D program.

Nokia engineers started researching digital phone exchanges in the 1970s, and this research led to an agreement with France's CIT-Alcatel—a world leader in telephone-switching technology—that granted Nokia access to advanced knowledge in the field. The move into telephone exchanges made sense for Nokia in the 1970s: switching markets were much larger than transmission-equipment markets (cables), and switches were becoming computer controlled and fully electronic, areas in which Nokia had accumulated expertise. There was also the belief that telephone operators preferred to do business with companies able to offer an integrated package of transmission and switching equipment.

Before the 1960s, CIT-Alcatel, Ericsson, and Siemens shared the telecommunications-infrastructure market in Finland.[7] This started changing slowly in the early 1960s. The first step in the transition was the birth, in 1962, of Televa Oy, a small R&D unit within the Finnish PTT created to focus on telecommunications. In the 1970s both Televa Oy and the Nokia electronics department conducted research on telephone switching, at a time when switching technology was evolving rapidly from purely electromechanical to crossbar and later to electronic and computer-controlled systems.

In the mid-1970s the Finnish Social Democratic Party attempted to reorganize the domestic electronics industry into a state-owned conglomerate. The plan failed, and Televa Oy, an integral part of the plan, ran into financial difficulties. Nokia intervened and lobbied for an arrangement with a smaller dose of state participation. Its intervention led to the birth of Telefenno Oy, a 50-50 joint venture between Televa Oy and Nokia established with the purpose of designing and marketing telephone exchanges. Telefenno Oy combined the switching know-how that Nokia had acquired through its CIT-Alcatel connection with that developed internally by Televa Oy. Nokia acquired a majority stake in Televa Oy three years later and ended up buying out the Finnish state in 1987.

Although early on Telefenno Oy conducted research on both analog and digital switching technologies, by the early 1980s the joint venture was focused on fully digitized switching systems and had started out-competing foreign suppliers. Telefenno Oy developed the DX 200 switch, which laid the foundation not only for Nokia's participation as a strong competitor in the world fixedline switching market but also for the company's entry into the design and manufacturing of cellular switches.

Nokia itself, however, entered the cellular business mainly through the supply of terminals. The Finnish precellular mobile-phone business took

3.2 The Nokia 101, introduced in 1992. It was one of the last handsets of the analog era. With its "candy bar" form factor, it claimed to be the world's most portable phone. Courtesy Nigel Linge and Andy Sutton.

off in the 1960s, when the military requested bids for radiophones. Various electronics manufacturers started conducting research in wireless communications, and Nokia introduced its first mobile phones in the mid-1960s. In the early days, the market was made up purely of public services, including police, firefighters, and railways, and it then expanded, in the early 1970s, to encompass a system of car phones for civil purposes.

In 1975 Nokia signed a cooperation agreement with Salora, Finland's leading radio- and television-equipment manufacturer. Under the terms of the partnership, Nokia would supply Salora with base stations and links, whereas Salora would provide Nokia with car telephones. Nokia and Salora set up Mobira in 1979, a company that would soon become an important player in the market for cellular terminals. In the early 1980s Nokia ended up fully acquiring Salora and thus absorbing the latter's mobile-phone expertise.

MOTOROLA

The brothers Paul Galvin (1895–1959) and Joseph Galvin (1899–1944) founded Motorola in Chicago in 1928 as the Galvin Manufacturing Co.[8] In that year, the brothers bought the bankrupt Stewart Battery Company's battery-eliminator manufacturing facilities. Battery eliminators, which enabled battery-powered radios to operate on household electricity, quickly became obsolete, and the brothers searched for other opportunities. They found them in a newly emerging consumer market: car radios for entertainment, an incipient business that was usually done on a custom basis. Motorola managed to take AM radio devices, which at the time were usually installed in the living room of a family home, and mass produce them for automobiles.[9]

The car-radio-for-entertainment business was not only new but also seasonal: workers here hired in the early spring and laid off at the end of the summer. The Galvin brothers realized that they needed to explore a new business line to smooth out revenue and employment fluctuations over the course of the year. In the late 1930s they started developing two-way mobile radio devices for police departments in the state of Illinois. This was a different market: whereas car radios for entertainment were a consumer business, two-way radio devices installed on police cars allowed police departments to track crimes in progress and more effectively pursue criminals on the run.[10]

In 1940, as the police car radio business was growing, Paul Galvin managed to recruit Dan Noble, a professor at the University of Connecticut who had already designed a prototype of an FM mobile-radio system for the Connecticut State Police. Noble soon became one of the pivotal figures in the evolution of what turned into Motorola's Mobile Communications (COMM) Division.

During World War II, Motorola entered a third market: it started designing and manufacturing portable radios for the US military, including the SCR-536 AM two-way radio, the "Handie-Talkie," and later the first portable FM two-way radio, the SCR-300 backpack radio also known as the "Walkie-Talkie." Both played an important role in facilitating the intervention of the American armed forces in the war. These products also helped Motorola become the leader in the market for commercial two-way mobile communications after the war.

As the end of the war approached, Motorola started designing and manufacturing products for civilian use. After the war, the company regained its leading position in car and home radios for entertainment and began manufacturing a new consumer product: television sets. In addition, the company explored a variety of new markets, including semiconductors, radio equipment for NASA space flights, and microwave telecommunications for businesses.

In the 1950s Motorola's COMM Division started working on a product line that would underpin the future growth of the company: (precellular) mobile phones. In the early 1960s AT&T started developing the Improved Mobile Telephone System (IMTS), a network of mobile-phone services that would offer nationwide coverage, would be fully automatic, and would enable users to interconnect with the public switched telephone network. Motorola competed with General Electric to supply mobile devices for IMTS and won the contract; by the mid-1960s the company had a 98 percent share of the IMTS market.

In 1968 the Federal Communications Commission (FCC) began entertaining the idea of assigning a block of frequencies to cellular-phone systems, with the idea that AT&T would be the only company authorized to provide such services. Motorola took the position that the radio common carriers should not be excluded from the new spectrum allocation. In mid-1972 the COMM Division proposed a cellular system to the FCC and started developing an initial version. In April 1973 Martin Cooper of Motorola telephoned Joel Engell of Bell Laboratories from Sixth Avenue in New York City using a handheld prototype DynaTAC (Dynamic Adaptive Total Area Coverage) telephone. This call is frequently described as the birth of the cellular phone, even though no full-fledged cellular networks were in operation at the time anywhere in the world.[11]

The famous Cooper "cellular call" to Engell was an engineering feat that Motorola pulled off in a matter of months. At a time when mobile phones were bulky contraptions set up in automobiles, Motorola was able to design and build a prototype handheld portable phone. Cooper himself has explained that the phone he used in 1973 combined mobility and portability in a way that was radically new at the time. He has also described the 1973 demonstrations as designed to show the FCC that Motorola had unparalleled expertise in wireless communications and to convince the FCC that AT&T should not be given a monopoly in cellular systems. To make the "cellular" demonstrations possible, Motorola engineers installed base stations operating at 900 MHz and adapted a telephone switch to connect the new portable phones to the AT&T phone network. With this setup, a call initiated by Motorola staff on the East Coast could reach anybody anywhere in the world.[12]

In 1977 the FCC granted a license to Motorola to install and demonstrate an experimental cellular system in the Washington-Baltimore area, and another one to Illinois Bell to set up a cellular system in Chicago. (A third experimental license was awarded to Millicom for the Raleigh-Durham area.) Motorola became the largest supplier of mobile telephones not only for its own system but for the Illinois demonstration as well.

COMPETING IN HANDSETS AND INFRASTRUCTURE

The markets for cellular infrastructure and cellular handsets were differentiated-product markets: thus, customers chose products not based on prices only but rather in light of the combination of prices and features that each product offered. But there was an additional factor that mattered: the long-term relationships between technology manufacturers and cellular carriers, usually established for the first time when both groups of players participated in standard-setting committees. Sometimes, long-term relationships between equipment vendors and operators were established even before the advent of cellular, when equipment suppliers provided infrastructure equipment to fixedline operators. Competition based on product prices and attributes has been labeled "market-based competition," and the impact of long-term relationships originally established in committees (or otherwise) has been described as "committee-based competition."[13]

Cellular products adhered to standards (and still do), and many of those standards were developed by committees in the context of standard-setting organizations. Thus, equipment vendors had incentives to participate in such organizations not only to learn about the technology being developed (and influence the development process) but also to establish the sort of connections with cellular carriers that would eventually lead to equipment sales. Carriers and equipment manufacturers participated in committees, and this led to the joint development of technology and to long-term relationships between the former and the latter.

Long-term relationships between technology manufacturers and carriers were more important when products were more complex and less important when markets were more fragmented. Thus, they played a more decisive role in infrastructure than in handsets because the latter were simpler products. They also played a role of lesser importance in the United States, where there were many regional markets (as opposed to just one of national scope), and where two cellular carriers competed in each regional market.

COMPETING IN NETWORK INFRASTRUCTURE

In the first decade of cellular history, six manufacturers dominated the global market for cellular-network equipment: Motorola and AT&T from the United States, Nokia from Finland, Ericsson from Sweden, Northern Telecom from Canada, and Nippon Electric Company (NEC) from Japan.[14] The type of cellular-equipment expertise these companies accumulated during the 1980s varied considerably. Ericsson managed to successfully produce equipment for a variety of standards, whereas AT&T,

Northern Telecom, and Nokia focused on one standard only (AMPS in the case of AT&T and Northern Telecom, and NMT in the Nokia case).

Table 3.1 shows the share of each standard in the total number of contracts awarded to each of the major firms for the provision of analog cellular systems up to 1991. The standards included in the table are the American Advanced Mobile Phone Service (AMPS), the Japanese Nippon Telegraph and Telephone High-Capacity System (NTT HCS), the British Total Access Communications System (TACS), and the Scandinavian Nordic Mobile Telephone (NMT) system. The table shows how diversified Ericsson's expertise was in comparison with, say, AT&T's, Nokia's, and Northern Telecom's.

FIRM CAPABILITIES IN NETWORK INFRASTRUCTURE

Two industrial capabilities helped firms win cellular-infrastructure contracts: radio-related R&D and know-how in fixedline switches.[15] Radio-related research was important in the design of cell-site equipment, especially base-station equipment. During the 1980s, radio technologies evolved at a steady pace, and companies that invested in radio-related R&D were able to manufacture equipment that was smaller, consumed less power, and transmitted a higher-quality signal.

Table 3.1 Shares of various analog standards in total contracts for key infrastructure manufacturers, 1991

	AMPS	NTT HCS	TACS	NMT	Total
AT&T	100	0	0	0	100
Ericsson	30	0	32	38	100
Motorola	75	0	23	2	100
NEC	47	13	40	0	100
Nokia	0	0	0	100	100
Northern Telecom	100	0	0	0	100

Source: US International Trade Commission, *Global Competitiveness of U.S. Advanced-Technology Industries: Cellular Communications* (Washington, DC: US International Trade Commission, 1993), 5–18.

The accumulated experience in developing, manufacturing, and marketing fixedline switching equipment was also crucial for competing in this market. Companies such as AT&T, Northern Telecom, Ericsson, Nokia, and NEC, which had such know-how, were able to transfer much of it to cellular switches. Synergies between fixedline and cellular technologies enabled these firms to develop high-quality, large-capacity cellular switches that were also competitive in price.

COMPETING IN NETWORK INFRASTRUCTURE: PRICES, ATTRIBUTES, AND LONG-TERM RELATIONSHIPS

In the analog cellular markets of the 1980s, there was substantial continuity between fixedline and analog infrastructure orders. In most countries, the analog service providers were the PTTs, and they tended to order cellular-infrastructure equipment from the same vendors that had previously supplied them with fixedline infrastructure. For example, NTT relied on NEC, the Scandinavian PTTs ordered from Ericsson and Nokia, Bell Canada selected Nortel, Deutsche Telekom bought switches from Siemens, France Télécom selected Matra for base stations and Alcatel for switches, and the Società Italiana per L'Esercizio delle Telecomunicazioni, the Italian PTT, ordered from Italtel.

To a lesser extent, the principle applied in the United States as well, where the Baby Bells—the Bell Operating Companies that arose from the 1984 AT&T breakup—tended to buy analog-cellular infrastructure from Western Electric (the manufacturing arm of AT&T, renamed Lucent in the mid-1990s). Western Electric had less success with cellular carriers that were not part of the Baby Bell system, since these companies had not purchased fixedline equipment from Western Electric in the precellular world. Western Electric had even less success with the radio common carriers, which were absolute newcomers to the telecommunications world and had never purchased fixedline infrastructure. When they entered American cellular markets, the radio common carriers acquired infrastructure equipment from Motorola and Ericsson.

The role of product prices and features was particularly important in the United States (relative to long-term relationships) because the second service provider in a regional market—the nonwireline cellular carrier—usually tended to buy equipment from a vendor other than that used by the wireline carrier. More generally, the greater fragmentation of the US market—many regional markets and two competitors in each market—contributed to enhancing the role of product prices and attributes. As the 1980s went by, four companies emerged as the top suppliers of analog-cellular infrastructure in the United States: Western Electric, Motorola, Nortel, and Ericsson.

Motorola and Ericsson were particularly successful in international AMPS and TACS infrastructure markets—that is, in countries other than those where the standards originated. In such countries, competition based on product prices and attributes was crucial, and local service providers tended to trust companies that had accumulated experience designing and manufacturing equipment for the standard in the countries of origin. Motorola and Ericsson were able to create a large global customer base from the early years of cellular telephony because they had been successful in developing equipment for the first (open) analog standards—NMT, AMPS, and TACS—when the standards were being developed. Many of the other firms—including Siemens, Alcatel, Matra, and Italtel in Europe, and Western Electric and Nortel in North America—prospered less well because they remained too focused on their existing customers, usually the national PTT in each country.

Figure 3.3 shows the global shares of network-infrastructure providers in 1990. The figure shows the role played at the time by three firms: the Swedish Ericsson and the American Motorola and AT&T. These three vendors combined had about 75 percent of the world cellular infrastructure market. Among the smaller firms (included in the "other" category) were the Japanese Toshiba, the Canadian Northern Telecom, the German Siemens, and the Finnish Nokia.

COMPETING IN CELLULAR TERMINALS

During the 1980s several companies developed expertise in the design and manufacturing of cellular terminals.[16] In 1990 the top five firms in this market were Motorola from the United States, Nokia from Finland, and Matsushita, Mitsubishi, and NEC from Japan. Figure 3.4 shows the world share of the main terminal manufacturers.

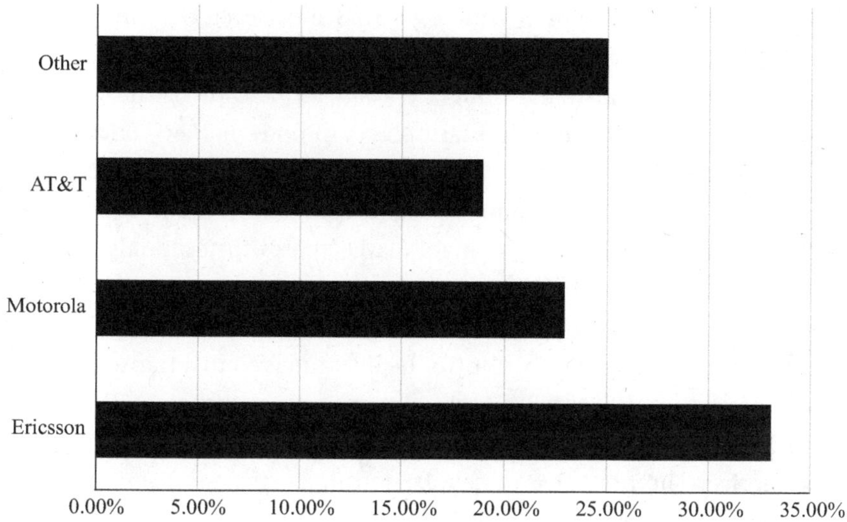

3.3 World sales of network infrastructure, 1990, shares in number of subscribers. *Source*: US International Trade Commission, *Global Competitiveness of U.S. Advanced-Technology Industries: Cellular Communications* (Washington, DC: US International Trade Commission, 1993), 2–8.

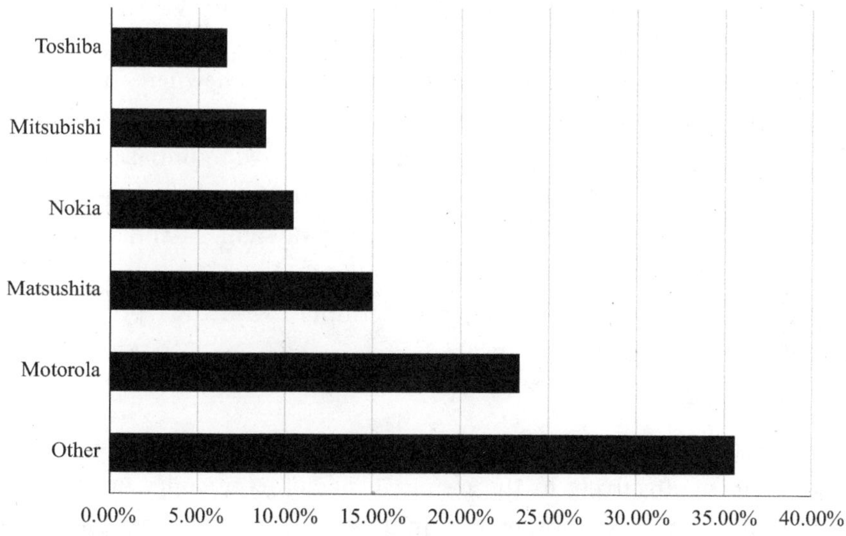

3.4 World sales of terminals, 1990, shares in units. *Source*: US International Trade Commission, *Global Competitiveness of U.S. Advanced-Technology Industries: Cellular Communications* (Washington, DC: US International Trade Commission, 1993), 2–9.

A comparison of figure 3.4 and figure 3.3 shows that the infrastructure market was more concentrated: the top three vendors had 75 percent of the infrastructure market but less than 50 percent of the terminals market. Motorola was an important player in both markets, Ericsson was considerably more dominant in infrastructure, and Nokia played a much more important role in terminals. Japanese firms such as Matsushita, Mitsubishi, and Toshiba had reasonable large shares in terminals but were absent from the infrastructure market.

Three types of cellular terminals were in use during the 1980s. All of them included a handset or control head connected to a transceiver unit with an antenna that could be in a separate location or built into the unit.[17] The quality and placement of the antenna affected the quality of the service. The first type of cellular terminal—the most common and least expensive—was a vehicle-based mobile telephone in which the handset or control head was mounted in the vicinity of the driver and the transceiver was placed either in the trunk or under the front seat. The second type, the most expensive and least common, was the portable, a lightweight, battery-powered unit that could be held in one hand. Portables were particularly well suited for cellular systems operating at the high end of cellular frequencies, which at the time was around 800 or 900 MHz. The third type, the transportable, was somewhere in between the other two: it had longer battery life and higher power than the portable, but it was also heavier and bulkier. All three terminal types were controlled by microprocessors and had full duplex transceivers—that is, capable of simultaneously sending and receiving signals—and a variety of standard features, including abbreviated dialing and last-number recall. Table 3.2 presents the three types of terminals with their prices in the United States around 1988.

FIRM CAPABILITIES IN CELLULAR TERMINALS

To compete effectively in the terminal market, companies needed two types of competencies: experience in designing and manufacturing (precellular) wireless communications devices and semiconductor expertise.[18] Firms with expertise in precellular mobile terminals and paging equipment were able, first, to produce phones that transmitted and received

Table 3.2 Types of mobile phones in the 1980s and their prices in the United States in 1988

	Price range (USD)	Average price (USD)
Vehicle-based mobile		$1,000
Transportable	$750–$3,000	$1,500
Hand-portable	$1,000–$4,000	$2,000

Source: US Department of Commerce, *A Competitive Assessment of the U.S. Cellular Radiotelephone Industry* (Washington, DC: US Department of Commerce, 1988), 27–29.

high-quality signals and, second, to quickly develop broad product lines for domestic and foreign customers.

Breadth of terminal offerings could be measured by the variety of models in a company's portfolio and the variety of standards for which a firm produced mobile units. By the early 1990s, three of the top five terminal manufacturers—Motorola, Nokia, and Mitsubishi—offered a broad range of models. Further, two of the top five—Motorola and Matsushita—designed and manufactured cellular phones for at least four different standards (the American AMPS, the British TACS, the Scandinavian NMT, and the Japanese NTT HCS), whereas the other three—Nokia, Mitsubishi, and NEC—did so for at least three standards.[19]

Firms with prior experience in wireless communications devices were also able to better anticipate consumer demand for new terminal features, which in turn gave them the ability to bring new phones to market before their competitors. Motorola, for example, used its experience with non-cellular mobile phones and pagers to anticipate the rise of vehicle-based cellular phones first and hand-portable cellular phones later. It managed to supply seven car-based models and two portable models in the US market in the first year after the first commercial cellular licenses were granted.[20]

The second capability that mattered for competition in the terminal market was semiconductor knowledge, especially in integrated-circuit design and manufacturing. Companies with expertise in this area knew how to develop the smaller, more efficient, and more powerful integrated circuits required to manufacture smaller and lighter phones. All the top terminal manufacturers were involved in the design of cellular-specific integrated circuits.

Semiconductor progress during the 1980s led to better cellular phones—smaller, lighter, and with longer talk time. The Motorola DynaTAC-800 F, for example, introduced in the US market in 1984, weighed about 850 grams, whereas phones sold by Toshiba and Motorola in the early 1990s (the MUX 500 and the MicroTAC LITE, respectively) weighted between 200 and 300 grams. In addition, portable-phone talk times improved by a factor of about three in the US market in the second half of the 1980s, from less than 50 minutes in the mid-1980s to around 150 minutes in the early 1990s.[21]

COMPETING IN CELLULAR TERMINALS: PRICES, ATTRIBUTES, AND LONG-TERM RELATIONSHIPS

During the 1980s product prices and features played a larger role in the cellular-terminal market than they did in cellular infrastructure. One reason for this is that terminals are simpler products than base stations and switches, and thus relationships between service providers and equipment vendors are less persistent—switching costs are lower for simpler products.

Long-term relationships mattered, however, in the analog-terminal market. This becomes evident when we compare the top terminal vendors in each market. Table 3.3 presents the market shares of the top three terminal vendors in America, Europe, and Japan in 1990.

The table highlights the distinction between the American and European markets, on the one hand, and the Japanese market, on the other. In the American and European markets, Motorola was the top vendor and Nokia played a significant role. In the Japanese market, none of this happened: in 1990 the top three vendors were Japanese companies. In addition, the Japanese market was considerably more concentrated, with the top three vendors accounting for about 75 percent of the market.

In Japan, NTT, which had been the key player in the supply of fixedline telephone services, exercised tight control over the analog standard-setting process. Rather than developing an open cellular standard, NTT chose to develop a proprietary standard. It also chose to interact exclusively with a select group of domestic suppliers, maintaining some degree of control over them. This explains why foreign vendors such as Motorola, Nokia, and

Table 3.3 Key terminal vendors in top cellular markets, shares in units, 1990

I. The US market		
Rank	**Company**	**Share (%)**
1	Motorola	23.06
2	Matsushita	16.50
3	Nokia	10.15
Top 3		49.71
II. The European market		
Rank	**Company**	**Share (%)**
1	Motorola	28.56
2	Nokia	15.25
3	NEC	11.43
Top 3		55.24
III. The Japanese market		
Rank	**Company**	**Share (%)**
1	NEC	26.09
2	Matsushita	24.09
3	Mitsubishi	24.09
Top 3		74.27

Source: US International Trade Commission, *Global Competitiveness of U.S. Advanced-Technology Industries: Cellular Communications* (Washington, DC: US International Trade Commission, 1993), 5-28.

Ericsson played a much smaller role in Japan than they did in the United States and Europe.

The rest of the chapter analyzes competition in cellular-network services. We focus on Britain first and on the United States later.

COMPETING IN NETWORK SERVICES IN BRITAIN: VODAFONE AND CELLNET

The driving force behind the creation of Vodafone, the only private competitor in the British cellular market during the 1980s, was a company named Racal, founded in 1950 to supply radio units to the military.[22] In the summer of 1982, as the deadline to submit applications for a second cellular network in Britain was approaching, the electronics entrepreneur Jan Stenbeck (1942–2002), who had recently founded the Millicom Group in Raleigh, North Carolina, approached Racal with the idea of setting up a joint venture to bid for the second license. The venture—in which Racal held an 80 percent ownership stake, Millicom 15 percent, and Hambros, an investment bank, 5 percent—was awarded the license in late 1982. The company operated as Racal Millicom until 1986, when it changed its name to Vodafone, a name that reflected the expectation that mobile phones would eventually be used for the transmission of both voice and data. Soon after its founding, Racal Millicom hired Christopher Gent (born 1948), who would later play a pivotal role in transforming Vodafone into one of the top cellular carriers in the world.

On January 1, 1985, the first cellular call in Britain was made on the Vodafone network, the first to officially launch in the country. The network was rolled out quickly to serve the demand coming from thriving "yuppies," who were starting to view cellular phones as both useful for business and indispensable status symbols. Since the terms of the license prevented Vodafone from selling its services directly to the public, the company created Vodac, a wholly owned subsidiary, to function as a service provider for the network. Vodafone targeted affluent business users first, and the early growth of the network—concentrated on large cities, polo fields, and yacht harbors—reflected this orientation.

Vodafone proved to be a more agile competitor than Cellnet, the cellular firm created through a partnership between Securicor and British Telecom (BT). Although Vodafone's initial target was a 30 percent market share, within a few years it reached 50 percent and became the largest cellular network in the world. Several factors made Vodafone a more potent competitor than Cellnet, one of which was network quality.[23] In the late 1980s Cellnet developed a poor reputation for network quality, mainly because it failed to invest in capacity as the number of subscribers increased at a rapid pace. Cellnet was forced to launch a large investment program to increase the number of both cells and channels in 1989, but by this time Vodafone was well on its way to becoming the market leader.

The network-quality differential reflected a difference in culture between the companies. Cellnet, although a joint venture between BT and Securicor, was dominated by BT, which up to the mid-1980s operated as a typical European PTT: it was bureaucratic in its procedures and at times disconnected from what customers wanted. Further, cellular services were more important for Racal than they were for BT. For BT, cellular was a somewhat peripheral activity, whereas Racal viewed it as the most important business decision it had ever made. Once Racal decided to enter the British cellular market, it committed fully to be successful in it. All these factors combined to enable Vodafone to catch up with, and eventually surpass, Cellnet in the British market.

Cellnet and Vodafone competed intensely on service quality but less so in prices.[24] Between 1985 and 1992, service prices for Cellnet and Vodafone were similar to one another and remained stable in nominal terms—although they did decline considerably in real terms, given the high inflation rates of the 1980s. One explanation for the price pattern is that the analog networks colluded tacitly, implementing a business model of high prices and profits targeted mainly at business customers.

COMPETING IN NETWORK SERVICES IN THE UNITED STATES

Unlike what happened in Britain, where there was a national duopoly from the moment cellular-phone services were launched in the country, in the United States there were regional duopolies. From the early years, many of the leading cellular carriers—and especially the Baby Bells—engaged in a process of acquisitions that led them to own stakes in cellular licenses in many regional markets.

COMPETING THROUGH ACQUISITIONS

The United States had many regional duopolistic markets in cellular services during the 1980s, and it is impossible to track them all. We examine competitive issues and strategies in American cellular markets by concentrating on two of the main competitors, PacTel and McCaw Cellular, in their interactions with each other and with other cellular players.[25]

Craig McCaw (born 1949), the son of a Seattle cable television and radio promoter, inherited a portion of his father's cable business in Centralia, a small town about eighty miles south of Seattle in the state of Washington. He quickly emerged as the family's business leader and in ten years increased the value of the cable assets to around USD 200 million. In the early 1970s McCaw obtained a license to provide (noncellular) mobile-phone and paging services in Centralia and, through the radio common carriers' trade association, became aware of the business potential of cellular-phone services. When cellular licenses were allocated by the FCC in the early 1980s, McCaw managed to collect several of them, then prepared to face the Baby Bells, which, in his view, would not be effective competitors in the cellular race. He was wrong: some of the Baby Bells developed entrepreneurial skills in the new cellular markets and became formidable rivals for McCaw.

The AT&T breakup in 1984 gave rise to the seven Regional Bell Operating Companies, or Baby Bells: Ameritech covered the Midwest; Bell Atlantic, the Mid-Atlantic states; BellSouth, the states on the southern portion of the Atlantic coast plus a number of states on the coast of the Gulf of Mexico; NYNEX, the New England states plus New York; Pacific Telesis, California and Nevada; Southwestern Bell, the central and southern portion of the country; and US West, the states of the Pacific Northwest. Each of the Baby Bells created a cellular subsidiary, which was awarded a cellular license for its region, the "wireline" license. The other cellular operator in each region received the "nonwireline" license.

Pacific Telesis was a holding company for Pacific Bell, Nevada Bell, and several other companies. It showed a strong entrepreneurial culture from the beginning. It organized the so-called PacTel Companies, including one to deliver cellular-network services, as separate subsidiaries with their own boards of directors. The cellular subsidiary, to which we will refer simply as PacTel, started operating as a cellular carrier in Los Angeles in June 1984, well before the nonwireline competitor, and organized an extensive network of resellers to sign up customers and sell or lease mobile terminals. (We will use Pacific Telesis for the holding company and PacTel for the cellular subsidiary, even though some sources also refer to the former as PacTel.) Although the cost of service was high (USD 45 per month plus 45 cents per minute), PacTel signed up seven thousand customers in the first two months of service.

Because cellular markets in the United States were regional, customers frequently traveled outside their home markets. Thus, PacTel's executives quickly realized that the company needed acquisitions to provide national coverage and become a major player in the new world of cellular. PacTel's first target was San Francisco, where the nonwireline cellular operator was Bay Area Cellular Telephone, a consortium of companies that included Communications Industries and McCaw Cellular. Communications Industries, a Dallas-based company, was interested in selling its share in the consortium, and PacTel offered to buy it for USD 10 per POP, a term roughly equivalent to "potential customer." McCaw Cellular, which was interested in acquiring cellular licenses from anyone ready to sell, interpreted PacTel's action as a declaration of war and prepared to respond in kind. An intense period of mergers and acquisitions ensued, in which PacTel, McCaw Cellular, and several other companies, including several of the other Baby Bells, participated. Table 3.4 summarizes the key transactions that led to a consolidation of cellular assets in the United States in the second half of the 1980s.

In 1986, after almost a year of legal maneuvering, PacTel acquired Communications Industries outright—a portfolio of cellular and paging franchises in San Francisco, Georgia, Texas, Missouri, Kentucky, Arizona, and Florida. The transaction was controversial because at that stage it was not even clear that the wireline cellular operators were allowed to acquire

Table 3.4 Consolidation in the American network-services market, 1986–1989

Year	Transaction	Comments
1986	PacTel acquired Communications Industries	PacTel acquired portions of cellular licenses in San Francisco, Phoenix, Saint Louis, Dallas, Tampa, and Jacksonville
1986	BellSouth acquired a stake in Mobile Communications Corp. of America	BellSouth acquired a portion of the cellular license in Los Angeles
1986	Southwestern Bell acquired Metromedia	Southwestern Bell acquired portions of the cellular licenses in New York, Boston, Chicago, and Washington, DC
1987	McCaw Cellular had its initial public offering	McCaw Cellular raised USD 2.39 billion
1989	McCaw Cellular acquired half of LIN Broadcasting	McCaw Cellular acquired portions of the cellular licenses in New York, Philadelphia, Houston, Dallas, and Los Angeles

portions of nonwireline licenses in geographic markets other than their own. (It was clear that such acquisitions were not allowed in their own markets.) Disregarding McCaw Cellular's persistent complaints, the FCC approved the transaction, which, paradoxically, ended up benefiting McCaw Cellular: after the PacTel acquisition of Communications Industries, the value of companies such as McCaw Cellular skyrocketed, which enabled McCaw Cellular to continue borrowing to fund its own cellular-franchise purchases.[26]

BellSouth responded in kind by acquiring a stake in Mobile Communications Corp. of America, which gave BellSouth a presence in Los Angeles, PacTel's home turf. Southwestern Bell, another of the Baby Bells, acquired Metromedia, the largest non-Bell wireless operator, a transaction that gave the acquirer a presence in several large markets, including New York, Chicago, Boston, and Washington, DC.[27] In the midst of the acquisition frenzy, McCaw sold all of its cable assets and used the funds to intensify the process of cellular-license acquisitions. In 1987 the McCaw Cellular initial public offering (IPO) raised a remarkable USD 2.39 billion.[28]

By the time of its IPO, McCaw Cellular controlled markets across the West, the South, and along the Eastern Seaboard but could not provide its customers with the ability to roam across cities without losing a call. Further, the company was absent from New York and Los Angeles, the two crown jewels of the American cellular marketplace. For this reason, it started exploring the potential acquisition of New York–based LIN Broadcasting, which owned television stations and portions of cellular licenses in New York, Philadelphia, Dallas, Houston, and Los Angeles. Through ingenious financial engineering, and with the help of an army of Wall Street investment bankers, McCaw Cellular fended off a counter bid by BellSouth, which was also interested in acquiring LIN Broadcasting. In 1989 McCaw Cellular acquired half of LIN for USD 3.4 billion to become the largest cellular player in the nation, more than twice as large as PacTel, the second largest.[29]

In the late 1980s and early 1990s, the acquisition process led to significant consolidation in the US cellular markets, to the point that, by 1993, eighteen companies accounted for most of the cellular assets in all major US cities. In that year, the seven Regional Bell Operating Companies, GTE, and McCaw Cellular were dominant. At the same time, ownership in each regional market was fragmented, with up to six operators sharing ownership of the cellular licenses in some of the largest markets.[30]

COMPETING IN PRICES AND FEATURES

The question whether the cellular duopolists operating in each American regional market during the 1980s competed or colluded is a complicated one. In the early 1990s, responding to a US Senate request, the General Accounting Office conducted a study of competitive conditions in cellular markets. Published in 1992, the report expressed concern about the fact that service prices were curiously similar across competitors in about two-thirds of markets.[31] Later in the 1990s, economists developed models suggesting that two features of American cellular markets may have facilitated tacit collusion: first, the same companies competed with one another in multiple markets, and second, companies that competed in one market co-owned a cellular license in another.[32]

Between its origins in 1983 and the early 1990s, the American cellular industry was one of rapidly rising volumes and declining average service prices. Even the General Accounting Office report had to admit that cellular had grown from about 92,000 subscribers in 1984 to about 7.6 million in 1991, "making it one of the fastest growing industries in the country."[33] In addition, a study prepared by the US International Trade Commission and published in 1993 showed that, between 1987 and 1992, nominal cellular-service prices in the United States declined, on average, by roughly 30 percent, and real service prices declined even more, by almost 40 percent.[34] That service prices declined on average does not imply, however, that each regional monopolist lowered its prices when a cellular competitor entered the market. The incumbents' responses varied across markets.[35]

In Atlanta, for example, the incumbent (Bell South Mobility) started operating in September 1984 and the competitor entered only in February 1987. The incumbent lowered prices on all plans in 1986, right before the competitor entered, and, in addition, increased the number of retail plans available in 1986 and raised the number of paid features available in 1987, coinciding with the competitor's entry.

In the Los Angeles market, the incumbent (PacTel) started operating in June 1984 and the competitor entered almost three years later, in

March 1987. The incumbent did not lower prices at all in response to the competitor's entry—in fact, the price for low-volume users went up. The incumbent, however, did make several changes in anticipation of the competitor's entry: in 1986 it increased the number of pricing plans available to consumers and raised the number of paid and free features available.

Cleveland offers an example of yet another type of cellular market in the 1980s. The incumbent (GTE Mobilnet) started operating in December 1984 and the competitor entered shortly thereafter, in June 1985. The incumbent lowered prices on all plans in 1986 but did not change the number of pricing plans available to consumers, and only increased the number of features available in 1987.

In several US regional markets, cellular incumbents lowered prices on at least one plan right before (or after) the entry of a competitor. In some markets, they did not change the prices of existing plans, but they increased the number of pricing plans and the number of features offered to consumers. Competition in American cellular markets did have an impact on the behavior of incumbents, but competitive responses were heterogeneous and often involved features other than prices.

More generally, duopolistic competition during the 1980s does not appear to have led to substantial price declines. In the United States, subscription tariffs fell modestly during the 1980s, and in other duopolistic markets, such as Britain and Sweden, they did not decline at all.[36]

COMPETITION, PRICES, AND ADOPTION

A somewhat different question is whether cellular-service prices were lower, on average, in countries where there was competition—Britain and the United States—than in countries where there was a monopoly (most other countries during the 1980s). Price comparisons across countries are complicated—it is well understood that a service that costs USD 100 per month in, say, one of the Organisation for Economic Co-operation and Development (OECD) countries is relatively much less expensive than the same service at the same price in a country where per capita income is much lower. That is why, when comparing cellular-service prices across countries, economists usually focus on the cost of a basket of services in a country as a percentage of that country's per capita income, or on the cost of a basket of services in purchasing-power-parity terms, a calculation that accounts for differences in living standards across countries.

Figure 3.5 presents the annual cost of a basket of cellular services in 1992 purchasing-power-parity terms, with countries ranked from highest to lowest.

There are nineteen countries in the figure. If we divide them roughly into three groups, the high-price group consists of six countries from Luxembourg to Japan; the midprice group consists of another six from Spain to Italy; and the low-price group consists of seven from the United States to Iceland. In this figure, the intensity of competition does not explain the level of prices. Britain, one of the competitive pioneers, is in the high-price group, whereas the United States, the other pioneer, is in the low-price group. Further, most of the members of the low-price group are countries in which there was a monopolistic cellular carrier, including the Nordic countries, Switzerland, and Iceland.

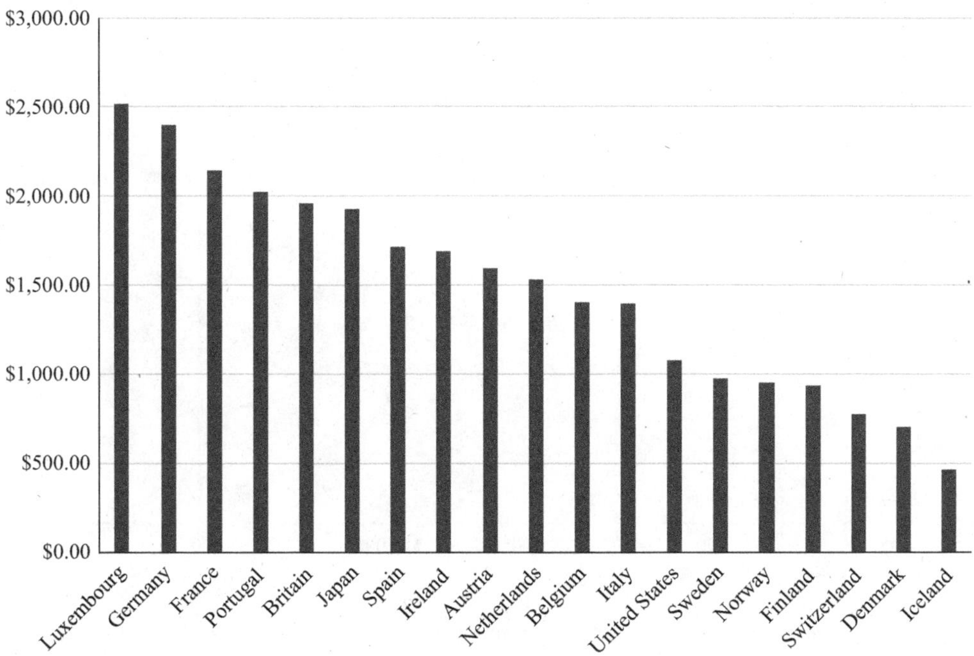

3.5 Cellular-service prices across OECD countries, 1992. *Source*: OECD, *OECD Communications Outlook 1993* (Paris: OECD, 1993), 85.

Yet a different question is whether the level of prices was correlated with the intensity of cellular adoption—that is, with cellular density. Figures 3.6 and 3.7 address this issue.

Figure 3.6 plots cellular-service prices and cellular density for the cellular pioneers—that is, for countries where cellular networks were launched in the late 1970s and early 1980s. Figure 3.7 shows the same relationship for countries that introduced cellular services in the mid-1980s.

The figures show that, in each group of countries, prices mattered. Among the pioneers, high-price countries such as Japan had low density in 1992 and, conversely, low-price countries such as Sweden had high density. The same was true for the countries that launched cellular service in the mid-1980s. Luxembourg, for example, had high prices and low

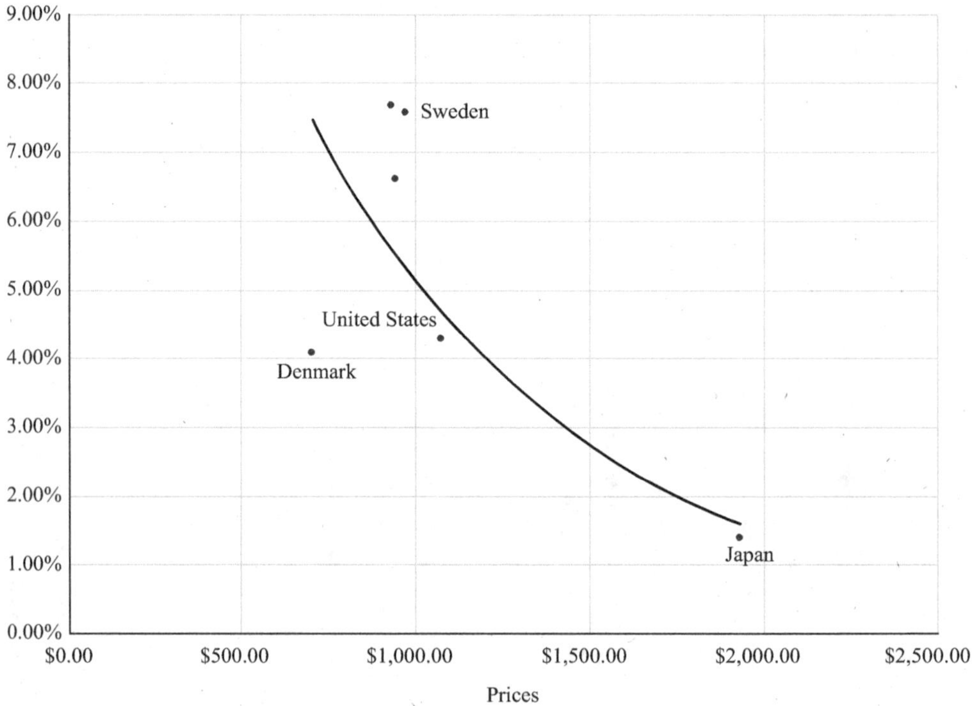

3.6 Cellular-service prices and cellular density among cellular pioneers, 1992. *Sources*: OECD, *OECD Communications Outlook 1993* (Paris: OECD, 1993), 85; and International Telecommunications Union database.

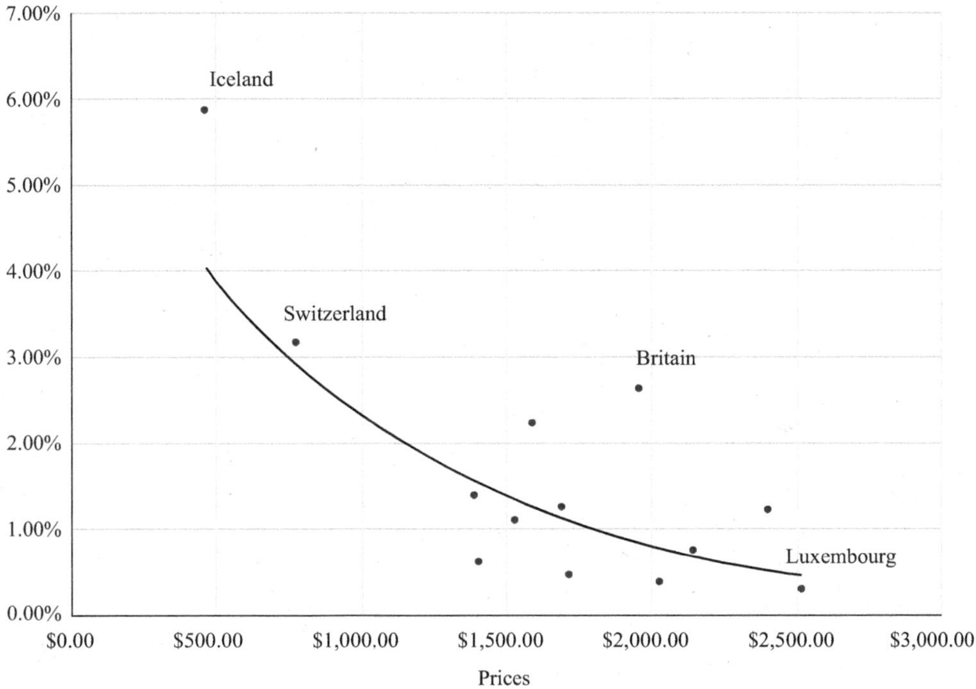

3.7 Cellular-service prices and cellular density among second-wave countries, 1992. *Sources*: OECD, *OECD Communications Outlook 1993* (Paris: OECD, 1993), 85; and International Telecommunications Union database.

density, while Iceland had low prices and high density in 1992. Among the countries of the second wave, Britain was an outlier: it had relatively high prices and relatively high density. In Britain, duopolistic competition did not lead to lower nominal service prices during the 1980s. However, it enhanced awareness of cellular phones and it increased cellular capacity, all of which accelerated the growth of cellular density.

II

THE SECOND GENERATION, THE 1990s

4

NEW TECHNOLOGIES, STANDARDS, CUSTOMERS, AND MARKETS IN THE WORLD OF 2G

4.1 A typical 2G cell tower. Courtesy of AT&T Archives and History Center.

In the transition from the 1980s to the 1990s, mobile phones underwent important transformations. First, cellular technology changed from analog to digital, which improved several aspects of mobile communications, including spectral efficiency and the privacy and security of conversations. While the analog cellular phones of the 1980s are usually referred to as first generation (1G), the digital phones of the 1990s are frequently labeled second generation (2G). The technology also transitioned from systems based on frequency division multiple access (FDMA) to ones that relied on time division multiple access (TDMA) or code division multiple access (CDMA).

Second, while cellular phones in the 1980s had been prestige items reserved for the wealthy and well-connected individuals who were able to afford them, in the 1990s they became consumer items relatively accessible to the masses. To some extent, this transformation was driven by a change in the network-services market: whereas in the 1980s cellular competition had been rare, in the 1990s it became common. In those countries—such as the United States and Britain—that had had duopolistic cellular markets during the 1980s, competition turned considerably more intense in the 1990s; and in those that had had monopolistic cellular markets in the 1980s, competition became the norm during the 1990s. The pressures of competition forced operators to offer pricing plans that made cellular service affordable for consumers for the first time in the history of mobile phones. The average price of cellular services in most countries declined at a much faster rate in the 1990s than it did during the 1980s. The fact that handset prices continued falling in the 1990s helped as well.

Third, while in the 1980s comparative hearings (beauty contests) had been the main approach governments had used for allocating spectrum, in the 1990s a new method was tried: spectrum auctions. In 1990 New Zealand pioneered the use of auctions for spectrum allocation, including allocation to cellular carriers.[1] (Economists like auctions because auctions tend to allocate assets to the buyers that value them the most.) In the mid-1990s the Federal Communications Commission (FCC) used auctions for the same purpose in the United States. Later, several countries in Europe implemented spectrum auctions for cellular services as well.

Fourth, during the 1990s, cellular communications expanded to many countries in Africa, Latin America, the Caribbean, Asia, Central and Eastern

Europe, and the Middle East. The 1990s also saw substantial growth in markets—such as China and South Korea—that would eventually rank among the largest in the world.

Finally, in the 1990s people started using cellular phones for purposes other than voice communications. During the decade, early forms of mobile-data use started appearing on the cellular horizon, first in Europe and later in other countries. The key manifestation of this trend was short messaging service (SMS), or texting.

We examine all of these processes—and related transformations—in chapters 4 through 6. This chapter focuses on new technologies, standards, subscribers, uses, and markets. Chapter 5 analyzes the interactions between national governments and private companies in several cellular markets during the 1990s, including the United States, Britain, several countries in continental Europe, China, and India. Chapter 6, finally, explores competition in product and service markets during the 2G era of cellular phones, which was also the first era of digital cellular.

NEW TECHNOLOGIES

In the transition from the 1980s to the 1990s, cellular systems went from being analog to being digital. This did not mean that, in the 1990s, cellular operators started using digital computers in certain sections of their networks, such as the switching system. Digital computers were already in use for that purpose in the 1980s.

What went from analog to digital was the mode of encoding the user's voice, somewhat like the difference between vinyl records and CDs. An analog signal, such as voice, is a continuous wave form that varies in frequency (pitch) and amplitude (volume or strength).[2] Analog transmission, which was at the core of cellular systems in the 1980s, involved the transmission of complex waveforms corresponding closely to the waveforms produced by the original sounds of human speech. In the cellular systems that appeared in the 1990s, by contrast, voice was digitized: the original complex waveform was reduced to an easily manipulated bitstream.[3] In short, digital cellular involved the transmission of zeros and ones.

The digitization process involved three steps. The first was analog-to-digital conversion, through which the original complex waveform was reduced to a numerical form. The second was modulation, which converted the digital data into a transmittable form. The third was the transmission itself. Digital cellular had several advantages relative to analog cellular: more traffic per cell, smaller cells, improved voice quality and network security, greater confidentiality, and a large range of mobile data services.[4]

The transition from the 1980s to the 1990s also witnessed the move from FDMA to either TDMA or CDMA. To increase the efficiency in spectrum usage, spectrum is divided into frequency bands, or channels, and then allocated to different users. FDMA, TDMA, and CDMA are approaches for doing this allocation.[5] All these methods are called "multiple access" methods because they all address the problem of granting access to the available spectrum band to multiple users.[6]

FDMA was the approach used in all analog cellular systems set up in the 1980s. When a user wanted to make a call, the user was assigned a channel about 25 kHz wide—enough to carry the human voice—and kept control of that channel for the duration of the call. In FDMA, each channel could be assigned to at most one user at a time. The advantage of the approach was that it required limited computing power and simple mobile terminals; the disadvantage, however, was that the available frequencies were used inefficiently.

TDMA, the method applied in the second-generation Global System for Mobile Communications (GSM) standard developed in Europe, took a different approach to the problem of allocating frequencies. In TDMA, the channels are wider and are broken down into time slots. Each user has available a channel that is about 200 kHz wide, but only for a fraction of the time. The advantage of TDMA over FDMA is that TDMA uses the available spectrum more efficiently, and the disadvantage is that it requires more complex terminals.

CDMA, also a second-generation technology, was pioneered by the American firm Qualcomm in the late 1980s and early 1990s, and it competed with TDMA to become the standard approach for allocating frequencies to users in the 1990s. With CDMA, all communications take place at the same time in the available frequency area, which generates interference. CDMA addresses the interference problem by assigning each user a code that distinguishes his or her signal from all others, which are treated as noise.

During the 1990s, when Qualcomm was spreading the CDMA gospel among equipment makers and network operators, the firm's public-relations team developed the "noisy cocktail party" analogy to explain multiple-access approaches in nontechnical terms.[7] A couple organizes a cocktail party at their large mansion, the analogy goes, and they invite dozens of guests. To facilitate communications among attendees, the hosts have three approaches available: FDMA, TDMA, and CDMA.

In an FDMA cocktail party, each one-on-one conversation takes place in a separate room (the assigned frequency). While two people are talking in, say, the living room, nobody is allowed to interrupt them—in fact, nobody else is even allowed to enter that room. Therefore, there is no possibility for interference, but the number of available rooms (frequencies) in the mansion (the portion of spectrum allocated to an operator) sets a limit to the number of conversations that can take place simultaneously.

If they follow the TDMA method, the hosts organize the cocktail party differently. In a TDMA party, many pairs of individuals talk to each other in the same room—the living room, for example. The first pair talks for, say, thirty seconds, and then the second pair gets their turn, and so on and so forth. There is no possibility of interference even when many pairs share the same room, since each pair only communicates during their allotted segment of time. The advantage of this approach is that it enables more conversations to take place in the mansion.

The CDMA approach, finally, is more akin to a real cocktail party, in which everybody shares the same space—the whole mansion, say. In this scenario, the background noise is much higher and there are plenty of opportunities for interference. The CDMA method solves the problem by forcing each pair of individuals to speak in a different language. Since the conversation of each pair of attendees is encoded in their own dialect, they are still able to communicate with each other despite the level of interference.

NEW STANDARDS

In the transition from 1G to 2G cellular phones, the analog standards of the 1980s were replaced with the digital standards of the 1990s. Europe went from multiple standards to one (GSM). The United States, by contrast, went from one standard (Advanced Mobile Phone Service [AMPS]) to several.

THE EUROPEAN 2G STANDARD GSM

The analog standards that prevailed in Europe, Japan, and the United States during the 1980s had several limitations. Since all of the standards were analog and relied on FDMA, some of the shortcomings were common to all standards. Others were specific to individual standards and geographies. An important problem in Europe was simply the fact that there were many standards. With a few exceptions, users could only use their mobile phones in their own country, which limited the scope of network effects.

Moreover, since different standards required different terminals and infrastructure equipment, many of the benefits that would have arisen from economies of scale and learning by doing in manufacturing were lost. Additionally, if in the future each country spent resources developing its own individual cellular technology, the duplication of research and development would result in wasted resources. All these factors suggested that it made sense to try to develop a common European standard.

A common standard requires that a common radio band be made available—a sufficiently wide portion of spectrum that has been allocated to cellular services in each country, and that is the same across countries. In Europe, this had been done in the late 1970s by the Conference Européene des Postes et Télécommunications (CEPT), an organization of the European postal, telegraph, and telephone (PTT) administrations that settled standardization issues in telecommunications and postal services. In 1978 the CEPT had reserved two 25 MHz blocks in the 900 MHz band for mobile communications.

In 1982, during the CEPT conference in Vienna, a new standardization working group was formed under the name of Groupe Spéciale Mobile, or GSM.[8] (The acronym GSM originally referred to Groupe Spéciale Mobile, but later came to stand for Global System for Mobile Communications.) At the end of that year, representatives from eleven European countries held the first GSM meeting in Stockholm.

The work of the GSM group generated increasing interest among the European telecommunications administrations, and as a consequence the number of participants and contributions rose. In 1985 it was decided that the group would be split into working parties, which would meet separately and would then report to GSM plenary meetings. In 1986 a permanent secretariat, the "Permanent Nucleus," was established in Paris. In 1987, the GSM group started inviting industry representatives to participate in the deliberations, and the next year, the European Telecommunications Standards Institute (ETSI) was formed. Most of the CEPT standardization activities, including those related to GSM, were transferred to ETSI.

GSM was built around open interfaces. Figure 4.2 presents the architecture of the GSM system.

As the figure shows, GSM has several components. First, there are a variety of mobile stations—some of them are handheld and others are mounted in cars. Second, there is the base station subsystem, which includes the base station transceivers and the base station controller. Third, there is the network and switching subsystem, including the mobile service switching center and the relevant customer databases: the home location register and the visitor location register. Finally, there is the operation subsystem, which deals with network operation and management, subscription management, and mobile-equipment management.

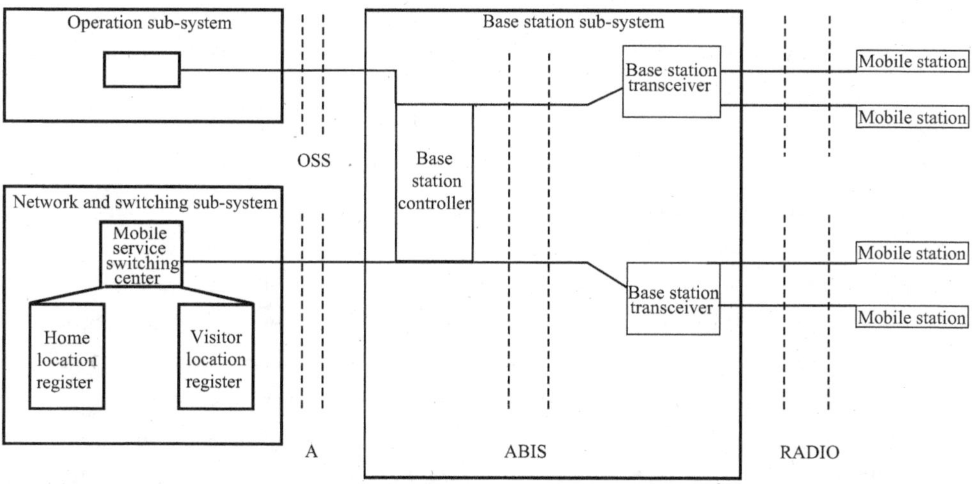

4.2 The architecture of the GSM system. *Source*: Adapted by the authors from J. Meurling and R. Jeans, *The Mobile Phone Book* (London: Ericsson Radio Systems, 1994), 104; and W. Webb, *Understanding Cellular Radio* (Boston: Artech House, 1998), 9–19.

The figure also identifies the main interfaces in the system: first, the radio (or air) interface between the mobile stations and the base stations; second, the Abis interface between the base stations and the base station controller; third, the A interface between the base station controller and the mobile switching center; and finally, the OSS interface between the base station controller and the operation subsystem. In the GSM system, all these interfaces were clearly defined and standardized, and all these specifications were in the public domain. In the future, network operators and equipment manufacturers willing to work within the GSM standard agreed to abide by these specifications, which meant that cellular carriers would be free to choose different equipment manufacturers to supply different components of the system. The GSM specifications were designed to ensure compatibility among all components designed by different manufacturers, and thus to foster competition among terminal and infrastructure suppliers.

The process of defining the GSM standard involved heated debates about cellular technology. GSM was based on TDMA, but there were many variants of the technology. French and German equipment manufacturers formed consortia that developed a broadband TDMA technology, a version of TDMA that relied on a 2 MHz channel width. The Nordic manufacturers—Nokia and Ericsson, primarily—created an alternative version, narrow-band TDMA. The Ericsson system, for example, had a channel width of 300 kHz, and the other Nordic systems were similar.

In early 1987 all the European countries participating in the standard-setting process except France and Germany voted in favor of narrow-band TDMA. In the spring of 1987, and after a period of intense negotiations and political maneuvering, the members of the GSM group agreed that the GSM standard would be based on narrow-band TDMA with 200 kHz channels. That year, at a meeting in Copenhagen, the European PTTs signed the GSM memorandum of understanding to formalize their commitment to cooperate on commercial and operational matters. The memorandum addressed matters such as time plans for procurement and deployment of systems, compatibility of numbering and routing plans, and tariff principles. In later years, the memorandum was signed by other operators and was amended to allow for cooperation with non-CEPT countries, which became necessary as GSM started to gain acceptance in countries outside Europe.

The first drafts of the GSM specifications became available in mid-1988, which allowed manufacturers to start developing products compliant with the standard. Although the original version of the GSM standard operated in the 900 MHz band, in 1990 a version of the standard adapted to the 1,800 MHz band was included in the scope of the GSM specifications. This version became known as Digital Cellular System (DCS) 1800 and was created in response to a request from the British telecommunications authorities, which were interested in licensing high-capacity systems in metropolitan areas under the name of Personal Communications Networks (PCNs).

Roaming was one of the essential features of the GSM standard. As shown in figure 4.2, the standard identified two databases—the home location register and the visitor location register—associated with each mobile switching center. The home location register was used for storing data for the subscribers who belonged to the area covered by a mobile switching center, whereas the visitor location register was used for temporarily storing data related to visiting subscribers from other areas.

From the early years, the roaming feature was crucial for the expansion of GSM. By making it possible for subscribers to use their mobile phones even when they were outside their home country, the roaming capability contributed to the success of GSM well beyond the European countries for which the standard was conceived.

THE GSM STANDARD IN A DIFFERENT FREQUENCY BAND: DCS 1800

In the late 1980s the British government determined that the competition between Cellnet and Vodafone had not been intense enough and, as part of the Phones on the Move initiative, it decided to allocate three additional licenses, the PCN licenses, in the 1,800 MHz band.[9]

Three consortia won these licenses, and two of them merged in 1992. The two that remained were One-2-One (owned by Mercury Telecommunications, in turn owned by Cable & Wireless) and Orange Personal Communications (in which Hutchison Telecommunications UK was an important shareholder). In 1990 Britain asked ETSI to develop a standard for these networks, which created a heated political debate. The countries that had signed the GSM memorandum of understanding viewed the 1,800 MHz networks as a threat, since the higher frequency band would enable PCN operators to build systems with higher-capacity cells. Further, some GSM stakeholders argued that the PCN license holders would free ride on the investments already made in the development of the GSM standard.

PCN had started in Britain, but it was viewed as a mobile option that would extend to other countries, and ETSI was not interested in developing a cellular standard that could potentially become an alternative to GSM. The stalemate between the British PCN operators and the GSM member countries was resolved by forcing the former to pay a substantial contribution to the latter, and by preventing the former from using the GSM name and trademark. ETSI developed a standard for the British PCN systems: it was based on GSM, and it was called DCS 1800. Approved by ETSI in January 1991, DCS 1800 was designed for systems that were deployed in densely populated areas, had small cells, and relied mainly on low-power handheld terminals. Eventually, the standard was adopted not only in Britain but also in many other European countries and beyond.

In most Western European countries, the transition to GSM in the first half of the 1990s coincided with the transition to competition in cellular network services. In the second half of the 1990s, many European countries introduced DCS 1800 networks, usually more than one, which intensified cellular competition, accelerated price declines, and led to an expansion of domestic cellular markets.

2G STANDARDS IN THE UNITED STATES: D-AMPS, N-AMPS, CDMAONE, AND PCS 1900

The process of defining a digital standard started earlier and took longer in Europe than it did in the United States.[10] Further, whereas in Europe the process was driven from the beginning by the idea of creating a single standard for the continent as a whole, in the United States the FCC allowed market forces to pick one or more digital standards. Additionally, whereas in Europe the contours of the GSM standard were defined mainly through the interaction of representatives from the domestic PTTs (and the equipment manufacturers were invited to participate late in the process), in the United States, by contrast, industry representatives were involved from the beginning in the formation of digital cellular standards.

The United States had a single standard (AMPS) throughout the 1980s. Thus, the interest in moving the country to a single standard, so strong in Europe, did not exist in the United States, and the decision to transition to digital technologies was driven by the need to accommodate more subscribers within the existing spectrum allocation. In the late 1980s this problem was particularly acute in densely populated metropolitan areas.

The initiative to start considering the transition from analog to digital originated in the Cellular Telecommunications Industry Association (CTIA), an industry trade group founded in 1984 by nineteen wireline and nonwireline cellular operators. During the 1980s, one of the CTIA's main responsibilities was to enable roaming across American regions, and a committee was created for this purpose. The committee set the requirements for the content and format of the data that needed to be exchanged when subscribers roamed outside their home markets, which in turn enabled the Telecommunications Industry Association (TIA) to develop an intersystem standard known as IS-41, widely used throughout the United States.

In 1988 the CTIA developed the so-called Users' Performance Requirements document, which would serve as the foundation for the creation of a digital standard. The TIA, which had been accredited by the American National Standards Institute, was then charged with the responsibility of developing the actual standard considering the Users' Performance Requirements. According to the requirements, systems adhering to the new standard would have to have a capacity at least ten times greater than was available at the time with analog cellular technology. In addition, the requirements envisioned a gradual transition from analog to cellular systems—this was different from the transition to digital in Europe, since brand-new networks were required for implementing the GSM standard.

At the same time, the FCC released its decision on cellular standards in the United States: rather than dictating a single standard for the country, the agency announced that cellular carriers would be allowed to introduce new cellular technologies at any time without prior regulatory approval. This was a radical departure from the approach adopted by the CEPT in Europe.

Internal debates took place at the CTIA regarding the virtues and limitations of TDMA versus FDMA technologies. In 1989 the CTIA determined that TDMA was the technology of the future and instructed the

TIA to work on specifying the technical details of the new standard, which became known as D-AMPS, IS-54, or simply TDMA. The first draft of the D-AMPS standard was published in mid-1990.

D-AMPS comprised three standards: one for mobile terminals, another one for base stations, and a third one for compatibility between terminals and base stations. Two features of these standards were particularly important. First, the standards gave operators the flexibility to convert as many analog channels into digital ones as they believed necessary to meet capacity requirements. Each analog channel converted to digital was able to accommodate three channels, and it was expected that operators would start by converting channels from analog to digital in densely populated metropolitan cells. Second, contrary to what happened in GSM, in which all interfaces were carefully specified, only the air interface was precisely defined in D-AMPS.

After the publication of the D-AMPS standard in 1990, some of the major operators started conducting technical validation trials, and the first D-AMPS systems were installed in 1992. The D-AMPS standard was not fully digital. In 1994 the specifications of a fully digital standard were published. It was named IS-136, and it was an enhancement of IS-54 that provided the functionality and performance of GSM.

Just as the CTIA was choosing TDMA (over FDMA) in 1989, alternatives to TDMA were starting to appear on the horizon. Qualcomm, a company based in San Diego that had expertise in military communications, started lobbying the cellular industry with proposals to use CDMA, rather than TDMA, for future cellular standards. In June 1989 Qualcomm made a CDMA presentation at a CTIA forum in Chicago. The CTIA's technology committee decided that the organization would continue to support IS-54 (TDMA) as the industry standard for digital cellular, but also recommended that the evolution of CDMA be monitored.

Qualcomm continued lobbying operators and equipment manufacturers. In November 1989 it showed industry observers that CDMA was a feasible technology by conducting a demonstration in San Diego with assistance from Pacific Telesis. It later carried out a more extensive field trial with NYNEX in Manhattan. In September 1990 the company published the first version of the air-interface specification for CDMA systems.

In the meantime, Motorola proposed a narrow-band analog standard, known as N-AMPS, which offered a solution to the AMPS capacity problem.

Motorola proposed reducing the bandwidth of AMPS channels from 30 kHz to 10 kHz, which would enable three times as many channels in the same band of spectrum. The CTIA was now faced with three alternatives for America's cellular future: D-AMPS (or TDMA), CDMA, and N-AMPS. In 1991 trials of the three systems were carried out to assess their virtues and limitations.

In January 1992 the CTIA's board of directors adopted a resolution that reaffirmed the role of D-AMPS (TDMA) as the industry standard but also opened the door for the consideration of a CDMA-based standard as an alternative. The resolution requested that TIA work with CTIA's technology committee to organize open forums focused on CDMA, and that TIA facilitate the development of a CDMA-based standard (without interrupting TDMA's progress). From then on, CTIA's standardization efforts focused on D-AMPS and CDMA, and Motorola's N-AMPS was left out of consideration.

Qualcomm's lobbying efforts paid off in June 1992, when CTIA's board of directors adopted a new resolution: the president of CTIA requested that TIA begin developing a CDMA standard, with the caveat that such activities should not interfere with efforts that were being made to define the TDMA standard. In practice, CTIA was formally opening the door for competition between two standards in the United States: one based on TDMA, formally known as IS-54, and another based on CDMA, which came to be known formally as IS-95 and informally as cdmaOne. In 1994 IS-95 was formally adopted as a second cellular standard in the United States. Qualcomm encouraged equipment manufacturers to obtain licenses to make CDMA terminals and infrastructure, and by mid-1996 it had convinced sixteen of them to join the CDMA Development Group.

Apart from D-AMPS and CDMA, a third 2G standard was adopted in the United States: a version of the European GSM standard adapted to the 1,900 MHz frequency band. This standard became known as PCS 1900, IS-661, and GSM 1900.[11]

THE JAPANESE 2G STANDARD PDC

The United States was not the only country that generated digital standards that competed with GSM: Japan did as well.[12] Aware that digital standards were being specified in the United States and Europe, the Japanese government started fostering the creation of a Japanese digital standard in the late 1980s. Originally called Japan Digital Cellular, it was later renamed Pacific Digital Cellular (PDC).

The PDC specifications were created by the Development Center for Radio Systems under the supervision of the Ministry of Posts and Telecommunications. The standard had a strong American flavor: it borrowed many parameters from D-AMPS, including its reliance on the TDMA approach, and Motorola and AT&T were invited to collaborate with Japanese firms in the process of developing the standard. PDC was specified for both the 800 and 1,500 MHz bands, and it was particularly well suited for densely populated areas. Just like D-AMPS, PDC contained precise specifications only for the air interface, which opened the door for vendors to define proprietary interfaces in other sections of the system. Unlike GSM and the American standards, PDC was never adopted outside the geographic region where it was created.

NEW METHODS FOR ALLOCATING SPECTRUM: AUCTIONS

In the 1980s beauty contests—more formally known as comparative hearings—were the method most governments used to allocate spectrum licenses to cellular carriers. This changed in the 1990s: New Zealand pioneered the use of auctions to allocate spectrum in the year 1990, and the United States followed in the mid-1990s.[13]

In 1985 the FCC began requesting permission of the US Congress to rely on auctions to allocate spectrum, but the request was denied. It was only in 1993 that Congress granted the FCC the authority to use spectrum auctions. In the late 1980s and early 1990s, the US government ran large budget deficits, and the antideficit political rhetoric reached its peak in 1993. Congress passed the Balanced Budget Bill, in which it allowed the FCC to conduct spectrum auctions and hence reduce the government's debt.

To allocate the Personal Communications Services (PCS) licenses, the FCC divided the country into fifty-one regions called major trading areas, then divided 60 MHz of spectrum in each area into two blocks of equal size, the A and B blocks. For at least two reasons, auctioning two blocks was considerably more complicated than auctioning a single block: first, two licenses for regions close together were worth more to bidders if won as a package than the sum of their values if won separately; further, the value that a license had for an operator depended, to some extent, on who else had won licenses in neighboring areas. Faced with these complications, the FCC had to decide whether to auction blocks sequentially or simultaneously and what auction format to use. It could go with a first-price sealed-bid auction, a Dutch (descending clock) auction, a second-price sealed-bid auction, or an English (ascending clock) auction. (In chapter 7, we discuss the pros and cons of various auction formats.)[14]

Once the FCC was authorized to conduct spectrum auctions, the major telecommunications firms hired academic experts to advise them, and so did the FCC. At the end of the consultation, the FCC decided to use a version of the simultaneous ascending auction format, also known as the simultaneous multiple-round format. The auction proceeded in rounds: In the first round, bidders submitted their sealed bids, and at the end of the round the results were made public, including the high bidders and the bids necessary to beat them. Participants were then given time to analyze the results, and then a new round of bidding started in which bidders were allowed to submit new bids. The format worked extremely well for the PCS A and B blocks.

NEW USERS

Up to the late 1980s, cellular phones had been used mostly by wealthy individuals for business purposes. During the 1990s, cellular use grew among consumers, who relied on their mobile phones to stay in touch with family and friends much as they used a fixedline phone.

In the United States, this transformation started in the late 1980s. A study conducted at the University of California at Berkeley in 1991 noted that the demographic profile of cellular users changed in the late 1980s. The mean household income among cellular customers dropped from USD 90,000 in 1988 to USD 44,000 in 1990, and the average age of subscribers fell from fifty-five to thirty-nine. Further, in the same time period, the proportion of women among cellular users rose from 10 percent to almost 20 percent. The study concluded that the industry had aggressively promoted cellular technology and had made inroads into the household market.[15]

The same study noted that the reasons people used cellular phones changed in the late 1980s. Before 1988, they were used mainly for business purposes. In the second quarter of 1990, however, only 45 percent of subscribers used their mobile phones for business only, whereas around 31 percent used them for both business and personal reasons, and the rest used them for personal communications only.[16]

Price declines were the key drivers of these changes. Mobile-phone prices collapsed during the 1980s in the United States and elsewhere. According to one study, the average price of a cellular phone in the United States declined from around USD 3,500 in 1984 to about USD 500 in 1989. In 1990 some service providers offered phones for free to potential customers to entice them to sign up for service.[17] Another study pointed out that, between 1988 and 1990, the typical price of a cellular phone in the United States plunged from over USD 2,000 to under USD 500.[18]

The price of cellular phones kept on falling during the 1990s. One of the reasons was technological change, which lowered the cost of electronic components. Another was the rise of the pan-European standard GSM, which made it possible for phone manufacturers to reap the benefits of large-scale production and learning by doing.

In Germany, for example, one of the first countries where a GSM cellular network was introduced, prices for phones that worked on the Netz-C analog standard fell by 30 percent in anticipation of the introduction of digital services. At a time when the cheapest analog phones were priced at around DM 3,200 (USD 2,100), Nokia introduced its first phones for one of the German digital networks at DM 2,700 (USD 1,780). Later in 1992, Motorola cut the prices of its GSM phones all over Europe; in Germany, they could be bought for DM 1,600 (USD 1,050), about half the price of the cheapest analog device.[19]

Between the late 1980s and the late 1990s, the cost of cellular service fell consistently as well. In the year 2000, the Organisation for Economic Co-operation and Development (OECD) published a study of price structures and trends for cellular service.[20] The study showed that, in OECD countries, the cost of a fixed basket of cellular services fell by 29 percent between 1992 and 1999. This decline captured only the decline in prices, holding the composition of the basket constant. During the 1990s, however, the composition of the typical bundle of cellular services did not remain constant because the makeup of the subscriber base changed radically.

Two factors explain why the cost of cellular service (to subscribers) declined during the 1990s. First, as a result of competition, service prices fell for any given level of usage. Second, the profile of the "typical" cellular user changed: cellular carriers in many OECD countries introduced innovative pricing plans, which allowed many consumers to access the service for social, rather than business and professional, purposes. These plans usually had lower fixed charges and higher variable charges than the plans designed for heavy cellular users. The new cellular users of the 1990s tended to use their mobile phones less often—and for shorter periods of time—than the typical users of the 1980s.

In the United States, cellular carriers started targeting the consumer market in the late 1980s. Figure 4.3 presents the evolution of the cellular monthly bill from late 1987 through late 2000 in the American market. The figure also shows the growth in the number of cellular subscribers.

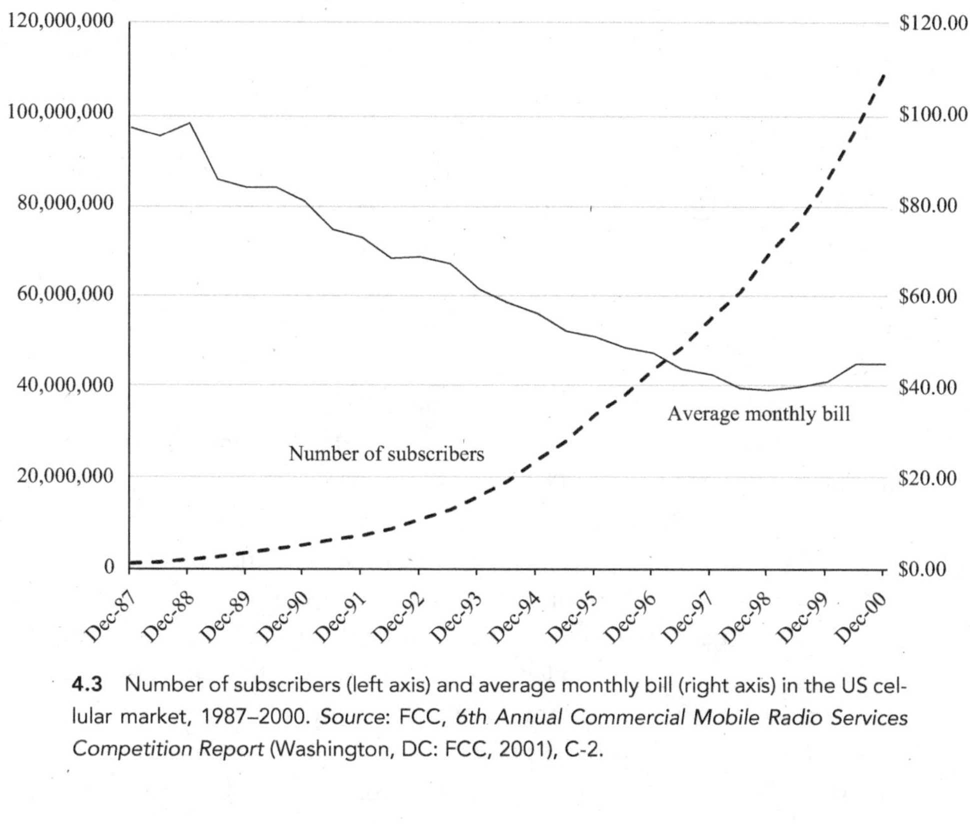

4.3 Number of subscribers (left axis) and average monthly bill (right axis) in the US cellular market, 1987–2000. *Source*: FCC, *6th Annual Commercial Mobile Radio Services Competition Report* (Washington, DC: FCC, 2001), C-2.

The average monthly bill for cellular services declined from about USD 97 in December 1987 to about USD 45 in December 2000, a fall of almost 55 percent. According to the FCC, this reflected not only a decline in prices for a fixed level of usage but also cellular penetration among segments with lower usage and thus lower bills.[21]

NEW MARKETS

During the 1980s, mobile phones were adopted in two types of countries. Some of them—including the United States, Japan, and a few European countries—were standard creators, but most of them were standard importers. The group of countries that imported a standard in the 1980s encompassed most Western European countries plus Canada, Australia, and New Zealand. It also included several countries in the Asia-Pacific region and several Middle Eastern countries. In Asia, two countries that would eventually become large markets also launched their first cellular systems in the 1980s, South Korea early in the decade and China at the end of the decade.

In Africa, mobile-phone adoption was scant during the 1980s, except for a few countries north of the Sahara Desert and in the southernmost extreme of the continent (South Africa). Adoption in Latin America was almost nonexistent, except for a few countries in Central America and the Caribbean and a few of the larger countries, such as Venezuela, Argentina, Chile, and Mexico, which imported a cellular standard toward the end of the 1980s.

The worldwide cellular-adoption picture changed radically during the 1990s. Table 4.1 presents a summary by cellular standard and region. (We deal with cellular adoption in Eastern Europe later in this section.)

Most Latin American countries introduced cellular systems for the first time in the early 1990s. All the early Latin American systems adhered to the AMPS standard, although some of them adopted AMPS/TDMA, and a few countries eventually adopted N-AMPS to provide greater capacity. In most cases, the system was run by a local operator in partnership with a foreign one. The American firm Millicom was particularly active in the early years.

Many Caribbean countries also introduced cellular systems in the first half of the 1990s with the participation of foreign operators, especially

Table 4.1 Cellular adoption in Latin America, the Caribbean, Africa, and the Asia-Pacific region between 1990 and 1995

		1990	1991	1992	1993	1994	1995
AMPS	**Latin America**	Guatemala, Peru	Bolivia, Uruguay	Brazil, El Salvador, Guyana, Paraguay	Belize, Ecuador, Nicaragua, Suriname		
	Caribbean	Barbados, Grenada, Saint Lucia, Saint Vincent	Guadeloupe, Jamaica, Martinique, Puerto Rico, Trinidad and Tobago	Montserrat	Aruba, Cuba		Anguilla
	Africa			Gabon		Angola, Burundi, Guinea	Zambia
	Asia Pacific			Cambodia, Vietnam	Bangladesh, Nauru	Guam, Samoa	
TACS	**Africa**			Gambia, Ghana, Mauritius, Nigeria	Kenya	Tanzania	
	Asia Pacific				Solomon Islands		
GSM	**Africa**					Cameroon	Ivory Coast, Seychelles, Namibia, Uganda
	Asia Pacific				Fiji		French Polynesia, Tahiti

Sources: G. Garrard, *Cellular Communications: Worldwide Market Development* (Boston and London: Artech House, 1998), and J. Funk, *Global Competition Between and Within Standards: The Case of Mobile Phones* (Houndmills, UK: Palgrave, 2002), 43-45

the British Cable & Wireless. The American AMPS standard was universally adopted in the Caribbean.

Most African countries launched cellular networks in the first half of the 1990s. Unlike what happened in Latin America and the Caribbean, where AMPS was pervasive, in Africa multiple standards were introduced. Most of the early systems in Africa adhered to European standards (either the analog Total Access Communications System [TACS] or the digital GSM), but AMPS was not completely absent from the African continent. Many small countries in the Asia-Pacific region also introduced cellular systems in the first half of the 1990s, most of them adhering to AMPS, GSM, or TACS.

Finally, in the first half of the 1990s, cellular systems were also launched for the first time in Central and Eastern Europe. Table 4.2 presents a summary.

In the late 1980s and early 1990s, the collapse of communism in Central and Eastern Europe created opportunities for the introduction of cellular

Table 4.2 Cellular adoption in Central and Eastern Europe in the 1990s

	Analog	GSM 1	GSM 2
Bulgaria	December 1993	September 1995	
Croatia	October 1990	March 1996	
Czech Republic	September 1991	July 1996	September 1996
Estonia	January 1991	September 1993	January 1995
Hungary	October 1990	March 1994	April 1994
Latvia	October 1991	January 1995	March 1997
Lithuania	February 1992	March 1995	October 1995
Poland	June 1992	September 1996	October 1996
Romania	May 1993	April 1997	June 1997
Serbia	January 1996		
Slovak Republic	September 1991	January 1997	February 1997
Slovenia	October 1990	July 1996	

Sources: H. Gruber, *The Economics of Mobile Telecommunications* (Cambridge: Cambridge University Press, 2005), 117–123; G. Garrard, *Cellular Communications: Worldwide Market Development* (Boston: Artech House, 1998), 380–381.

services. After the USSR dissolved in 1991, the countries that had belonged to the Soviet bloc inherited a telephone infrastructure that was generally viewed as insufficient to meet demand.[22] Fixedline density was considerably below the levels achieved in OECD countries. The newly independent countries soon understood that a sound telecommunications infrastructure could play an important role in facilitating economic development. Since expanding the fixedline infrastructure required massive investments, cellular phones offered an interim approach to address the unmet demand for telecommunications service. Although cellular service was expensive, it could be introduced quickly, and it would be enough to meet the needs of government officials and people traveling for business.

A few patterns emerge from an analysis of the early cellular systems in Central and Eastern Europe. For one, only European standards were used in these countries (NMT [Nordic Mobile Telephone] 450 and later GSM). Moreover, all these countries relied on NMT 450 technology when they set up their first cellular systems in the early 1990s, even though many Western observers considered such technology obsolete at the time. There were, however, at least two valid reasons for doing so. One was that Russian military equipment used the 900 MHz band and, even after the withdrawal of the Russian army from Central Europe, it was not clear how quickly that band would become available for cellular services. The other was that the early systems in Central Europe were set up with a focus on coverage rather than capacity, and thus a 450 MHz system (with larger cells) was more appropriate than one operating on the 900 MHz section of the spectrum.[23]

In addition, all the Central and Eastern European systems were set up by the domestic PTT in collaboration with a foreign partner—usually US West and Bell Atlantic from the United States, and the PTTs from the Scandinavian countries. Finally, all of the Central and Eastern European countries eventually transitioned to a GSM system, usually with the help of a foreign partner.

NEW USES: SMS

At the time of writing, people are used to having access to a wide array of data services through their cellular phones. The original "killer data app" for mobile phones, however, was neither Internet access nor email but rather SMS—the ability to send and receive text messages on cellular phones.[24] Originally limited to 160 characters, text messages were feasible much earlier than they were popular: the first text message was reportedly sent from a personal computer to a mobile terminal in 1992, but SMS did not really take off until the late 1990s.[25]

One reason for the lag in the diffusion of SMS technology was that, up to the late 1990s, cellular carriers thought of themselves as being in the voice-telephony business, whereas SMS was viewed as a component of the paging business. It did not make much sense for them to spend too many resources promoting SMS against what was a strongly competitive paging industry in both Europe and the United States. In the late 1990s, however, the frontier between the paging and the cellular industries became blurred. In addition, there was a confluence of factors that contributed to raising the visibility of SMS starting around 1998, including the evolution of mobile-handset technology, the expansion of the commercial Internet, the popularity of prepaid cards for cellular service, and the increasing popularity of mobile-commerce applications.

Mobile handsets evolved considerably during the 1990s. Older handsets were often "receive only," so that they could be used only for receiving "broadcast" data services: users were able to read stock market quotes or sports results on their terminals. By contrast, handsets introduced in the late 1990s allowed users to write and send their own messages. This capability enabled cellular subscribers to rely on SMS in circumstances when there was an advantage to using text rather than voice or in addition to voice, such as in situations when text enhanced the privacy of the communication or when it was difficult to hear a voice conversation.

The advent of the commercial Internet in the mid-1990s also contributed to enhancing the value of SMS, and it did so through network effects.[26] Since it was feasible to send a text message from a personal computer to a mobile handset (and back), after 1995 the number of potential SMS users grew not only because the number of cellular subscribers increased but also because the number of people connected to the Internet did. Several mobile carriers in Europe and the United States set up SMS on their websites, so that any user of the carrier's website was able to send a text message via the Internet to any of the carrier's subscribers.

4.4 Vodafone advertisement promoting SMS, a key feature of 2G cellular phones. *Source*: Martin Campbell-Kelly.

The popularity of SMS increased in the late 1990s also because network operators started offering a range of e-commerce applications via text messaging, including mobile banking and mobile payments. In addition, SMS became an important tool for mobile workers to communicate with their offices.

The main reason why SMS grew in the late 1990s was that cellular-phone use rose among the youth. Innovative tariff options allowed consumers, including many young people, to obtain access to mobile phones, and SMS became particularly popular among them. Young people started transferring and adapting to mobile texting many of the abbreviations, acronyms, and text-based emoticons used in Internet messaging. The technological constraints of mobile texting gave rise to a new and unique SMS argot, and young mobile-phone users took pride in being able to master the SMS jargon.[27]

THE TRANSITION TO DIGITAL CELLULAR: TAKING STOCK

The 1990s represented a new era in the history of cellular phones. During the 1990s, the nature and dynamics of mobile-phone markets changed.

Some of the transitions—from analog to digital cellular, and from FDMA to TDMA and CDMA—happened under the hood, so to speak. Although users were mostly unaware of them, these changes improved the mobile-phone experience considerably. The short-term implications of changes in cellular technology were less significant than the long-term ones, since new technologies laid the foundation for the mobile Internet experience that arrived not in the 1990s but in the 2000s.

During the 1990s, cellular phones became gadgets that consumers were able to have access to. This transformation came about because the 1990s were the era in which many cellular markets felt the impact of competition for the first time, and competition drove product and service prices down. In addition, semiconductor progress lowered the cost of many components, which further contributed to lower cellular product prices.

The 1990s were also the time when many countries—especially in Latin America, the Caribbean, Africa, Asia, and Central and Eastern Europe—obtained access to cellular-phone services for the first time. In short, the 1990s were the decade when cellular phones diffused from the wealthy to the masses, and from a limited set of countries to the far ends of the world.

5

CELLULAR SYSTEMS IN THE WORLD OF 2G: BRITAIN, THE UNITED STATES, CONTINENTAL EUROPE, CHINA, AND INDIA

5.1 A phone for the Orange network. Britain's fourth GSM network was launched in spring 1993 with the catchy slogan "The future's bright, the future's orange." To initiate first-time customers, the launch pack included a Nokia 2140 handset and comprehensive user guides. Courtesy Nigel Linge and Andy Sutton.

National governments changed their approach to cellular markets in the 1990s. During the 1980s, most of the governments that allowed for the introduction of cellular phones viewed the state-owned telecommunications operator as the only entity entitled to provide cellular services. There were a few exceptions, and the main ones were Britain and the United States.

The British government was unique in Europe, not only because it encouraged competition between two carriers in a national cellular market from the beginning but also because it took systematic measures to protect the private operator from potentially anticompetitive behavior on the part of the state provider. The US government also fostered cellular competition from the start, although it did so through a collection of regional duopolies.[1]

In the 1990s this situation changed in three ways. First, the countries that had introduced cellular services under a duopolistic market structure in the 1980s took steps to enhance competition by licensing additional carriers in the 1990s. Second, during the 1990s, most of the countries where cellular markets operated in monopolistic fashion during the 1980s decided that the time had come to allow for competition. Third, most of the countries that did not introduce cellular phones during the 1980s authorized the launch of cellular networks at some point during the 1990s. These transformations can be labeled "enhanced competition," "competition for the first time," and "cellular for the first time," and in this chapter we analyze all three.

In the 1990s cellular phones became a worldwide phenomenon, and it is impossible to track in detail all national markets in a book of modest length. Thus, we analyze in detail a small number of markets on each continent. Generally, we focus on national markets that were, or would become, "large" in the overall scheme of the mobile-phone industry—in the sense that they accounted for, or would eventually account for, large fractions of worldwide users. We also cover a number of countries that are interesting from a comparative perspective, because they enable us to examine the impact of different telecommunications regulatory regimes.

In this chapter, we study the British and the American markets as cases of "enhanced competition." As examples of "competition for the first time," we focus on three countries in Europe (Germany, Italy, and

Portugal) and one in Asia (China). Finally, we also examine India as a case of "cellular for the first time." Other large markets, including Japan and South Korea, are covered in subsequent parts of this book.

As a prelude to this chapter, table 5.1 summarizes the history of cellular networks in Western Europe in the 1980s and 1990s. As the table shows, in the first half of the 1990s most governments licensed competing digital networks based on the European Global System for Mobile Communications (GSM) standard. The table also shows that many of them allowed for the entry of additional competitors later on, all of which operated under the Digital Cellular System (DCS) 1800 standard, the version of the GSM standard originally specified for Personal Communications Networks (PCNs) in Britain.

ENHANCED COMPETITION: BRITAIN AND THE UNITED STATES

The British case of enhanced competition is important for several reasons. Britain pioneered competition not only among analog cellular operators in the 1980s but also among digital operators in the early 1990s. Moreover, competition between digital cellular operators in Britain took an institutional form that would soon become pervasive all over the world: whereas one of the operators was usually the state provider, the competing operator was selected among many consortia, each one linking a powerful domestic company with experienced foreign cellular carriers.

The American case is interesting for two reasons. First, the United States was the largest cellular market in the world in the 1990s. Second, the US government relied on auctions to allocate spectrum for cellular services starting in late 1994. It was not the only country to use auctions in the 1990s—New Zealand, Ireland, Italy, and Greece used them too—but it was among the first to do so.

THE BRITISH MARKET: PCNS, GSM, AND INTERNATIONAL CONSORTIA

During the 1980s, competition between the two licensed operators—Cellnet and Vodafone—stimulated rapid growth in the British cellular market, even though the service prices charged by each of the duopolists remained relatively stable and similar to those charged by the other.[2] In

Table 5.1 Start of service for cellular systems in Western European countries, 1G and 2G

	Service started (1st and 2nd operators)					
	Analog		GSM		DCS 1800	
Country	1st	2nd	1st	2nd	1st	2nd
Sweden	1981	1981	1992	1992	1996	1996
Britain	1985	1985	1992	1994	1993	1994
France	1985	1989	1992	1992	1996	
Denmark	1982		1992	1992	1997	1997
Finland	1982		1992	1992	1995	
Portugal	1989		1992	1992	1998	
Germany	1985		1992	1992	1994	1997
Norway	1981		1993	1993		
Greece			1993	1993	1998	
Luxembourg	1985		1993			
Spain	1982		1995	1995	1999	
Netherlands	1985		1994	1995	1998	1998
Italy	1985		1992	1995	1998	1999
Austria	1984		1993	1996	1997	1998
Belgium	1987		1993	1996	1998	
Iceland	1986		1994			
Switzerland	1987		1993			
Ireland	1985		1993			

Sources: H. Gruber, *The Economics of Mobile Telecommunications* (Cambridge: Cambridge University Press, 2005), 65–143; G. Garrard, *Cellular Communications: Worldwide Market Development* (Boston: Artech House, 1998), 95–312; J. Funk, *Global Competition between and within Standards: The Case of Mobile Phones* (Houndmills, UK: Palgrave, 2002), 43–45.

early 1989 the Department of Trade and Industry published the "Phones on the Move" document, which argued that, although the analog cellular market had grown at a brisk pace in the 1980s, it had remained restricted mostly to businesspeople. It was necessary to implement new cellular technologies that would usher in a consumer cellular market in Britain. The Department of Trade and Industry invited suggestions on how such market could be developed, simply specifying the portion of spectrum where the new technology would operate.

The new services to be delivered were called Personal Communications Networks, and they were not supposed to compete directly with the existing cellular services, at least at the beginning. Although the PCN standard would be based on GSM, the "Phones on the Move" document suggested that the PCN systems would have several distinguishing features: they would operate on a higher frequency band (around 1,800 MHz as opposed to 900 MHz, where GSM operated), would have smaller cells than GSM, and would be ideally suited for light, hand-portable terminals. The document was conceived as a consultation paper and, as such, left several features of the PCN systems undefined. For example, it raised the question whether PCN services should include the call-handover feature but did not take a firm stance about it, leaving it open for public comment.[3]

In 1989, the British government issued three PCN licenses, all of them to consortia that linked a well-known British company with foreign partners. Cable & Wireless, Motorola, and Telefónica participated in the Mercury consortium; British Aerospace, Pacific Telesis, and Millicom were the main players in the Microtel partnership; and STC, US West, Thorn EMI, and Deutsche Bundespost partnered to form the Unitel consortium.

The government requested that the European Telecommunications Standards Institute (ETSI)—the organization that had been responsible for defining the GSM standard in 1989—develop a standard based on GSM for the implementation of PCN services in Britain. In 1991 ETSI published the specifications of such standard: it was called DCS 1800 and was a derivative of the GSM standard designed for low-power, handheld terminals. DCS 1800 was developed to deliver mass-market mobile-phone services in densely populated areas through small cells and was conceived to interoperate smoothly with GSM.[4] Before the PCN launch, Mercury and Unitel merged into a firm called Mercury PCN, and the ownership structure of Microtel changed, with British Aerospace selling to Hutchison Whampoa, a Hong Kong–based conglomerate.

Almost four years went by between when the licenses were awarded in 1989 and the time the PCN networks started operating. Mercury PCN launched its Mercury One-2-One network in 1993, and Hutchison inaugurated its Orange service the following year. The delay between the award of the PCN licenses and the start of service allowed the incumbents, Vodafone and Cellnet, to build up their GSM systems: Vodafone started GSM service in 1992 and Cellnet followed in 1994. Thus, by 1994 there were four providers of mobile-phone services in Britain—Cellnet, Vodafone, One-2-One, and Orange—and six networks: two Total Access Communications System (TACS), two GSM, and two PCN.

Although GSM and DCS 1800 were not originally conceived to compete with each other, the frontier between GSM and DCS 1800 systems quickly faded.[5] Understanding how this happened requires us to take a brief look at the cellular market in Germany, where a DCS 1800 license was granted to the E-Plus consortium. E-Plus wanted to compete with the two German GSM operators by offering nationwide coverage but realized that it would be extremely costly to offer such coverage with the DCS 1800 standard as originally defined. Since this standard was developed for cells with small radius (microcells), numerous base stations would be required to cover the whole German territory. Thus, E-Plus requested that ETSI modify the DCS 1800 standard to allow for larger cells. In response, the GSM operators asked ETSI to modify the GSM standard, so that GSM would be competitive with DCS 1800 through small cells in densely populated areas. This competitive dynamic of encroachment between GSM and DCS 1800 carriers was evident not only in Germany but also in Britain: even before the first PCN services were launched in Britain (on the DCS 1800 standard), Vodafone, which had a GSM license but no PCN license, announced its plans to introduce microcells to attract what it viewed as the typical PCN customers.

The entry of the PCN operators had a significant impact on the price of cellular services in the country. Between early 1987 and late 1992, Vodafone and Cellnet had the same rental fee of twenty-five pounds per month, and the same peak and off-peak charges, which varied depending on whether users were in the London area. Over that six-year period, for example, both companies charged thirty-three pence per minute for calls during peak times in and around the city of London. Starting in mid-1992, and anticipating the entry of the PCN operators, Vodafone and Cellnet started introducing packages for low-traffic users, with lower fixed charges and higher variable charges. They also lowered some of their tariffs and eliminated the London premium on some programs. After its launch in April 1994, Orange started charging users by the second (as opposed to in thirty-second or one-minute increments, as the incumbents did), which put additional pressure on Cellnet and Vodafone.[6] From the beginning, PCN operators and the incumbent cellular carriers tended to provide similar services, and whatever differences existed between them early on disappeared in 1996, when the incumbent cellular providers were allocated spectrum in the same spectrum band where PCN services operated.

Price cuts and the proliferation of tariff plans tailored to specific groups of users had a powerful impact on the market, with the number of subscribers growing from about 5.7 million in 1995 to about 43.5 million in the year 2000. By that year, Britain had become the fifth-largest cellular market in the world, and it ranked sixth in the world in cellular density, which increased from about ten in 1995 to almost seventy-four in 2000.

Figure 5.2 compares cellular density in Britain with that in the cellular pioneers (the United States, Japan, and Sweden) between 1979 and 2000.

Several facts stand out. First, in all countries—with the exception of the United States—cellular density grew faster from the mid-1990s, which points to the impact of enhanced competition. (Density started growing faster in the United States as well, but later than in the other countries included in the figure.) Second, the British performance during the 1990s was exceptional, in the sense that, by the year 2000, it had caught up with, and surpassed, all of the pioneering countries in cellular density. This is remarkable because Britain introduced cellular phones only in 1985. The British performance provides evidence of the success of a model of radical telecommunications reform in which competition was encouraged early (in both the analog and the digital periods) and an independent telecommunications regulator was established at the very beginning to monitor the competitive process.

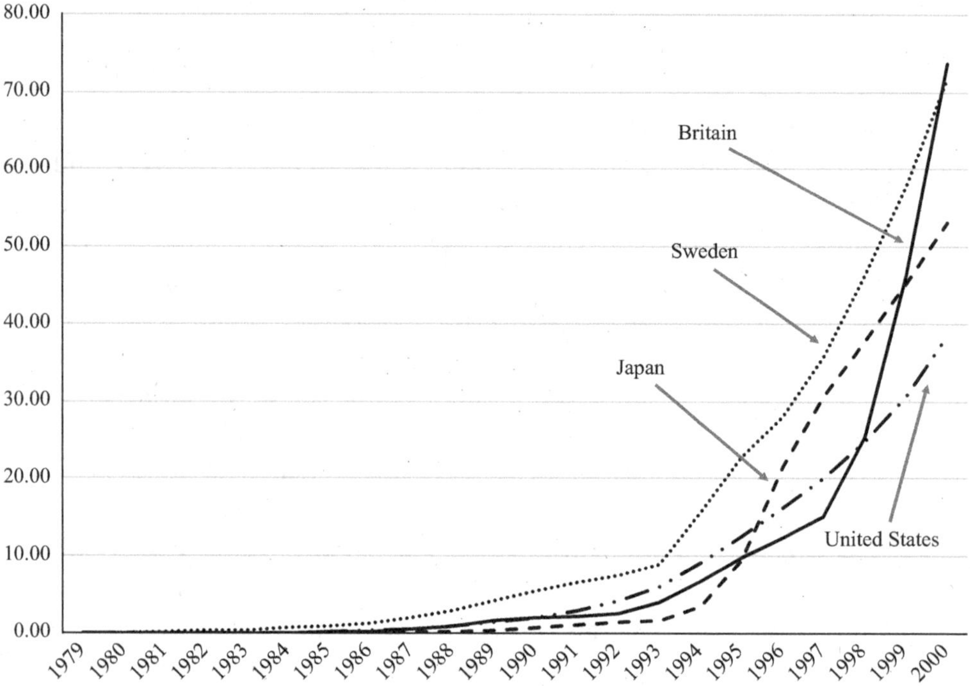

5.2 Cellular density in Britain, the United States, Japan, and Sweden, 1979–2000. *Source*: International Telecommunications Union (ITU) database.

THE AMERICAN MARKET: PERSONAL COMMUNICATIONS SERVICES AND SPECTRUM AUCTIONS

In Europe, the switch from analog to digital cellular facilitated the introduction of a single standard, GSM, for the continent as a whole. (Several European countries also awarded DCS 1800 licenses, but DCS 1800 was a derivative of GSM and interoperated well with it.) In the United States the situation was different, since the country already had a single standard (Advanced Mobile Phone Service [AMPS]) during the 1980s.[7] In Europe, the introduction of GSM was also an opportunity to proceed with the liberalization of cellular markets in most countries, whereas in the United States there was already cellular competition in the 1980s. Thus, there was no perceived "need" in the United States to rush the transition from analog to digital cellular, which explains why the adoption of digital cellular happened later in the United States than it did in Europe and Japan. The Federal Communications Commission (FCC) did not sanction a single standard for digital cellular services, and three technological alternatives arose in the early 1990s—D-AMPS and cdmaOne were the main options, and a third, N-AMPS, was considered but then discarded. In addition, during the 1990s the United States also witnessed the launch of cellular networks that adhered to a version of the European GSM standard.

During the 1990s, the United States followed the British example and attempted to create a consumer market for cellular services via Personal Communications Services (PCS), the American equivalent to the British PCN. Whereas Britain granted three PCN licenses in 1989, it was only in 1995 that the first PCS licenses were awarded in the United States. For the allocation of PCS licenses, the FCC changed the regional licensing units from metropolitan statistical areas and rural statistical areas to fifty-one larger entities known as major trading areas, subdivided into 493 basic trading areas. It also changed the licensing approach from beauty contests to auctions.

It was decided that there would be two licenses, the A and B blocks, in each major trading area. In addition, each basic trading area would have four licenses—blocks C through F. In theory, any town in the country could be covered by as many as six PCS operators. Blocks A and B were auctioned off in 1994–1995; block C, in 1995–1996; and blocks D through F, in 1996–1997. The auctions, combined, raised about $10 billion.

Following its market-oriented philosophy, the FCC did not sanction a national standard for PCS services. PCS licensees adopted three different cellular standards: two digital systems developed for the American market, D-AMPS and cdmaOne, and a version of the European GSM standard adapted to the frequencies allocated to cellular in the United States. Toward the end of the 1990s, the frontier between PCS services and digital cellular faded and, additionally, there was a trend toward the formation of national (as opposed to regional) cellular networks. The fact that competition intensified in the late 1990s likely explains why, as figure 5.2 shows, cellular density started growing faster around that time in the United States, later than in many European countries. We provide more details on these trends in the next chapter when we discuss the history of AirTouch.

By the year 2000, the United States was the largest cellular market in the world. The number of subscribers rose from 33.8 million in 1995 to 109.5 million in the year 2000. Cellular density in the United States increased from about thirteen in 1995 to almost thirty-nine in 2000, a level that, although much higher than just a few years earlier, was low by Organisation for Economic Co-operation and Development standards.

COMPETITION FOR THE FIRST TIME IN EUROPE: GERMANY, ITALY, AND PORTUGAL

It was during the 1990s that most European countries experienced competitive markets in cellular network services for the first time. The introduction of competition in cellular services in Europe in the early 1990s did not come out of the blue. There was a movement in the late 1980s within the European Commission toward the liberalization of telecommunications markets, a goal that was resisted by many governments and their postal, telegraph, and telephone (PTT) administrations. Introducing competitive markets in cellular network services first was, to some extent, a compromise: by embracing cellular competition, governments were able to align themselves with the liberalization directives emanating from the commission without undermining what at the time were important sources of revenue for the PTTs. Some PTTs eventually warmed up to the idea of competition in cellular services, which they came to view as an experiment that would allow them to prepare for the wider liberalization that they were bound to face anyway.

We have included three case studies in this chapter: Germany, Italy, and Portugal. Germany was one of the first countries in continental Europe to have two digital networks and a private consortium running one of them: two competing GSM networks started operating at about the same time, in mid-1992. It was also one of the first countries with an independent telecommunications regulator in 1989.

Italy was different from Germany in two ways. First, there was a lag between the launch of the first GSM network, run by the PTT, in 1992 and the inauguration of the second network, operated by a private competitor, in 1995. Second, an independent telecommunications regulator was not established in Italy until the late 1990s. Portugal, finally, did not have cellular phones for most of the 1980s. However, once liberalization and competition under an independent regulator were introduced in the late 1980s and early 1990s, the national cellular market grew at a rapid pace.

THE GERMAN MARKET: BEAUTY CONTESTS AND GSM COMPETITION

The analog phase of cellular phones in Germany demonstrated both engineering achievement and limited market acceptance.[8] Netz-C, the cellular system that Siemens developed for the PTT, was technically advanced but also proprietary, and thus closed. As a consequence, there was no competition among equipment suppliers, and handsets remained expensive throughout the 1980s. Moreover, the telecommunications monopoly of the Deutsche Bundespost (DBP) was sanctioned by the constitution, and the idea of introducing competition in the supply of cellular services was anathema. Cellular tariffs were set high, among the highest in Europe, and by the late 1980s Germany was considerably behind the Nordic pioneers in cellular density.

In 1988 the government started entertaining the possibility of introducing competitive markets in telecommunications especially in sectors, such as mobile phones, that had a limited impact on DBP's revenues. Entry of a cellular competitor was planned to coincide with the introduction of GSM, the new European standard. The 1989 reform legislation created three distinct public corporations—one for telecommunications, another for postal services, and a third for financial services. It was determined that DBP Telekom, the public telecommunications corporation, would operate one of the GSM networks, the D1 system, and that a private competitor would be granted a license for the other GSM network, the D2 system. The Bundesministerium für Post and Telekommunikation (BMPT), which up to 1989 had had both operational and regulatory responsibilities, became an independent regulator in charge of adjudicating disputes among the competitors.[9]

The German government used a beauty contest to allocate the D2 license. Beauty contests, also known as administrative hearings or criteria contests, encompassed several methodologies and institutional arrangements. Some included transparent and well-defined quantifiable criteria such as spectrum fees and commitments to geographic coverage. Others resembled black boxes and often seemed to rely on discretionary decisions. The beauty contest for the D2 system attracted ten consortia, almost all of them combinations of German companies and telecommunications operators from the United States and Europe. The winner was Mannesmann

Mobilfunk, a consortium that linked Mannesmann, an old-style German industrial company, with Pacific Telesis from the United States, Cable & Wireless from Britain, and Lyonnaise des Eaux from France.

The D1 and D2 systems started providing services in mid-1992. DBP Telekom charged Mannesmann two types of fees—interconnect and leased-line fees. Interconnect fees were those charged for the right to terminate a mobile call on the fixedline telephone network. Such fees depended on distance: the farther a mobile call was carried by the public switched telephone network after it was handed over by the cellular network, the higher the interconnect fee. To avoid high interconnect charges, the cellular operator had the option to carry the call as far as possible on its own transmission network. Doing so, however, had its own cost: to carry the calls as far as possible, the operator often had to lease transmission lines from the fixedline network operator, for which it paid lease-line fees. (In Britain, for example, Cellnet and Vodafone were the largest users of transmission capacity for many years.) Cellular operators had to assess relative prices—interconnect fees versus leased-line fees—and, on that basis, decide whether they were better off handing mobile calls over to the fixedline network as soon as possible.[10] Disputes between the competitors arose almost immediately regarding the interconnect fees and leased-line fees that DBP Telekom charged Mannesmann, and BMPT, the regulator, had to intervene and set caps on such fees.

In 1992, before the D1 and D2 digital systems started providing services, BMPT announced a tender for a DCS 1800 license, a competition that attracted two consortia. The license was won by E-Plus, a partnership that linked the German firms Veba and Thyssen with Bell South and Vodafone. Subscription fees declined considerably around 1992, when the market transitioned from a monopoly to a duopoly, and they declined again around 1994, when E-Plus entered.[11]

In 1997 a second DCS 1800 license was awarded to the Viag Interkom consortium, which linked the German firm Viag with British Telecom and Telenor from Norway. At this point, five cellular networks were operating in Germany: one of them was analog (Netz-C) and the other four—two GSM and two DCS 1800—were digital.

After the entry of E-Plus, the market experienced steady growth. Between 1995 and 2000, cellular subscriptions rose from 3.7 million to

48.2 million, and cellular density jumped from about five to about fifty-nine. By the year 2000, Germany had become the fourth-largest cellular market in the world.

THE ITALIAN MARKET: AUCTIONS, CONFLICT WITH THE EUROPEAN UNION, AND GSM COMPETITION

In the 1980s, Italtel, the main supplier of telecommunications equipment in Italy, developed a proprietary analog standard for the Società Italiana per L'Esercizio delle Telecomunicazioni (SIP), the domestic PTT. The Italian analog system retained the concept of calling areas, and call handover was possible only within areas. Further, because of the proprietary nature of the standard, there were only a few terminal suppliers, and SIP was the only entity from which subscribers could rent mobile devices. The number of subscribers grew slowly until 1990, when the government installed a second analog network that adhered to TACS, the British standard.[12] Even though subscriber growth accelerated, by 1991 Italy lagged behind the Nordic pioneers in cellular density.

In 1990 the Italian Ministry of Post and Telecommunications announced that it would start the legal process required to grant a GSM license to a private operator. (It was assumed that the PTT would operate its own GSM network.) From then on, the awarding of this license became entangled in the dynamics of Italian politics, including a debate regarding the termination of SIP's telecommunications monopoly, and several years went by before a private operator was able to launch a competitive GSM network in Italy. An independent telecommunications regulator—the Autorità per le Garanzie nelle Comunicazioni—was established only in 1997 and started its operations in late 1998.[13]

Bids for a GSM license auction were invited from private consortia in late 1993. The process had two steps: potential bidders had first to convince the government that they had the resources to build and operate a cellular network, and only then were they allowed to submit bids in the auction. Four consortia—all of them partnerships between established Italian companies and foreign telecommunications operators—submitted bids in a first round. In a second round, two of the consortia, Omnitel and Pronto, merged to form Omnitel Pronto Italia (OPI), which combined the strengths of several foreign cellular powerhouses, including Mannesmann from Germany, AirTouch and Bell Atlantic from the United States, and Telia from Sweden. OPI was awarded the GSM license.

The delay in granting a GSM license to a private operator allowed SIP, the domestic PTT, to both build up its TACS network and launch its own GSM network in 1992, far earlier than its private competitor. In mid-1995 the cellular-phone division of the PTT was spun off as Telecom Italia Mobile, and later that year OPI, the private competitor, launched its competitive GSM system. From then on, subscriber growth accelerated for several reasons. Both operators introduced innovative tariff packages to capture new groups of consumers. In addition, prepaid cards became the instrument of choice in Italy for obtaining access to cellular services. By 1999, prepaid mobile services accounted for about 75 percent of the Italian cellular subscriber base, a percentage that was much higher than in any other European country. In the late 1990s the granting of two DCS 1800 licenses intensified competition even further.

The Italian government was confronted by the European Commission regarding issues affecting the cellular market. One issue was that OPI had to make a substantial payment to the government to receive its cellular license, whereas the public operator, SIP (and later Telecom Italia Mobile), received its license for free. In late 1994 the commission argued that this asymmetry was anticompetitive, but the Italian government refused to address the problem. A year later, the commission gave the government an ultimatum under Article 90 of the Treaty of Rome, which deals with anticompetitive practices. Although the Italian government agreed to a deal that gave OPI a number of advantages, including reduced interconnection charges, the political disputes between the commission and the government continued for years.[14]

Although the introduction of cellular competition in Italy was delayed by bureaucratic wrangling, subscriber growth accelerated once the competing GSM network was launched in 1995. Between 1995 and 2000, the number of subscribers grew from about 3.9 million to around 42.2 million and cellular density rose from around seven to about seventy-four. By the year 2000, Italy had become the sixth-largest cellular market in the world.

THE PORTUGUESE MARKET: RADICAL TELECOMMUNICATIONS REFORM, GSM COMPETITION, AND RAPID GROWTH

The Portuguese experience with cellular phones during the 1980s was almost nonexistent.[15] Portugal was one of the poorest countries in Europe at the time and had limited fixedline telecommunications infrastructure. The country inaugurated an analog cellular system at the end of the 1980s simply because, in the context of a trade deal, Germany offered a Netz-C system at a subsidized price and Portugal accepted it.

In early 1989 the state telecommunications operator launched the network in Lisbon. Since Portugal modified the original Siemens system, the terminals used there were even more expensive than those used in Germany. This, combined with high service tariffs, led to slow growth in the Portuguese market in the first couple of years after launch.

The Portuguese cellular market started changing almost immediately, however, because in April 1989 the government decided to restructure the entire telecommunications industry. In October 1989 a new telecommunications law was passed that introduced changes almost as radical as those implemented in Britain in the first half of the decade.[16]

Up to that point, the two public operators—Telefones de Lisboa e Porto and Correios e Telecomunicações de Portugal—had functioned as both operators and regulators. The 1989 Basic Telecom Law established that, going forward, telecommunications operations would be segregated from telecommunications regulation. An independent regulator—the Instituto das Comunicações de Portugal—was created or, more precisely, activated: the entity had existed since 1981 but had played no role in telecommunications regulation in the country.

The 1989 law established that, even though it was the responsibility of the state to guarantee the availability of basic telecommunications services, private operators would be allowed to supply complementary and value-added services, as long as they obtained the proper licenses. This set the foundation for liberalizing the mobile market.

An invitation to apply for a GSM license in Portugal was issued in April 1991, and it attracted about a dozen foreign cellular operators, including Pacific Telesis, US West, Cable & Wireless, and BT. All of them formed consortia with domestic Portuguese companies, and a consortium named Telecel, which included Pacific Telesis, won the beauty contest. Of all the competing consortia, it was the most optimistic regarding the potential of the Portuguese cellular market, and that tilted the balance in its favor. While Telecel was in the process of planning its GSM network, the two state operators formed a joint venture—Telecomunicações Moveis Nacionais—to launch their own network. Both GSM networks—Telecomunicações Moveis Nacionais's and Telecel's—were inaugurated in October 1992.

Competition between the cellular networks drove terminal prices and service tariffs down, and the market soared almost from the beginning. Cellular prepaid subscriptions helped accelerate market growth. By the time a PCN system—such as the ones launched earlier in Britain and elsewhere—was set up in 1998, the Portuguese cellular market had already had several years of steady growth. Between 1995 and 2000, the number of cellular subscribers rose from about 341,000 to around 4.7 million, and cellular density grew from about three to around sixty-four.

Figure 5.3 presents the evolution of cellular density in Britain, Germany, Italy, and Portugal.

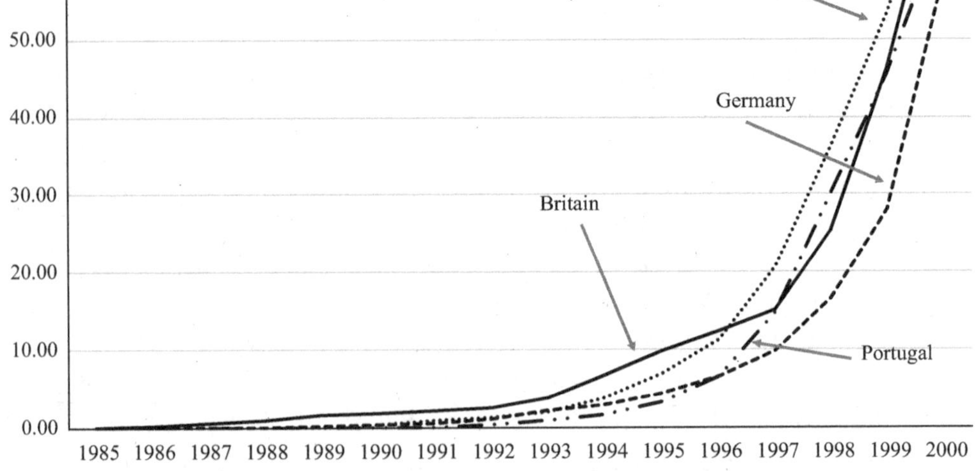

5.3 Cellular density in Britain, Germany, Italy, and Portugal, 1985–2000. *Source*: ITU database.

The figure highlights several important facts. In all these countries, with the exception of Germany, cellular density grew faster starting in the mid-1990s. This suggests that competition—first competing GSM networks, and later competition from DCS 1800 licensees—had an impact on market growth. Moreover, in Germany, growth accelerated later, in the late 1990s.

The figure also shows that Italy matched the British performance in cellular density, and Portugal almost did. This is impressive, since Britain was one of the best performers among Organisation for Economic Co-operation and Development countries during the 1980s and 1990s. This suggests that Portugal's model of radical telecommunications reform, which included competing cellular carriers under monitoring by an independent regulator, worked well. It also highlights the importance of the European Commission's intervention to ensure a level playing field in the Italian cellular market, and also the role played by prepaid cellular subscriptions in fostering growth in both Italy and Portugal.

A CASE OF COMPETITION FOR THE FIRST TIME IN ASIA: CHINA

China introduced competition in cellular services for the first time during the 1990s. The Chinese model of competitive markets, however, was different from the Western European model. In Western Europe, the private operators that received licenses to compete in domestic cellular markets during the 1990s were generally international consortia linking domestic firms with foreign cellular operators. In China, by contrast, cellular competition was, to coin a term, "bureaucratic" competition—it originated in an alliance formed by several powerful ministries that came together to establish a telecommunications company to compete with the Chinese PTT.

Since its creation in 1949, the Ministry of Posts and Telecommunications exercised control over postal and telecommunications services in China.[17] The minister was responsible to the premier of the State Council, the country's chief administrative authority. Until the mid-1990s, the powers of the ministry in telecommunications were all-encompassing: it controlled the operations of all postal and telecommunications enterprises, and it was responsible for development plans, technical standards, service policies, and regulations.

Even though the Ministry of Posts and Telecommunications was ultimately in charge of everything telecommunications-related in China for decades, the sector had a complex hierarchical structure. Below the ministry, and under its direct control, were thirty post and telecommunications administrations, which were responsible for overseeing telecommunications operations in China's twenty-one provinces, five autonomous regions, and four special municipalities (Beijing, Shanghai, Guangzhou, and Tianjin). Below the regional administrations, telecommunications operations were run by a variety of organizations, including enterprises devoted to equipment production, research institutes, and postal and telecommunications offices.

In late 1987 the Ministry of Posts and Telecommunications introduced the first analog cellular phones in Guangzhou, in the Guangdong Province. These early cellular terminals, which operated on the British TACS standard, were extremely expensive, and only high-level businesspeople, high-ranking government officials, and certain sectors of the military had access to cellular services.

Between 1987 and 1994, all thirty regional telecommunications administrations set up cellular systems in the geographic areas under their authority, each run by the regional telecommunications operator. Although the Ministry of Posts and Telecommunications did not mandate a national standard, in practice most of the regional systems adhered to TACS, which became a de facto standard. Given the high terminal prices and service fees, cellular became an important source of revenue for the regional telecommunications operators (and the regional post and telecommunications authorities that supervised them), even if adoption remained limited.

In the early 1990s three factors created pressure on the Chinese government to break the telecommunications monopoly. First, the demand for telecommunications services from consumers and businesses was rising rapidly with the steady growth of the economy, and the services provided by the Ministry of Posts and Telecommunications were increasingly viewed as subpar and insufficient. Second, on the international front, China's trading partners were putting increasing pressure on the country to open its telecommunications sector to foreign investment and competition, especially in light of China's intentions to join the World Trade Organization.

Third, several Chinese ministries were growing increasingly unhappy with the Ministry of Posts and Telecommunications' performance and monopolistic position. The Ministry of Electronic Industry, for example, which was the leading supplier of telecommunications equipment in China, had formerly benefited from government research grants and equipment orders, but in recent years it had become progressively excluded from access to telecommunications contracts. Other organizations were also dissatisfied: the growing gap between the demand for telecommunications services by several other ministries and the quantity and quality of services delivered by the state monopoly had led those ministries to set up networks for their own use. Among them were the Ministries of Railways, Electrical Power, Petroleum, and Coal Industry, as well as several banks and airlines. Many of these private networks were technically advanced, and the ministries that owned them resented being prevented from selling telecommunications services in the market.

In 1994 the rebellious ministries set up an entity that weakened the monopolistic role of the Ministry of Posts and Telecommunications: it was called China United Communications Corporation (China Unicom). The idea behind China Unicom was that it would provide telecommunications services to the public by combining the expertise of the Ministry of Electronic Industry in manufacturing telecommunications equipment with the other ministries' know-how in operating private networks. The proposal to create China Unicom led to confrontations at the State Council between the rebellious ministries and the Ministry of Posts and Telecommunications, but the State Council ended up authorizing the creation of China Unicom.

In 1994, the year China Unicom was created, the State Council ordered the Ministry of Posts and Telecommunications to modify its institutional structure so that, in the future, operations would be separated from regulation. Going forward, the ministry would become, in theory, an independent regulator and would relinquish operational responsibilities to an organization named China Telecom. Such an entity was indeed created in 1994, but the change did not bring about the consequences the State Council envisioned. The separation between the Ministry of Posts and Telecommunications and China Telecom was more nominal than real, not only because China Telecom continued to be managed by the ministry but also because the ministry did not function as an independent regulator. Between 1994 and 1998, the ministry created as many obstacles as possible to prevent the growth of China Unicom. For example, it forced

Unicom to pay for the full cost of building gateways between its cellular networks and China Telecom's fixedline telephone network.

Although the Ministry of Posts and Telecommunications discriminated consistently against Unicom between 1994 and 1998, Unicom managed to grow its cellular networks with the support of its founding ministries and the central Chinese government. Competition between China Telecom and China Unicom in the network services market resulted in declining tariffs, shorter waiting lists, and improved quality of service. It also resulted in the adoption of advanced network technology. Before Unicom entered the Chinese cellular market with its digital GSM networks in 1995, China Telecom was still relying on the analog TACS networks set up in 1987, even though GSM technology had been available since the early 1990s. Once Unicom started setting up GSM systems, China Telecom was forced to do the same.

Important institutional changes occurred in 1998, when the central government launched a large-scale reform. The number of ministries was cut from forty to twenty-nine—some were abolished while others were forced to merge. One of the driving forces behind the restructuring was, once again, the pressure exerted by the rebellious ministries, which pushed for the complete separation of the Ministry of Posts and Telecommunications and China Telecom. In March 1998 the government established a new Ministry of Information Industry, which combined the old Ministry of Posts and Telecommunications with its former rival, the Ministry of Electronic Industry.

The creation of the Ministry of Information Industry meant that, for the first time, telecommunications regulation would be separated from operations. Because it was created through the combination of opposing ministries, the new ministry behaved like a truly independent regulator that did not discriminate against China Unicom. The ministry reversed many of the policies that had handicapped China Unicom, and even forced China Telecom to provide roaming services to customers of China Unicom who were traveling in areas not served by Unicom.

Competition in cellular services and equipment drove prices down and facilitated adoption. Between 1995 and 2000, cellular subscriptions rose from around 3.6 million to about 85.3 million, and cellular density grew from about 0.3 to almost 7.0. By the year 2000, China had become the second-largest cellular market in the world.

5.4 A disguised cellular mast in the picturesque town of Wuzhen, China. *Source*: Wikimedia.

Important changes happened in the transition from the 1990s to the 2000s that laid the foundation for continued cellular growth in the twenty-first century.[18] For one, China Unicom started setting up CDMA networks in several regions. Although GSM networks continued to account for most of China Unicom's customers and revenues, CDMA networks attracted an increasing number of subscribers as the 2000s went by. Further, a new telecommunications technology called personal handy system (PHS) was approved, which soon started competing with traditional cellular services. PHS allowed fixedline customers to roam within their local phone area, using their handsets as if they were mobile phones. PHS was attractive to many customers because it was cheaper than traditional cellular and operated under calling-party-pays principles.

In 2000 the cellular services of China Telecom were spun off into a new corporation named China Mobile, and China Telecom started offering PHS service to its fixedline subscribers. Moreover, in 2002 a new fixedline operator was created under the name of China Netcom, and it also offered PHS service to its customers.

Thus, as of 2002 there were two entities in China providing traditional cellular services (China Unicom and China Mobile) and two offering fixed-line service (China Telecom and China Netcom). Given that the two fixed-line operators also offered PHS services to their customers, from 2002 there were effectively four providers of cellular or quasi-cellular services in the country.

A CASE OF CELLULAR FOR THE FIRST TIME IN ASIA: INDIA

The evolution of India's cellular market offers an interesting contrast with that of China's. First, whereas in China cellular phones were introduced in 1987, in India there were no cellular phones until the mid-1990s. Second, China had monopolistic cellular-phone markets in operation between 1987 and the mid-1990s, while India's cellular markets were competitive from the beginning. Third, in China cellular competition came about as bureaucratic competition, but India followed the Western European pattern of launching competitive markets in cellular network services by facilitating the entry of international consortia. Finally, both countries established an independent telecommunications regulator in the late 1990s—hence China had cellular markets that operated without a regulator for about a decade, whereas India had an independent regulator almost from the beginning. In short, India's cellular market had a strong British flavor.

Following India's independence from Great Britain in 1947, telecommunications in the country fell under the authority of the Ministry of Post and Telegraph, the national PTT. The provision of telecommunications services and the production of telecommunications equipment were government monopolies until the 1980s.[19] Telephones spread at a slow pace in the four decades after India's independence. By 1990, fixed-line phone density in the country was about 0.58, among the lowest in the world. In the 1980s Prime Minister Rajiv Gandhi started introducing changes in Indian telecommunications. Britain had separated postal services from telecommunications in 1981, and Gandhi followed the British example by creating the Ministry of Communications in 1986.

Gandhi planned to create a government-owned company to run the telephone business, again following the British example, but the bureaucracy of the Ministry of Communications opposed the change and convinced him to corporatize two businesses first as an experiment—telephone services in the metropolitan areas and long-distance service. Two government-owned telecommunications corporations were created in 1986, Mahanagar Telephone Nigam Limited to manage telecommunications in Bombay and Delhi, and Videsh Sanchar Nigam Limited to run international services. The rest of telecommunications services in India remained under the authority of the Ministry of Communications, with services run by the Department of Telecommunications. (The ministry also exercised considerable influence over Mahanagar Telephone Nigam Limited.) Despite the policy changes implemented in the 1980s, India did not introduce cellular phones during that decade. A car-based mobile-phone system was set up in New Delhi in 1986 but, by mid-1988, it had fewer than one hundred subscribers.

In 1990 a balance-of-payments crisis served as the catalyst for the introduction of more radical changes in Indian telecommunications. The following year, the Indian government entered into an agreement with the World Bank and the International Monetary Fund, both of which provided relief tied to certain liberalization conditions. The government created the so-called Athreya Committee, which made several recommendations toward the restructuring of Indian telecommunications: going forward, the policy-making role of the Department of Telecommunications should be separated from its operational role; private operators should be allowed to provide certain telecommunications services, including mobile-phone services; and an independent regulatory authority would have to be created to oversee these changes and act as a referee. The Ministry of Communications refused to allow the entry of private companies into any of the fixedline services but agreed that there was a role to be played by such enterprises in the provision of mobile-phone and other services.

In early 1992, and in line with the Athreya recommendations, the Department of Telecommunications invited bids in an auction for the supply of cellular services in the four metropolitan areas of Bombay, Delhi, Calcutta, and Madras. Two licenses would be awarded in each metropolitan area. Although the winners were announced toward the end of that year, licenses were not granted immediately because four of the losing bidders contested the process in court and protracted litigation ensued. Cellular service in the metro areas finally began in 1995. Because license fees were relatively modest, the operators survived, although five of the eight sold out within a few years.

In 1994, when the legal battles surrounding the 1992 cellular-licensing process were still in full swing, the government issued a National Telecom Policy (known as NTP-94). The policy acknowledged that the government did not have enough resources to make the investments required to address the unmet demand for telecommunications services, and thus would have to encourage the participation of private operators. The licensing model that had been implemented for cellular services in the metro areas would now be extended to other regions of the country, and it would be applied to both cellular and fixedline services. NTP-94 invited Indian private businesses, in partnership with foreign operators, to submit bids to provide fixedline and cellular services in twenty-one geographic circles that roughly corresponded to Indian states. Once again, to ensure that competition would discipline the carriers, two cellular licenses would be awarded in each circle.

Bidders had to show they possessed the capital required to build the telecommunications networks, and foreign operators would be invited to participate but only as minority partners in consortia led by Indian companies. License fees were to be paid over a ten-year period, and licenses would be awarded to the consortia submitting the highest bids.

The winners of the 1995 cellular and fixedline licenses were all Indian businesses, including some of the largest conglomerates, such as the Tatas. Most of the licenses, however, went to smaller Indian companies such as Bharti and BPL, both manufacturers of telecommunications equipment. Foreign operators that participated in the winning consortia as minority partners included AT&T from the United States, Nippon Telegraph and Telephone from Japan, Telstra from Australia, Bell Canada, the Swiss PTT, France Télécom, and private companies from the United States, Hong Kong, and Thailand. All of the consortia set up cellular networks that adhered to the European GSM standard.[20]

The Ministry of Communications—which made the rules, interpreted them, and acted as the arbiter of all telecommunications-related disputes—made life complicated for the newly licensed cellular carriers. It set a low ceiling on how much they were allowed to charge their subscribers, low enough to undermine the financial viability of the operators. Further, the Department of Telecommunications imposed an interconnection regime that handicapped the private cellular carriers: they had to pay a fee both when calls originated in the fixedline network and terminated in the cellular network, and when they went in the opposite direction. As a result of these conditions, by early 1997 six cellular operators were in default of their license-fee payments.

An institutional change gave cellular carriers some breathing room: in 1997 the government established the Telecom Regulatory Authority of India, which became the operators' key ally. The confrontations between the old-style bureaucracy and the newly created independent regulator started immediately. One of the first disputes involved the nature of interconnection agreements between the private operators and the PTT. Between 1997 and 1999, the Telecom Regulatory Authority expressed a strong preference for replacing the interconnection regime that had handicapped the cellular carriers with a calling-party-pays regime: in such a regime, the user making the call would pay a charge, regardless of whether the call started in the fixedline or the cellular network. The authority determined that the new interconnection regime would be formally introduced in November

1999. In this case, and others, the authority ruled in favor of the operators, but the bureaucracy of the Ministry of Communications consistently appealed those rulings at the High Court of Delhi, which tended to reverse them.

Despite the persistent opposition of the Ministry of Communications to the liberalization process, cellular subscriptions grew in India from about 77,000 in 1995 to about 3.6 million in the year 2000, with cellular density rising to 0.34. In any case, by the year 2000, India was the supreme example of a country where the growth of cellular density had been severely slowed by bureaucratic intervention in favor of the status quo.

At the end of the 1990s, the government made a change of direction and introduced the New Telecom Policy 1999.[21] This piece of legislation laid the foundation for faster growth in cellular density in the twenty-first century. Two measures included in the new policy were particularly consequential. First, it allowed the cellular operators to migrate from a licensing regime based on fixed annual fees to one based on charges proportional to revenues, which made it easier for the carriers to be consistently profitable. Second, the Department of Telecommunications and Mahanagar Telephone Nigam Limited were allowed to enter the cellular business; they introduced CDMA networks and a low-price service that intensified competition among operators.

The New Telecom Policy 1999 made cellular service profitable and unleashed a period of consolidation among cellular carriers. Five of the fifteen operators that had obtained licenses before 1999 sold out or forfeited their licenses, and two—the Birlas and the Tatas—merged their cellular operations to form a new company named Idea. At the end of a process of mergers and acquisitions, five companies—Bharti, Hutchison, Reliance, Escorts, and Idea—held forty-five out of a total of fifty-five cellular licenses.

THE TRANSITION TO DIGITAL CELLULAR: SUMMARY

The cellular world underwent three kinds of transformations in the 1990s. First, countries with duopolistic network service markets—including Britain and the United States—enhanced the competitive environment by awarding new licenses. Second, markets with a monopolistic supplier of analog network services—including Germany, Italy, and Portugal in Europe and China in Asia—became competitive when they transitioned to digital cellular services. Finally, another group of countries—including India—started to catch up and launched cellular services for the first time.

The process of introducing network-services competition where previously there was none was fraught. In Europe and elsewhere, the transition to competitive cellular markets coincided with a more comprehensive process of telecommunications liberalization. In many countries where the domestic PTT had acted as a monopolistic supplier of telecommunications services for decades, several stakeholders—in governmental agencies especially—had a rather pessimistic view of cellular competition and of telecommunications liberalization more broadly. Conflicts between representatives of the old order and those of the new order often generated delays in the liberalization process: in Italy, for example, a competing GSM network was launched only after a delay of several years.

Creating, or enhancing, competition in cellular network services demanded more than just granting licenses to new operators. Newly licensed private operators frequently competed with the cellular subsidiary of the domestic PTT, and this generated problems. In the early years, most calls initiated on cellular phones were destined for a fixedline user, and the fixedline network was owned and operated by the domestic PTT. This raised the issue of the level of the interconnect fee—that is, the fee that the private cellular operators paid the PTT for calls terminated on the fixedline network. The PTT had the capacity to raise the costs of rivals, which turned the interconnect fee into a potentially problematic subject.

To deal with problems such as the level of the interconnect fee, most governments at some point during the 1990s segregated the operational and regulatory functions of the domestic PTTs. Most countries created an independent telecommunications regulator, which ensured that private cellular competitors were not handicapped by some of the PTT's tactics.

Countries tended to perform better when they introduced competition *and* established an independent regulator early in the process. Britain and, later, Portugal are the best examples of this model. Italy, however, delayed the introduction of competition until the mid-1990s and the establishment of an independent regulator until the late 1990s, and nonetheless was able to catch up. Plausible explanations for its rapid growth in cellular density are that the European Commission played the role a regulator would have played had it existed, and that prepaid subscriptions contributed to speed up cellular growth in the late 1990s.

Competitive cellular network services tended to raise cellular density—but they did so through a different channel in the 1990s relative to the 1980s. During the 1980s, duopolistic network-services markets experienced either small subscription-tariff declines (the United States) or no decline at all (Sweden and Britain).[22] Thus, in the 1980s, competitive markets in network services increased cellular subscriptions per capita by raising capacity without lowering prices. In the 1990s, by contrast, competition (or enhanced competition) in network services raised cellular density not only by raising capacity but also by lowering subscription tariffs considerably. This was the case in the United States, Britain, Sweden, and elsewhere.[23]

During the 1990s, there were important developments in the two countries that would eventually become the largest cellular markets in the world—China and India. China introduced cellular phones earlier, in 1987, and had monopolistic cellular-service markets until 1994. In that year, competitive markets in network services were allowed—not by facilitating the entry of consortia of local and international operators but rather in the form of a telecommunications entity, China Unicom, sponsored (and owned in part) by several powerful state ministries. Cellular density grew faster in China after competitive markets were introduced, and especially after a truly independent regulator was established in 1998.

In India, cellular services were introduced only in the mid-1990s. Competitive markets were permitted from the beginning, but the traditional telecommunications bureaucracy managed to handicap the private cellular operators until an independent regulator was introduced in 1997. The cellular market in India grew faster after the regulator started intervening in favor of the private carriers.

By the year 2000, China's cellular market was much larger than India's—China had 85.3 million subscribers against India's 3.6 million. Further, China's cellular density was 6.84 against India's 0.34.

During the first decade of the twenty-first century, however, India was able to catch up with China. Figure 5.5 shows cellular density over

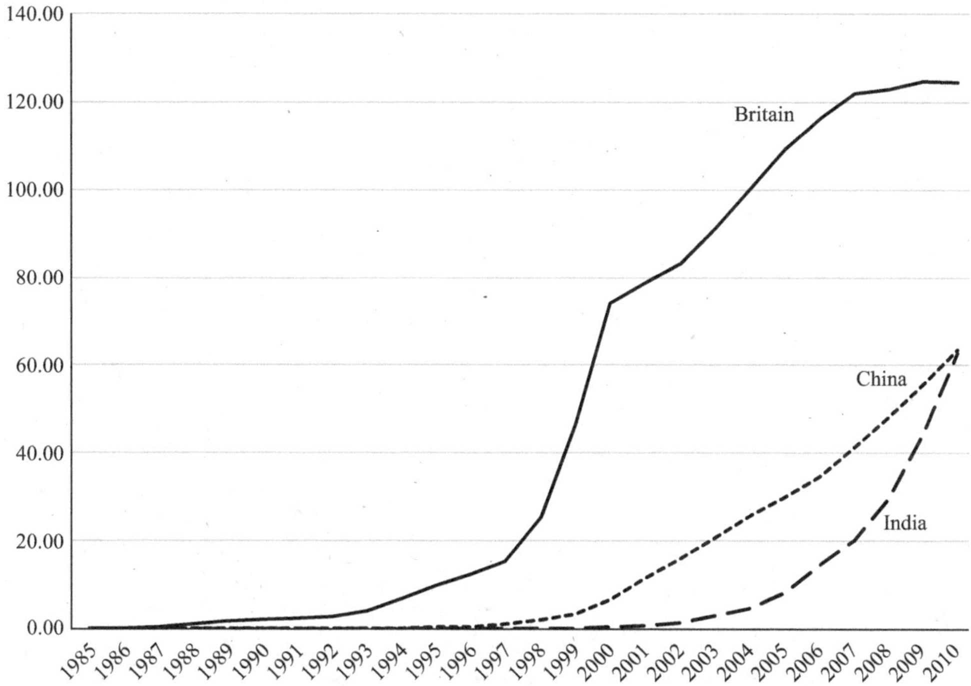

5.5 Cellular density in Britain, China, and India, 1985–2010. *Source*: ITU database.

time for India and China between 1985 and 2010. Britain is included for comparison.

In China, growth in cellular density accelerated slightly after China Unicom was created in 1994, and much more noticeably after an independent regulator was established in 1998. In India, growth in cellular density only accelerated in the late 1990s and early 2000s, after the creation of an independent telecommunications regulator that protected the private cellular carriers from the old telecommunications bureaucracy. The figure shows that cellular density grew much faster in India than in China during the first decade of the twenty-first century, to the point that, by 2010, both countries had roughly the same density.

Figure 5.6 presents a different perspective on the comparison between India, China, and Britain. The figure tracks cellular density at one to fifteen years since the launch of cellular networks in each country. By plotting trends this way, we can "control" for the fact that cellular started much later in India than in either Britain or China.

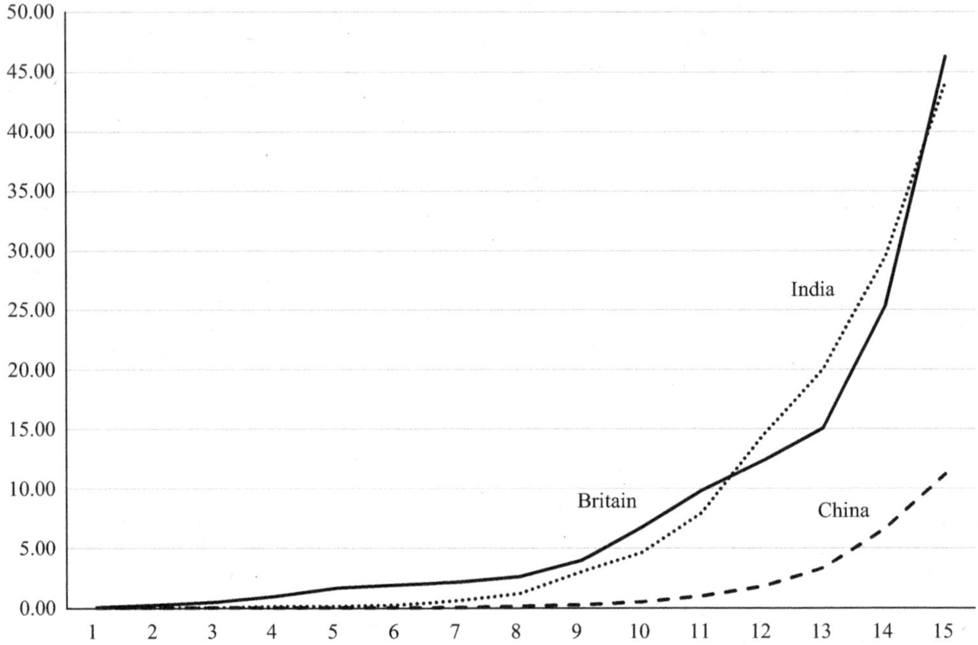

5.6 Cellular density in Britain, China, and India, 1–15 years after launch of service.
Source: Calculated by authors from ITU database.

The figure shows a remarkable fact: during the first fifteen years of cellular service, India performed as well as Britain, and China lagged considerably behind. The British model of fostering competitive markets from the beginning of service under monitoring by an independent regulator—a model that India tried to follow, although imperfectly—appears to have generated as good a performance in India as it did in Britain years earlier, at least during the first fifteen years of cellular service in each country.

6

COMPETING IN 2G CELLULAR MARKETS

6.1 The Ericsson LX100 phone, 1997. A typical lightweight cell phone marketed by AT&T wireless services. Courtesy of AT&T Archives and History Center.

During the 1990s, competition became more intense in cellular markets all over the world. In this chapter we examine the impact of competition in the markets for cellular network services, infrastructure, and phones.

We first track the history of AirTouch, a company that was born in 1984 under the PacTel name during the AT&T divestiture. The history of AirTouch allows us to take another look at some of the fundamental episodes in the history of cellular markets in the late 1980s and the 1990s.

PacTel (later called AirTouch) participated in the contest for the first private Global System for Mobile Communications (GSM) license in Germany and in the competition for one of the Personal Communications Networks (PCN) licenses in Britain. After Personal Communications Services (PCS) operators were licensed in the United States, PacTel participated in the process of consolidation that led to the rise of cellular giants such as Verizon Wireless. Further, it played a crucial role in supporting Qualcomm and the code division multiple access (CDMA) standard in the United States and beyond. And finally, AirTouch and Vodafone were the core actors in one of the most important mergers in the history of the cellular industry—the transaction that created Vodafone AirTouch. Covering the history of PacTel/AirTouch allows us to examine the connections between three pivotal companies in the history of the industry: PacTel/AirTouch itself, Qualcomm, and Vodafone.

The history of AirTouch sets the scene to analyze how competition in cellular services evolved during the 1990s. In this decade, two elements were new: the first, already emphasized, was that many markets went from monopoly to competition; and the second, closely linked to the first, was that the second licensee in many countries was usually not a local company acting in isolation but rather a consortium that linked a local company with foreign operators. In this sense, the industry became internationalized. As additional examples of this internationalization, in this chapter we examine the forays of American and Western European cellular carriers in the countries that used to belong to the Communist bloc.

Finally, we study competition in the supply of cellular infrastructure and cellular handsets. We analyze the role that product prices and attributes—and long-term relationships between equipment manufacturers and cellular carriers—played in these markets.[1]

AIRTOUCH: A CASE STUDY IN DOMESTIC AND INTERNATIONAL COMPETITION

Pacific Telesis was born during the AT&T breakup of 1984.[2] After the divestiture, the Regional Bell Operating Company based in California and Nevada became known as the Pacific Telesis Group. The Pacific Telesis cellular subsidiary, also known as PacTel, started operating a cellular network in Los Angeles in mid-1984. (We use the Pacific Telesis name for the holding company and PacTel for the cellular subsidiary, even though some sources also refer to the former as PacTel.)

PacTel participated in the frenzy of cellular transactions that took place in the American cellular market in the second half of the 1980s. Its main acquisition was Communications Industries, which owned shares of licenses in Phoenix, Saint Louis, Dallas–Fort Worth, Jacksonville, and Tampa, plus a stake in the San Francisco nonwireline license. While it expanded its cellular footprint in the United States, PacTel made international forays (with mixed fortunes) in China, Spain, Japan, and Britain.

The liberalization of the German market in the late 1980s was a watershed for PacTel as an international cellular operator. By the time Germany started considering granting a second license for digital cellular services, the D2 license, the country was the third-largest economy in the world, and its mobile market had great commercial potential. PacTel showed political savvy in aligning itself with Mannesmann, an old-style industrial company that led the consortium that won the digital license in 1989. A few days later, a consortium that linked PacTel with British Aerospace was awarded one of the PCN licenses in Britain.

The reunification of Germany in 1990 enhanced the value of the D2 license awarded to the Mannesmann consortium, since the German government decided to extend the license to East Germany. By the early 1990s, PacTel was facing the challenge of helping to launch digital cellular networks in two of the most promising markets in the world—Britain and a reunified Germany.

In its international ventures, PacTel was one among many American and Western European competitors. In the transition from the 1980s to the 1990s, the Soviet Union dissolved, and Central and Eastern Europe started undergoing a significant economic transformation. US West set up the first cellular system in Budapest, Hungary, in 1990, which was also the first cellular system in Eastern Europe. US West also joined Bell Atlantic to launch a network in Czechoslovakia in 1991 and participated in the construction of the first cellular systems in various Russian cities in the early 1990s.[3] (We discuss American and Western European forays in Central and Eastern European cellular markets later in this chapter.) Around the same time, Bell South made incursions in South America; Southwestern Bell acquired a major stake in Telmex, the Mexican telecommunications operator; and Bell Atlantic and Ameritech combined to buy New Zealand Telecom.[4] The Baby Bells, in short, used the experience they had gained while competing in duopoly cellular markets in the United States during the 1980s to acquire stakes in companies operating in major telecommunications markets—especially cellular markets—in other parts of the world during the 1990s.

Needing capital to fund its domestic and international expansion, PacTel started entertaining the idea of an initial public offering (IPO). In 1992 AT&T and McCaw Cellular struck a deal according to which AT&T would acquire one-third of McCaw Cellular Communications for USD 3.8 billion and would have the option to buy the rest later.[5] This transaction, which shocked the American wireless market, pushed cellular stocks up and solidified the idea of a PacTel IPO, which took place in late 1993, with the spin-off completed in April 1994. PacTel raised USD 1.57 billion and acquired a new name—AirTouch—in the process.[6] By the time PacTel became AirTouch, it was already an international wireless operator, with systems up and running (or about to be launched) in Germany, Portugal, Spain, Sweden, Japan, South Korea, and Thailand. It ended up selling its stake in the British PCN license, since it reckoned that dealing with the German and British markets simultaneously would strain its resources.

PCS LICENSES, QUALCOMM, AND THE ORIGINS OF VODAFONE AIRTOUCH

Soon, AirTouch faced new competition in the United States. The licensing policy implemented by the Federal Communications Commission (FCC) had generated a collection of regional cellular duopolies in the 1980s, and by the early 1990s studies suggested that cellular service prices had not declined as much as the agency expected they would. The FCC decided that additional competition would put pressure on the duopolists and that awarding PCS licenses would create such competition. The FCC also changed the spectrum allocation method from lotteries to auctions.

Many telecommunications companies were interested in the PCS licenses. Sprint and MCI, the long-distance carriers, had sold their first-generation cellular assets but were now interested in testing the waters again. Further, there was a new breed of cellular providers, such as Fleet Call (later renamed Nextel), that were also trying to grow their portfolios of wireless assets. Nextel was founded in 1987 by two telecommunications lawyers, Morgan O'Brien and Chris Rogers, and two investment bankers, Brian McAuley and Peter Reinheimer, as a dispatch service provider for commercial vehicles, and it accumulated a stock of dispatch frequencies around the country. (A fleet-dispatch service is one where a central operator, the dispatcher, assigns routes and stops to a fleet of commercial vehicles.) In 1991 the company received permission from the FCC to offer cellular services on those channels, despite the opposition of the more traditional cellular operators.[7]

The PCS license allocation process generated a frenzy of deals among the operators. In 1994 AirTouch and US West agreed to merge their cellular operations in the United States.[8] In 1995 AT&T acquired the totality of McCaw Cellular to form AT&T Wireless Services, and Bell Atlantic Mobile and NYNEX Mobile merged to set up Bell Atlantic NYNEX Mobile.[9] Most importantly, AirTouch, US West, Bell Atlantic, and NYNEX formed a joint venture named PCS PrimeCo to operate wireless networks in the PCS frequencies.[10]

During the 1994–1995 auctions, it became clear that three bidders—Sprint/APC, AT&T, and PCS PrimeCo—were attempting to acquire enough frequencies to set up national wireless networks, whereas the rest were pursuing regional strategies. After several months (and more than a hundred rounds) of bidding, the three companies with national strategies ended up obtaining a large portion of the frequencies: Sprint/APC spent USD 2.1 billion to acquire twenty-nine licenses covering 145 million potential subscribers; AT&T invested USD 1.68 billion to obtain twenty-one licenses covering 107 million potential subscribers; and PCS PrimeCo spent USD 1.1 billion to purchase eleven licenses covering 57 million potential subscribers. (In the wireless world, the potential subscribers associated with an area are usually the number of people located in that area.)[11]

The PCS auctions were important not only for AirTouch, which acquired multiple licenses through its PCS PrimeCo joint venture, but also for Qualcomm. By mid-1995, time division multiple access was quickly becoming the dominant digital standard in the American market. Qualcomm was heavily promoting the virtues of its CDMA technology but was still unable to show a single CDMA commercial network in operation. The tide turned in June 1995, when PCS PrimeCo announced that its PCS networks would adhere to the CDMA standard.[12] Soon thereafter, Sprint also expressed a preference for CDMA for its PCS systems. The decisions made by these large PCS operators—and especially by Sprint/APC, which was the largest—legitimized CDMA as a digital technology in the American market. By 1997, more than half of the PCS operators were implementing the CDMA standard on their networks.[13]

The choice of CDMA by PCS PrimeCo was the culmination of many years of CDMA support by AirTouch (and its predecessor PacTel). In the early 1990s, it was PacTel that came to the rescue of Qualcomm when the latter was running out of funds to develop the CDMA technology. It was also PacTel that convinced other Baby Bells, such as NYNEX, of the virtues of CDMA as a digital standard for cellular phones. And further, AirTouch not only promoted CDMA within the PCS PrimeCo alliance but also joined forces with the South Korean firm POSCO to acquire the second digital license in South Korea (and to then launch a CDMA network in that country).

During the 1990s, Vodafone and AirTouch crossed paths in many markets and ended up collaborating in cellular consortia in countries such as Germany, Italy, and Egypt. By 1996, executives of both companies were already discussing the possibility of combining their European cellular assets, but nothing came of those negotiations.[14]

In 1997 competition in the American market became more intense, a process that triggered the eventual merger of AirTouch and Vodafone. In that year, many of the PCS networks came online for the first time, and innovative marketing programs were introduced to capture new cellular customers. In 1998 AT&T changed competitive conditions in the US market when it announced its Digital One Rate plan, which allowed customers to pay one flat rate for a prespecified number of minutes of talk, without roaming charges anywhere in the country.[15] It was an aggressive move—the first one of its kind—toward implementing the concept of a truly national network in the American cellular market. Because it represented a substantial price cut, it generated all sorts of competitive responses. Soon after AT&T announced its national-rate plan, Bell Atlantic and GTE decided to merge, and rumors spread about the potential acquisition of AirTouch by Bell Atlantic.

After Bell Atlantic made an offer for AirTouch, Vodafone improved the offer by 25 percent and, after long negotiations, AirTouch chose Vodafone. The combination, completed in mid-1999, gave rise to Vodafone AirTouch, a company with a USD 110 billion market capitalization. At the time, it was Britain's largest telephone firm, and the third-largest company on the London Stock Exchange.[16]

ORIGINS OF VERIZON WIRELESS AND EXPANSION OF VODAFONE AIRTOUCH

The combination created a firm with a powerful international presence but did not address the competitive challenges AirTouch was facing in the US market: in the late 1990s AirTouch was losing customers to AT&T and Sprint, both of which were pursuing a strategy of creating a national cellular network with no roaming charges. A new corporate combination was required in order to compete with the likes of AT&T and Sprint in the United States, and such a combination was created in May 2000: Vodafone AirTouch and Bell Atlantic–GTE reached an agreement to form a new business, of which the former would own 45 percent and the latter the remaining 55 percent. It was called Verizon Wireless—it had about 28 percent of the American cellular market at the time, and it was the largest cellular operator in the country.[17]

The birth of Verizon Wireless in the year 2000, in turn, put additional pressure on many cellular operators in the United States. SBC and Bell South announced that they would combine their domestic wireless assets under the Cingular name. AT&T Wireless had its IPO soon thereafter, and it was completely spun off in 2001. The announcement of the Digital One Rate plan by AT&T in 1998 generated a sequence of corporate transactions, many of which gave rise to organizations—including Verizon and Cingular—that would be major players in the American cellular market for many years to come.

At the end of the decade, Vodafone Airtouch was involved in one additional corporate transaction of significance. By the year 2000, the German firm Mannesmann had almost completely transformed itself from an old-style industrial company into a cellular company. It had also made telecommunications investments in Austria, Italy, and France. When E-Plus won the Digital Cellular System (DCS) 1800 license in Germany, Vodafone, which was a member of the E-Plus consortium, invaded Mannesmann's turf as a competitor. In the late 1990s Mannesmann announced that it was in conversations with Hutchison Whampoa to buy Orange and become Vodafone's competitor on Vodafone's own turf. It was a tit-for-tat response to Vodafone's entry into the German cellular market.

Vodafone AirTouch responded by attempting a hostile takeover of Mannesmann. The pressure that Vodafone put on Mannesmann's shareholders finally bore fruit and, in early 2000, executives from both firms announced that they had reached a friendly agreement to combine their companies.[18] Vodafone continued its shopping spree with cellular acquisitions in Europe and Asia. By 2001, Vodafone AirTouch had become the largest telecommunications company in the world. At that stage, the company changed its name again, this time to simply Vodafone.

AMERICAN AND WESTERN EUROPEAN CELLULAR FORAYS IN CENTRAL AND EASTERN EUROPE

AirTouch was not the only American cellular carrier that pursued opportunities in foreign markets, although it was one of the most aggressive. Other US operators also explored emerging cellular markets, and a few of them ventured into what was, in the late 1980s and early 1990s, the new frontier for capitalism: the countries of Central and Eastern Europe that up to that point had belonged to the Communist bloc. In 1989 all of these countries started transitioning toward a market economy. Among American carriers, US West was the cellular pioneer in the formerly Communist countries, whereas AirTouch was more risk-averse and decided to wait until later. When they arrived in the formerly Communist countries,

US cellular operators faced competition from their Western European peers. To provide a flavor of cellular competition in Central and Eastern Europe, we examine developments in Hungary and Russia.

HUNGARY

In Hungary, the state-run Hungarian Post had been in charge of all postal and telecommunications activities since 1964.[19] In 1988 the government decided to allow foreign investment in the domestic telecommunications sector.

In many ways, the transition to a liberalized telecommunications sector followed a pattern observed in many Western European countries. In the late 1980s the commercial and regulatory activities of the government were segregated, with regulatory responsibilities transferred from the Hungarian Post to the Ministry of Transport, Telecommunications, and Construction. A 1989 act split the Hungarian Post into three separate service providers: the Hungarian Telecommunications Company, later renamed MATAV; the Hungarian Broadcasting Company; and the Hungarian Post Company. The Hungarian Telecommunications Company became a joint-stock corporation, owned initially by government agencies.

In late 1990 the first cellular network in the country—and in Central and Eastern Europe—was launched. The system, which adhered to the Nordic Mobile Telephone (NMT) 450 standard developed in Scandinavia, was operated by Westel, a joint venture between the American carrier US West, which owned 49 percent, and the Hungarian Telecommunications Company, which owned 51 percent.

In the joint venture, US West was responsible for marketing and technical support, and the Hungarian postal, telegraph, and telephone (PTT) administration was in charge of setting up the network. The Americans wanted to promote the network, but the Hungarians believed marketing was not necessary—and in this regard they were right: before launch, the waiting list had several thousand names on it.[20] By April 1996, the network had almost sixty-four thousand subscribers.

Hungary was not only the first ex-Communist country to inaugurate an analog cellular network but also the first to allow for the introduction of competing digital systems. In September 1993 the government launched a contest to set up GSM cellular networks. The tender had its hiccups. Originally, the government identified technical ability and low tariffs as key criteria for winning, but later decided to auction the licenses. A consortium that included the German DeTeMobil and British Telecom as partners submitted the highest bid (USD 48 million) but did not win, which generated plenty of controversy.

The winning consortia were Westel 900, an organization that was separate from the enterprise that ran the NMT 450 system but had the same corporate parents, and PANNON GSM, which included several Scandinavian operators and several Hungarian shareholders. Westel 900 launched its system in April 1994, and PANNON did a month later. By April 1996, the former had 129,000 subscribers and the latter had 93,000.

RUSSIA

In September 1990 the Russian government announced plans to award cellular licenses in all towns with a population of at least one hundred thousand. Separate licenses would be awarded for each city and a variant of the NMT 450 analog standard would be introduced first. Later, digital cellular licenses would be granted, preferably for systems adhering to the European GSM standard.[21] Foreign investors would be welcome and, under certain conditions, would even be allowed to take majority stakes in joint ventures with local partners.

US West was the pioneer in Russia, as it was in several other Central and Eastern European countries. In late 1990 it announced that it had reached an agreement with several local partners to launch an NMT 450 system in Leningrad, which would soon become Saint Petersburg. The network started service under the Delta Telecom name in late 1991. US West was also first in Moscow. Here it formed a joint venture with Millicom and the Ministry of Posts and Telecommunications to set up an NMT 450 system that started service in December 1991 under the Moscow Cellular Communications name.

The cellular-standard situation quickly became complex in Russia. In 1992 the ministry requested tenders for GSM systems in five cities and soon extended the invitation to an additional seven. US West bid for all twelve with a local telecommunications partner and acquired ten of them. It failed to obtain the two most valuable ones, Moscow and Saint

Petersburg, which ended up in the hands of Western European carriers (the German DeTeMobil, Telecom Finland, the Norwegian Telenor, and the Swedish Telia). The ministry continued granting NMT 450 and GSM licenses in smaller cities. Most of the NMT 450 licenses went to local telecommunications companies, whereas US West and a few large European carriers, such as Telecom Finland and DeTeMobil, obtained most of the GSM licenses.

The Russian cellular landscape became even more convoluted when, unexpectedly, the ministry started awarding Advanced Mobile Phone Service (AMPS) licenses as well. One of the early AMPS licenses was obtained by VimpelCom, a Russian-American joint venture.

SUMMARY

American and Western European cellular carriers participated in competitions for cellular licenses not only in Western Europe but also in the countries that, up to the late 1980s, belonged to the Communist bloc. Many of these countries inaugurated cellular service for the first time in the early 1990s, and all of them adopted the NMT 450 analog standard for their very first cellular systems. Later in the 1990s, all of them awarded GSM licenses, thus fostering the transition to digital cellular systems.

If AirTouch was involved in, and won, several of the Western European cellular contests, the American pioneer in Central and Eastern Europe was US West, which partnered with the domestic PTT to set up an analog network in Hungary and with Bell Atlantic to launch an analog system in Czechoslovakia. In addition, US West participated in consortia that were awarded the GSM licenses in Hungary, Poland, and Czechoslovakia. The company was also the American cellular pioneer in Russia, where it obtained multiple NMT 450 and GSM licenses in joint ventures with local partners. In their forays in Central and Eastern Europe, the American cellular carriers usually confronted telecommunications rivals from Germany and the Scandinavian countries.

Figure 6.2 shows the evolution of cellular density in several Central and Eastern European countries, with Britain included for comparison.

The figure shows that, as expected, all of the ex-Communist countries lagged behind Britain during the 1990s. All of them, however, caught up at some point during the 2000s, and the Czech Republic did early in that

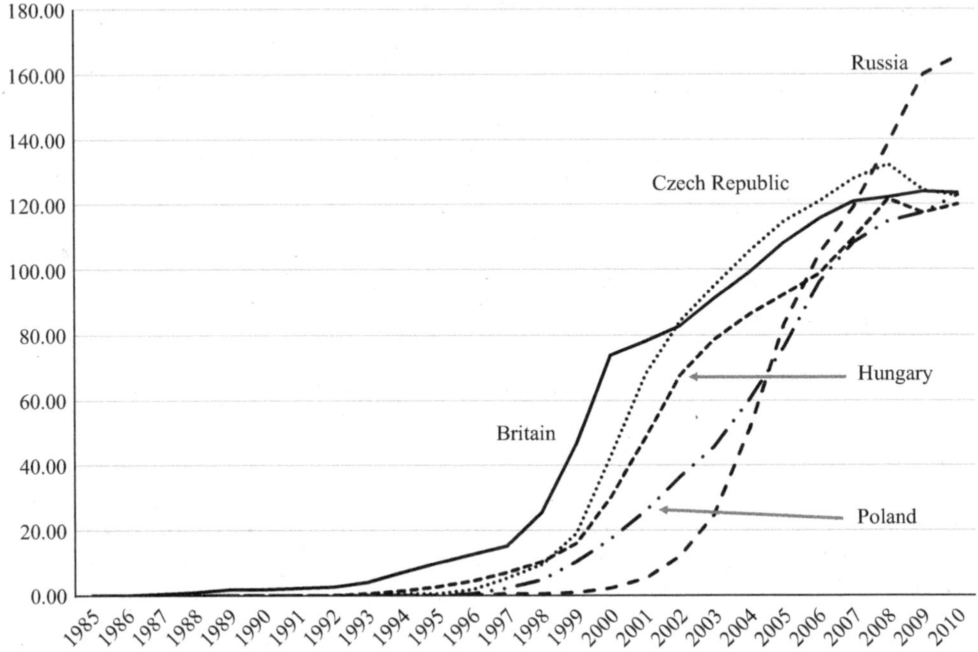

6.2 Cellular density in several Central and Eastern European countries, 1985–2010. *Source*: International Telecommunications Union database.

decade. This is remarkable because, as we emphasized earlier, Britain was for many years one of the best performers among Organisation for Economic Co-operation and Development (OECD) countries.

THE CELLULAR GOLD RUSH: COMPETITION IN NETWORK SERVICES

Competition transformed market structures all over the world in the 1990s.[22] Figure 6.3 presents this evolution: it shows the number of OECD countries that had one, two, three, and four cellular operators each year between 1990 and 2000.

The extent of the transformation is impressive. Although the figure does not show it, the OECD as a whole went from 35 cellular operators to 105 in a decade. The number of monopolistic markets declined from twenty-two to zero; the number of duopolistic markets peaked at seventeen in 1997 and then declined; the number of markets with three operators

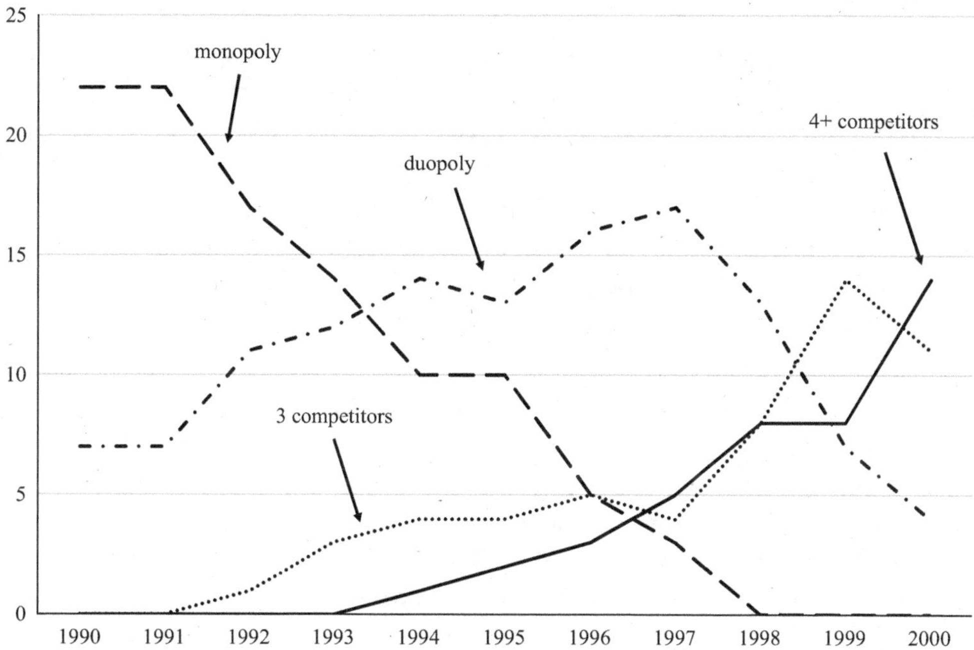

6.3 Market structure in OECD countries, number of countries, 1990–2000. *Source*: Organisation de coopération et de développement économiques, *Téléphonie mobile: Structures et tendances en matière de Prix* (Paris: Organisation de coopération et de développement économiques, 2000), 87.

peaked at fourteen in 1999 and started falling thereafter; and by the year 2000, there were fourteen markets with at least four cellular operators. Of the twenty-nine markets covered in the figure, by the year 2000, none were monopolistic, four were duopolistic, eleven had three operators, and fourteen had four or more.[23]

The second trend in the 1990s was toward the internationalization of the cellular industry, in the sense that, for the first time in the history of the industry, foreign operators made systematic attempts to establish a presence in national cellular markets. The industry became internationalized for two reasons. First, it was during the 1990s that many countries decided to grant additional cellular licenses, either to allow for the entry of competing operators in markets that had hitherto been monopolistic or to enhance competition in markets that up to that point had operated as duopolies. Second, many domestic companies that had no experience with cellular competition formed consortia with foreign companies—especially from the United States and Britain—that had years of experience operating in competitive cellular markets.

The cellular "gold rush" started when it became clear that the German government would grant the D2 license to a private operator: leaving Britain aside, this marked the beginning of cellular competition in Europe. Foreign operators participated in contests such as the one for the D2 license by forming consortia with domestic companies. On paper, companies such as the Baby Bells could only play the role of minority partners in international cellular consortia based in Europe. The choice, however, was clear: they could either hold a minority stake in the partnerships or miss out on the opportunities that were opening up on the European continent and elsewhere. Further, given their experience in operating cellular networks in competitive environments such as the American duopolistic regional markets, they were often able to exert more influence in the context of the consortia than was justified on the basis of their minority stake.

The leading firm in a consortium was not one of the foreign operators but rather the local partner. All the consortia that bid for the D2 license, for example, were formed at the initiative of German companies.

6.4 The Cellnet National Control Centre, Slough, UK, 1996. Courtesy The National Museum of Computing, UK.

Although there was no recipe for a "perfect consortium," the successful partnerships shared several features. They had a leading local partner that was politically acceptable and likely to win a bidding contest given political conditions in the local market. At least one of the partners was a company with experience in setting up and running a cellular network. And finally, the partners, in combination, were able to show that they had the means to support the capital investment required to build a network.

By the end of the 1990s, the process of internationalization through consortia had led to a world in which many of the main operators had their cellular assets heavily diversified across countries. As an example, table 6.1 presents Verizon's international holdings in 2001. The table shows that Verizon had assets in Western and Eastern Europe, Asia, and North, Central, and South America.

COMPETING IN DIGITAL INFRASTRUCTURE

Carriers chose cellular-infrastructure products on the basis of prices and features, but also in light of the long-term relationships they had established with equipment manufacturers.[24]

COMPETING IN GSM INFRASTRUCTURE

Long-term relationships between technology manufacturers and cellular carriers played a fundamental role in the market for GSM infrastructure. Most of the early orders came from the domestic PTTs, which were responsible for setting up the first GSM networks, and most of these orders were awarded on the basis of factors that went beyond product prices and attributes. Companies such as Ericsson, Nokia, Siemens, Alcatel, PKI, Matra, and Motorola were involved in the GSM standard-setting process and thus were beneficiaries of the early orders for GSM infrastructure. Firms such as Lucent, Nortel, and the Japanese electronics manufacturers, which did not participate in the GSM standard-setting committees, failed to capture any of the early orders.[25]

Among the companies that participated in the GSM standard-setting process, orders were allocated on the basis of preexisting relationships between service providers and equipment suppliers. Out of the twenty-seven carriers that were offering—or had offered—both analog and digital

Table 6.1 Verizon's holdings in foreign cellular operators as of 2001

Country	Subscribers (millions)	Venture	Ownership (%)
Italy	15.000	Omnitel Pronto Italia	23.1
Mexico	1.700	Iusacell	37.2
Czech Republic	2.110	EuroTel Praha	24.5
Slovakia	0.493	EuroTel Bratislava	24.5
Greece	1.645	STET Hellas	20.0
Indonesia	0.765	Excelcomindo	23.1
Japan	3.840	Tu-Ka	2.7
New Zealand	1.200	Telecom New Zealand	24.9
Philippines	0.026	BayanTel	19.4
Argentina	0.950	CTI Holdings	59.5
Canada	2.156	TELUS Corporation	22.0
Venezuela	1.700	CANTV	28.5
Taiwan	5.100	Taiwan Cellular Corporation	13.5
Dominican Republic	0.393	CODETEL	100.0

Source: FCC, *6th Annual Commercial Mobile Services Competition Report* (Washington, DC: FCC, 2001), C-9, C-10.

cellular service by 1998, twenty-five purchased their initial digital infrastructure equipment from one of their analog suppliers. Ericsson and Motorola were particularly successful in the early years of the GSM standard, since they had accumulated experience with analog equipment and had also developed an international network of customers.

Although long-term relationships shaped the GSM infrastructure market, product prices and attributes played an important role when the second and third GSM service providers in each country placed their orders. The second GSM network operator tended to order infrastructure from a vendor that was different from that used by the first operator—it was a way to emphasize product (or service) differentiation. Many of the carriers that came second (and beyond) in the GSM market had never offered analog cellular services, and thus had no established relationships with equipment vendors. Among these "new" operators, Nokia became a preferred supplier mainly because, despite its relatively short experience in telecommunications switching markets, it was able to develop an excellent digital switch for mobile systems.

As time went by, GSM carriers tended to acquire most of their infrastructure from companies that were able to supply both switches and base stations. This consolidated the role of vendors such as Ericsson and Nokia, and also weakened the market position of Motorola, which never managed to develop expertise in the switch market. Suppliers such as Alcatel, Siemens, PKI, and Matra remained too focused on their local PTTs to become important players in this market.

Table 6.2 shows the market shares in cellular infrastructure (switches and base stations) for the largest thirty-three GSM networks in Europe as of December 1996, with shares based on subscriber numbers. The table also shows shares in GSM handsets, discussed later in this chapter.

The table highlights several facts. Ericsson managed to achieve high shares in both switching systems and base stations. Nokia did too, although it was weaker in switches, an area in which it had accumulated relatively

Table 6.2 Market shares of main suppliers in GSM switching systems, GSM base stations, and GSM handsets, 1996–1997

GSM switching systems		**GSM base stations**		**GSM handsets**	
Supplier	**Market share (%)**	**Supplier**	**Market share (%)**	**Supplier**	**Market share (%)**
Ericsson	48	Ericsson	37	Ericsson	25
Siemens	21	Nokia	22	Nokia	24
Nokia	14	Motorola	13	Motorola	20
Alcatel	10	Alcatel	10	Siemens	9
Other	7	Other	18	Other	22

Note: The shares of GSM switching systems and GSM base stations are calculated for the thirty-three largest networks in Europe as of December 1996; the shares of GSM handsets are calculated for January 1996–January 1997. Switching-systems and base-stations shares are based on the number of subscribers.

Source: R. Bekkers, *Mobile Telecommunications Standards: GSM, UMTS, TETRA, and ERMES* (Boston: Artech House, 2001), 329, 334.

less expertise. Motorola's situation was similar to Nokia's, although Motorola was even weaker in switching systems. Finally, companies such as Siemens were able to retain reasonably large shares in the early years only because their GSM networks covered large domestic markets.

Outside Europe, long-term relationships played an important role among those carriers that set up the first GSM network in each country, usually the domestic PTTs, since there was a tendency for most of them to order GSM infrastructure from the same vendor that had supplied them with analog equipment. Motorola and Ericsson, for example, captured about 80 percent of the GSM infrastructure market in China, mainly because they had been successful suppliers of analog infrastructure in that country.

COMPETING IN DIGITAL INFRASTRUCTURE IN THE UNITED STATES AND JAPAN

Product prices and attributes were more important in the United States than in Europe. There were three digital standards in the United States in the 1990s: D-AMPS, cdmaOne, and GSM 1900. In the process of transitioning from analog to digital, the providers of analog cellular services upgraded to either D-AMPS or cdmaOne, whereas the new service providers—the ones that entered the market for the first time as PCS providers—implemented the three of them.

Infrastructure equipment suppliers tended to develop the digital technologies that their analog customers expected to use once they transitioned to digital cellular. Lucent, for example, developed both cdmaOne and D-AMPS infrastructure, Ericsson focused on D-AMPS, and Motorola became a major provider of cdmaOne equipment. By 1997, Lucent and Nortel were the top providers of cellular infrastructure in the United States, with shares of 35 percent and 20 percent, respectively, followed by Ericsson, Motorola, and Nokia, in that order. This suggests that continuity in vendor-carrier relationships did play a role in the American market, even if it was a less important role than in Europe: 73 percent of digital service providers in the United States relied on one of their analog-equipment vendors, whereas 93 percent of GSM providers did.

Long-term relationships between equipment manufacturers and cellular carriers played an important role when the American digital standards were exported to other countries. Motorola, for example, became the leading supplier of cdmaOne infrastructure equipment outside the United States, to a good extent because it had developed a web of relationships with international customers during the analog era.

In Japan, long-term relationships played a larger role than in either Europe or the United States. This was so because Nippon Telegraph and Telephone (NTT) was heavily involved in the creation first of a fixed-line telecommunications system and then of an analog cellular system in cooperation with various subcontractors. Motorola managed to penetrate the Japanese market, with help from the American government, when a mobile service provider, Daini Denden Inc. (DDI) Cellular, attempted to differentiate itself from NTT DoCoMo. DDI launched a Total Access Communications System (TACS) network in 1989 with Motorola equipment, which later led to orders of digital infrastructure as well.

Nippon Ido Tushin (IDO), another carrier, initially used the NTT standard. Later, however, it implemented a TACS network with Motorola infrastructure, which also led to orders of digital infrastructure. Table 6.3 shows the cellular service providers in Japan, the date when they started service under various standards, and the identity of the company that supplied the cellular infrastructure for each. In the table, NEC stands for Nippon Electric Company.

The table shows, first, that Japanese suppliers dominated the domestic market during the analog period. It also shows that Motorola and Ericsson grew as infrastructure suppliers from the early 1990s on.

COMPETING IN HANDSETS

Product prices and features tended to play a larger role in the cellular handset market than in the cellular infrastructure market. This was so because handsets are simpler products than base stations and switches, and thus relationships between service providers and handset manufacturers are less persistent.

COMPETING IN GSM HANDSETS

Participating in the standard-setting process was a condition for competing in the GSM handset market. Because they did not participate in the GSM standard-setting process, the Japanese handset manufacturers failed

Table 6.3 Cellular carriers, service start dates, and infrastructure suppliers in Japan, 1979–1999

Service provider	Standard	Start date	Major base station supplier	Major switch suppliers
NTT	NTT	1979	NEC	NEC, Fujitsu
DoCoMo	PDC	1993	NEC, Ericsson, Mitsubishi, Fujitsu	NEC, Fujitsu
IDO	NTT	1988	NEC	NEC, Fujitsu
IDO	TACS	1991	Motorola	NEC
IDO	PDC	1995	NEC, Fujitsu	Fujitsu, NEC
IDO	cdmaOne	1999	Motorola	NEC
Cellular Group	TACS	1989	Motorola	NEC
Cellular Group	PDC	1994	Motorola	NEC
Cellular Group	cdmaOne	1998	Motorola	NEC
J-Phone	PDC	1994	Ericsson	Ericsson
Tsuka Cellular	PDC	1994	Motorola	NEC
Tsuka Cellular	PDC	1994	NEC	NEC
Digital Tsuka	PDC	1996	Ericsson, NEC	Ericsson, NEC

Source: J. Funk, *Global Competition between and within Standards: The Case of Mobile Phones* (Houndmills, UK: Palgrave, 2002), 130.

to develop GSM know-how and patents, which in turn limited their role in the GSM handset market, at least in the early years.

Product prices and attributes were important for GSM handsets because the GSM standard was adopted widely (in Europe and beyond), which meant that volumes rose quickly, more so than in the AMPS case. With rapidly rising volumes, manufacturers such as Motorola, Nokia, and Ericsson implemented a strategy of introducing phones in various market segments and national markets simultaneously, while offering steep price discounts. Firms such as Alcatel and Siemens, which by the mid-1990s were still too focused on their own domestic markets, did not implement this approach until about 1996, and the Japanese only did in 1999.

The advantage of introducing GSM handsets in different price segments simultaneously was that it allowed the three leaders—Motorola,

Nokia, and Ericsson—to subsidize their low-end phones while making steady profits on the high end of the market. Among the three, Nokia performed best in the long run because it focused on the look and feel of the phone rather than merely on cost, which allowed the company to price some of its phones with large profit margins. By 1996–1997, Ericsson, Nokia, and Motorola shared the lead in the market for GSM terminals.

COMPETING IN DIGITAL HANDSETS IN THE UNITED STATES

Product prices and features also played a larger role in the US digital handset market than long-term relationships between equipment vendors and cellular carriers. Relationships had an impact, however, early on in the development of the digital cellular market, and they benefited mainly Ericsson (in the D-AMPS market) and Qualcomm and Sony (in the CDMA market).

In the early years of digital cellular in the United States, Ericsson and Motorola ruled the market for digital handsets (under the D-AMPS standard). Ericsson had played an important role in the process of standard formation in the United States—it had exerted substantial influence on the Cellular Telephone Industry Association in the adoption of the D-AMPS standard for the American digital market.

The US market changed in the mid-1990s, when PCS licenses were awarded. About 50 percent of all PCS firms chose cdmaOne as the cellular standard for their networks, which reshuffled the positions of various handset manufacturers: it benefited those that had accumulated CDMA expertise and harmed, in relative terms, those that had focused exclusively on other standards.

Early on, long-term relationships mattered in CDMA, just as they had for D-AMPS. The two companies that benefited the most from long-term relationships in CDMA technology were Qualcomm and Sony, which had a joint venture that ruled the American cdmaOne market until around 1998. Qualcomm and Sony dominated the CDMA market because they controlled the standard and the chipset that embedded the standard's key specifications.

Around 1998, Qualcomm started lowering the price of its chips and its patent royalty fees to expand the CDMA market, which allowed firms such as Nokia, Oki, and Motorola to start competing in the American CDMA handset market. Once these companies started invading that market, their superior strength in competition based on prices and features allowed them to quickly displace Qualcomm and Sony from their dominant positions. Both Motorola and Nokia gained market share in the CDMA handset market. By the late 1990s, Nokia was dominant, to a good extent because of its global development platforms, through which the company created many handset models with extreme efficiency. In addition, because Nokia relied on common designs for different markets, it was able to extract volume discounts from its suppliers, which in turn increased its margins. By the time cellular markets started the transition from 2G to 3G, Nokia was the world leader in sales of cellular handsets.

Figure 6.5 shows the market shares of the core vendors of digital handsets in the US market between 1995 and 1998.

The figure highlights the decline in Ericsson's share after 1996, which coincided with the relative decline of D-AMPS and the relative rise of CDMA in the US market. The figure also shows the increase in the share of Nokia

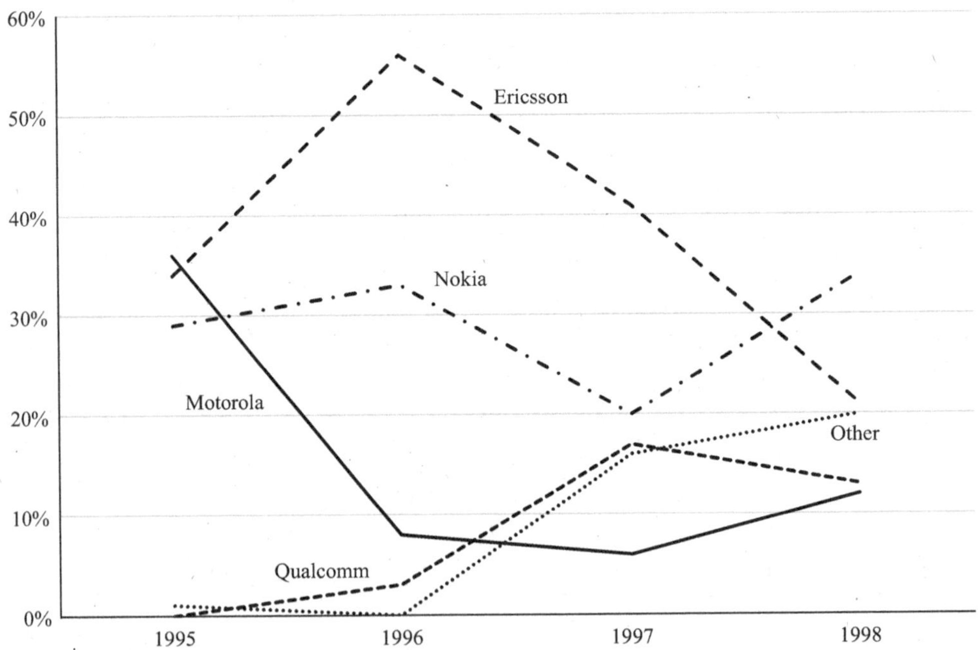

6.5 Shares of the US digital handset market, 1995–1998. *Source*: J. Funk, *Global Competition between and within Standards: The Case of Mobile Phones* (Houndmills, UK: Palgrave, 2002), 170.

and Motorola (and the decline in Qualcomm's share), which reflected the rising importance of the first two companies in the CDMA market.

COMPETING IN DIGITAL HANDSETS IN JAPAN

The Japanese market for digital handsets was the only one in which long-term relationships between equipment vendors and cellular carriers were at least as important as product prices and attributes. This was so because NTT exercised tight control over both the 1G and 2G standard-setting processes in Japan, which enabled it to interact exclusively with a select group of suppliers (and to maintain a good measure of control over them).

Another characteristic of the Japanese market that made product prices and attributes less important than elsewhere was that users tended to focus on the weight and size of phones (to the detriment of other features). A single-segment handset market developed in which only these features, plus battery life, mattered.

NTT had a monopoly on the Japanese cellular market between 1979 and the late 1980s, when DDI and IDO entered. Rather than developing an open standard for analog cellular phones, or adhering to one of the existing ones, NTT chose to develop its own (proprietary) standard. The four Japanese firms that supplied products for the standard during the 1980s were Matsushita, NEC, Fujitsu, and Mitsubishi.

In the digital era, NTT followed the proprietary approach and developed its own standard, Pacific Digital Cellular (PDC), first by itself and later in cooperation with its exclusive suppliers. When DoCoMo, NTT's cellular subsidiary, was spun off in 1992, the approach did not change, and most of the engineers who had worked on the development of PDC went to the newly formed cellular company. The Japanese suppliers were treated preferentially by DoCoMo, which granted them access to confidential information on the PDC standard in exchange for a commitment from them to delay delivery of PDC handsets to service providers other than DoCoMo. Such preferential treatment allowed the select group of suppliers to develop phones for the PDC market that were better than those manufactured by competitors.

The domination of the PDC market by DoCoMo and its suppliers during the 1990s encouraged competing carriers to look for alternatives, and two of them launched cdmaOne networks in the late 1990s. From 1998 on, competition based on product prices and features took a more central role in the domestic cellular market as information on the PDC standard became more widely diffused. This eliminated the advantage that the select group of suppliers had had as a consequence of their proximity to, and collaboration with, NTT DoCoMo.

COMPETING IN 2G MARKETS: TAKING STOCK

The transition from analog to digital cellular—that is, from the 1980s to the 1990s—brought about a number of profound transformations in the cellular industry. Some of them were technological and happened behind the scenes. Others were more visible, and they changed the user experience with mobile phones dramatically.

Competition among cellular carriers in national markets all over the world had the strongest impact. New licenses were awarded, and new service providers introduced innovative pricing plans with the goal of capturing new customers. The new competitive conditions led to lower service prices in many national markets, which in turn helped transform the demographic characteristics of cellular users. Markets that up to that point had been dominated by celebrities and company executives became consumer markets for the first time in the history of the industry, and they did so at different speeds in different countries.

In the markets for cellular infrastructure and handsets, product prices and attributes played an important role, but so did long-term relationships between equipment manufacturers and cellular carriers. Relationships were particularly important in the more complex products—switching systems and base stations as opposed to the phones themselves. During the 1980s, three companies managed to rise to the top ranks of the industry as suppliers of both infrastructure and handsets: Ericsson, Motorola, and Nokia. The three of them benefited, to some extent, from long-term relationships with cellular carriers. But more importantly, they managed to develop expertise and reputation in the manufacturing of products that adhered to open standards while the standards were being developed. This served the three companies well when cellular markets exploded all over the world, which started happening during the 1990s.

III

THE THIRD GENERATION, THE 2000s

7

NEW TECHNOLOGIES, STANDARDS, CUSTOMERS, AND MARKETS IN THE WORLD OF 3G

7.1 Imagining the future. With the arrival of 3G and Internet access on mobile phones, new possibilities were being explored. These "concept" phones announced by Ericsson in 2000 envisioned applications such as electronic payment systems and advanced user interfaces. The phones were never manufactured. Courtesy The National Museum of Computing, UK.

The cellular industry underwent important changes in the transition from the 1990s to the 2000s. In the 2000s new cellular standards were introduced. The third-generation (3G) standards shifted the focus of mobile-phone users from voice to data. This transition, which had started during the 1990s, accelerated in the 2000s. If during the 1990s short messaging service was the killer data app on cellular phones, in the 2000s mobile email and mobile Internet access supplemented text messaging.

The main 3G standards were wideband code division multiple access (WCDMA) and cdma2000. WCDMA was called "wideband" CDMA because it was based on CDMA as a multiple-access method but had a channel width of 5 MHz or more. By contrast, the 2G cdmaOne standard had a channel width of about 1.5 MHz (and is sometimes labeled "narrowband" CDMA).[1] A third cellular standard, time division – synchronous code division multiple access (TD-SCDMA), was developed in China starting in the late 1990s but was introduced only in 2009 (and only in the Chinese market).

Leaving China aside, around four hundred 3G cellular networks were launched in the world between 2001 and 2009. WCDMA accounted for about 74 percent of those networks and cdma2000 for the rest. (In these calculations, we are not considering 3G evolutions—namely, High-Speed Downlink Packet Access [HSDPA] and cdma2000 EVDO [Evolution, Data Only].) WCDMA and cdma2000 were not evenly distributed around the world. During most of the period, Western Europe accounted for more than 50 percent of all WCDMA networks launched, while cdma2000 networks were less geographically concentrated: by 2009, 30 percent of all cdma2000 networks had been launched in Latin America, 26 percent in the Asia-Pacific region, 12.5 percent in North America, 11.5 percent in Eastern Europe, and 10.5 percent in Africa.[2]

In the 2000s handsets changed—first from voice-only phones to feature phones, and later from feature phones to smartphones. The shift from voice-only phones to feature phones started in the mid-1990s. Up to the mid-1990s, mobile phones were used almost exclusively for voice communications. The second half of the 1990s witnessed a wave of new single-function consumer devices such as media players, digital cameras, and Global Positioning System (GPS) devices, as well as infrastructure to support them. For a small increment to manufacturing cost, it became feasible to incorporate some of these devices into the mobile phone, which thus became a "feature phone," an integrated bundle of special-purpose devices with capabilities that were frozen at the time of manufacture.

The term *smartphone* was used for the first time in 1995 and represented a new way of thinking about mobile phones. Smartphones were no longer single- or multipurpose devices but rather universal handheld computers that incorporated a telephone. A core feature of smartphones was their capacity to support application programs, later called apps, which enabled devices to perform tasks that had not been envisaged when they were manufactured. In the first decade of the new millennium, mobile phones became classic universal computers.

In the 2000s competition intensified in several areas, four of which were particularly important: cellular standards, device types, mobile operating systems, and approaches for accessing the mobile Internet. WCDMA and cdma2000 waged the standards battle.

The second battle involved device types. In the late 1990s and early 2000s there were four types of handheld devices—personal digital assistants (PDAs), voice-only mobile phones, feature phones, and the early smartphones—and it was unclear which one would eventually become dominant among consumers.[3] PDAs had mobile operating systems and were used mostly to manage personal information, although they eventually incorporated electronic mail and other applications. Some of them, including those created by Palm, were pen based, whereas others, including the BlackBerry handhelds from Research in Motion (RIM), were keypad based. Voice-only mobile phones were purely for making and receiving voice calls. Feature phones had simple operating systems and incorporated some data capabilities such as Internet access and PC synchronization. They often included digital cameras, video recorders and players, and MP3 players. Smartphones, finally, had more sophisticated operating systems than those of feature phones and were used both for voice communications and for running apps. Smartphones eventually became dominant. However, neither the transition from voice-only to feature phones nor the transition from the latter to smartphones happened overnight. It was only around 2012 that smartphones represented a larger proportion of new mobile-phone models introduced than feature phones;[4] and it was only in 2013 that smartphones, for the first time, accounted for the majority (54 percent) of worldwide mobile-phone units sold.[5]

The third battle was among smartphone operating systems. Most of them were originally developed for PDAs and later served as the operating systems for the early smartphones. Before 2007, the competitors in this market were Symbian, Palm OS, Microsoft Windows Mobile, and RIM's BlackBerry OS. Symbian was the leader in the global market, and BlackBerry OS was dominant in the United States. In late 2007, immediately after the launch of Apple's iPhone, Symbian had about 62 percent of the world smartphone operating-system market (in devices sold), with BlackBerry and Windows Mobile following far behind (with 11 percent and 12 percent, respectively).[6] In the United States, even as late as the fourth quarter of 2009, after the launch of the first Google Android devices, RIM was the market leader with a 42 percent share, followed by Apple, Microsoft, Palm, and Google.[7]

The fourth battle was among approaches for accessing the Internet through mobile devices. In the early 2000s there were two alternatives: i-mode, a proprietary approach pioneered by Japan's DoCoMo, and the Wireless Application Protocol (WAP), an open standard sponsored by several cellular players. Here, i-mode was the clear winner, even though WAP had widespread support across the industry.

Finally, in the first decade of the twenty-first century, the cellular industry underwent a geographic shift. Whereas at the end of the 1990s the United States was the largest cellular market in the world and many Western European countries were still among the top ten, by 2010 China and India had surpassed the United States. Moreover, by 2010, countries such as Russia, Indonesia, and Brazil had displaced the Western European countries from the top-ten list.

NEW STANDARDS: WCDMA, CDMA2000, AND TD-SCDMA

Three types of entities participated in the process of defining the 3G standards: standard-setting organizations, operators and equipment makers, and cooperatives. The standard-setting organizations included the European Telecommunications Standards Institute (ETSI), the Association of Radio Industries and Business from Japan, the Telecommunications Industry Association from the United States, and the International Telecommunications Union (ITU). Because of their individual histories, the interests of these institutions often failed to align with one another. The same was true about operators and equipment manufacturers, most of which had made substantial investments in the development and adoption of a specific 2G standard (either Global System for Mobile Communications [GSM] or cdmaOne). Cooperatives, finally, were organizations such as the Third Generation Partnership Project (3GPP), the Third Generation Partnership Project 2 (3GPP2), and the Operators Harmonization Group, which included standard bodies and private companies.[8]

EUROPE AND JAPAN

European researchers started working on the 3G standards at the beginning of the 1990s, even before GSM was in commercial operation. Much of the European research that led to the 3G standards was done in the context of large-scale European Community programs. From 1995 on, European researchers developed two proposals for an air interface for 3G cellular systems, one centered on wideband time division multiple access (TDMA) and another on wideband CDMA. All these projects, which were under the umbrella of the Universal Mobile Telecommunications System (UMTS), were rather theoretical and failed to draw the attention of network operators and equipment manufacturers.[9]

In the mid-1990s the European Commission set up the UMTS Task Force and later the UMTS Forum, but operators ignored them. The attitudes of stakeholders in Europe started changing only when they realized that Japan was taking the lead in 3G. The Japanese had several reasons to invest in developing 3G standards: in the 1990s cellular networks in Japan were facing capacity shortages and, moreover, Japan regretted the negligible role its companies played in the world market for 2G infrastructure.

Several Japanese organizations made progress in defining 3G standards, including the Association of Radio Industries and Business and the Ministry of Posts and Telecommunications. In 1997 Nippon Telegraph and Telephone (NTT) placed an order with several domestic and foreign vendors for an experimental WCDMA network. This was especially important for European companies such as Ericsson and Nokia, two of the vendors selected by NTT, since up to that point Ericsson had neglected the CDMA standard.

When it became evident that Japan was taking the lead in the process of defining the new cellular standards, European GSM operators and equipment suppliers decided that the time for action had come. Two ideas started to take shape in Europe: first, UMTS would not be a complete break with the past but rather an evolution relative to GSM; and second, ITU would end up adopting not just one 3G standard but many.

EUROPE AND THE UNITED STATES

In the second half of the 1990s, American companies and standard bodies became involved as well. American stakeholders developed multiple paths to migrate from 2G to 3G. Several companies—including Lucent, Nortel, Motorola, and especially Qualcomm—announced that they would work on specifying a 3G standard based on the cdmaOne 2G standard. This work was later transferred to the so-called CDMA Development Group, of which Samsung, Nokia, OKI, Philips, and Sony were also members. Another group of companies started working on a path from 2G to 3G based on D-AMPS. (Digital AMPS, or D-AMPS, was the digital version of the Advanced Mobile Phone Service standard that superseded the analog version in the United States in the 1990s.) Among these were Ericsson, Lucent, Nokia, Nortel, Alcatel, and several operators that had committed to D-AMPS during the 1990s. These firms formed an entity named the Universal Wireless Communications Consortium.

In 1997 ETSI announced the formal launch of a process for selecting an air interface for UMTS. Five different proposals were presented and voted on, and ETSI chose a hybrid interface that combined elements of the two most popular ones—one based on WCDMA and another on TD/CDMA. This happened at about the same time that the Japanese Association of Radio Industries and Business selected WCDMA as the air interface for the Japanese standard and NTT issued a tender for a WCDMA prototype network to the largest providers of network infrastructure.

In the late 1990s organizations and companies that supported a specific transition path to 3G cellular standards formed two cooperatives, 3GPP and 3GPP2. 3GPP was set up in late 1998 and included members of three types: standard-setting bodies, market-representation members, and individual companies. Representatives of ETSI and the American National Standards Institute met in mid-1998, and ETSI refused to include non-ETSI technologies in 3GPP. (The American National Standards Institute is a nonprofit entity that oversees the development of standards in the United States.) In response, the American National Standards Institute recommended that an alternative cooperative be created, 3GPP2, to focus on non-ETSI technologies such as CDMA. Several standard-setting bodies became members of 3GPP2, even though some of them were already members of 3GPP. Table 7.1 shows the entities that participated in 3GPP and 3GPP2.

ITU, WCDMA, AND CDMA2000

Early in the 1990s, ITU had entertained the idea of having one 3G cellular standard for the world as a whole. Work on what was expected to become the (single) 3G cellular standard started as early as the mid-1980s, with frequency bands reserved in 1992 and 1995. As the decade went by, however, it became obvious that a single, worldwide 3G standard would

Table 7.1 Membership in 3GPP and 3GPP2

Organization	**Region**	**3GPP**	**3GPP2**
I. Standard-setting bodies			
Association of Radio Industries and Business	Japan	Yes	Yes
Telecommunications Industry Association	United States		Yes
China Wireless Telecommunication Standard	China	Yes	Yes
European Telecommunications Standards Institute	Europe	Yes	
T1 Committee	United States	Yes	
Telecommunications Technology Association	South Korea	Yes	Yes
Telecommunications Technology Committee	Japan	Yes	Yes
II. Market-representation members			
GSM Association		Yes	
UMTS Forum		Yes	
Global Mobile Suppliers Association		Yes	
Universal Wireless Communications Consortium		Yes	

Source: R. Bekkers, *Mobile Telecommunications Standards: GSM, UMTS, TETRA, and ERMES* (Boston: Artech House, 2001).

not be feasible. In 1996 ITU declared that its goal would be more modest: rather than trying to attain a single standard for the world, it would encourage stakeholders to minimize differences among global 3G standards. Even the goal of developing similar 3G standards (rather than one) appeared unattainable in the late 1990s, since two important private actors—Ericsson on the European (GSM) side and Qualcomm on the American (CDMA) side—had been involved in patent-infringement litigation against each other since the mid-1990s.

In the late 1990s two crucial developments helped pave the way for a world with 3G standards that were different from, but also able to coexist with, one another. First, Ericsson and Qualcomm reached an agreement that settled their existing litigation and opened the door for cross-licensing deals. Second, network carriers formed the Operators Harmonization Group. This entity laid the foundation for the rise of the two main 3G standards: WCDMA, the successor to GSM, and cdma2000, the heir to cdmaOne. Table 7.2 presents their key features.

The table shows that, from the perspective of the air interface, WCDMA and GSM were different (even though WCDMA was considered GSM's heir). WCDMA abandoned the TDMA approach used in GSM and adopted a version of the CDMA technology that Qualcomm and others had pioneered.[10]

CHINA AND TD-SCDMA

WCDMA and cdma2000 were the main 3G standards, and the ones that achieved the widest diffusion around the world (if diffusion is measured

Table 7.2 Major 3G cellular standards

	WCDMA	**cdma2000**
Standard body	3GPP Release 99	3GPP2
Frequency bands	850/900 MHz, 1.8/1.9/2.1 GHz	450/850 MHz, 1.7/1.9/2.1 GHz
Peak data rate	384–2048 Kbps	307 Kbps
Typical user rate	150–300 Kbps	120–200 Kbps
Multiple access	CDMA	CDMA

Source: A. Ghosh et al., *Fundamentals of LTE* (Upper Saddle River, NJ: Prentice Hall, 2011), 12.

by number of networks launched). A third 3G standard, TD-SCDMA, was developed in China starting in the late 1990s but was not deployed until 2009, and it was deployed only in China.[11]

TD-SCDMA was developed jointly by Datang, a Chinese equipment manufacturer, and Siemens. In May 2000 TD-SCDMA was accepted by ITU as one of the 3G standards, together with WCDMA and cdma2000. The next year, in May 2001, TD-SCDMA was accepted by 3GPP.

During the 2000s, there was uncertainty in China as to which of the 3G standards, if any, would become the official one in the country, and at some point it was thought that the Ministry of Information Industry would impose TD-SCDMA as the only standard. This did not happen. There were differences of opinion among Chinese policy makers regarding the wisdom of adopting a domestic standard as the only official 3G standard, and to hedge its bets, the government also granted licenses to internationally accepted 3G standards.[12] Thus, the country ended up with three different 3G networks, one on each standard. China Telecom built a cdma2000 network, China Unicom a WCDMA system, and China Mobile a TD-SCDMA network.[13] It was widely believed that TD-SCDMA had slower data rates than the other two.

Subscribers of the three Chinese carriers started migrating to 3G services in the late 2000s, but they did so more slowly, in relative terms, on the China Mobile network. By mid-2013 about one-third of all Chinese cellular subscribers were on 3G networks.[14] By this time, many countries around the world were already starting to migrate to fourth-generation (4G) cellular standards.

EXTENDING THE LIFE OF THE OLD STANDARDS

Although by the early 2000s the core features of the main 3G standards were already defined, the 2G networks were not dismantled and replaced overnight. They were upgraded, and the upgraded 2G systems—usually referred to as 2.5G networks—competed with the new, 3G standards.[15] Generally, 2.5G networks in the GSM tradition were those that included at least one of the following: High-Speed Circuit-Switched Data (HSCSD), General Packet Radio Services (GPRS), or Enhanced Data Rates for GSM Evolution (EDGE). Table 7.3 presents the 2G standards with their "evolutions." The table includes not only GSM and IS-95 (cdmaOne) but also

Table 7.3 Major 2G cellular standards and their evolutions

	GSM	IS-95	IS-54/IS-136
Year of introduction	1990	1993	1991
Frequency bands	850/900 MHz, 1.8/1.9 GHz	850 MHz/ 1.9 GHz	850 MHz/1.9 GHz
Multiple access	TDMA/FDMA	CDMA	TDMA/FDMA
Data evolution	GPRS, EDGE	IS-95-B	CDPD
Peak data rate	GPRS: 107 Kbps; EDGE: 384 Kbps	IS-95-B: 115 Kbps	12 Kbps
Typical user rate	GPRS: 20–40 Kbps; EDGE: 80–120 Kbps	IS-95-B: <64 Kbps	9.6 Kbps

Source: A. Ghosh et al., *Fundamentals of LTE* (Upper Saddle River, NJ: Prentice Hall, 2011), 7.

IS-54/IS-136 (the D-AMPS standard used in the United States). In the table, FDMA stands for frequency division multiple access, the multiple-access method used in the first-generation standards.

Low data rates were *the* problem in 2G systems. A basic GSM network was able to deliver only a 9.6 Kbps user data rate, not good enough to surf the Web. HSCSD offered an easy approach to speed the process up: with HSCSD, the system used up to four time slots, rather than one, for a data connection. With each time slot supporting a speed of either 9.6 Kbps or 14.4 Kbps, the total data rate was the number of time slots times the speed of one slot. HSCSD was easy and cheap to implement, since it involved only software upgrades to the network, but it also had its weaknesses. Since it was circuit switched (as opposed to packet switched), it used spectrum inefficiently, because it allocated the used time slots constantly, even when nothing was being transmitted.

Most handset manufacturers preferred GPRS, which allowed data rates to be raised to a theoretical maximum of 107–115 Kbps. Typical user rates for GPRS, however, tended to be in the 20–40 Kbps range. Leaving data rates aside, the crucial difference between HSCSD and GPRS was that, while the former was circuit switched, the latter was packet switched, and thus did not allocate resources continuously but rather only when data were being transmitted. GPRS, which operators started deploying in 2001, was especially suitable for email and Web surfing, but less so for real-time applications. It was more expensive to implement than HSCSD but became a necessary upgrade as data traffic grew during the 2000s.

EDGE was the third 2.5G improvement to GSM systems. The key innovation behind it was a new modulation scheme that increased the data rates of GPRS networks by up to threefold. It was reasonably easy to implement, as it only required a software upgrade to a system's base stations. Peak data rates for EDGE systems were as high as 384 Kbps, although typical user rates were in the 80–120 Kbps range.

HSCSD, GPRS, and EDGE were upgrades created for GSM systems. The IS-95 (CDMA) standard was able to deliver 14.4 Kbps data rates. An upgrade, called IS-95-B, managed to reach typical user data rates of around 64 Kbps, but many operators skipped it and moved directly to cdma2000.

All of these improvements to 2G systems, which were developed in the late 1990s and became available in the early 2000s, had a significant impact on network operators. Since it was much easier, and less expensive, to implement the 2.5G upgrades than to set up new 3G systems, many operators started rethinking their decision to fully move from 2G to 3G networks. Eventually, the pressure of rising data traffic forced operators to build 3G networks, but the advent of the 2.5 upgrades slowed down the adoption of WCDMA and cdma2000.

Figure 7.2 presents the cumulative number of EDGE, 3G (WCDMA and cdma2000), and 3.5G (HSDPA and cdma2000 EVDO) networks launched in the world between 2001 and 2009. (We discuss 3G evolutions, also known as 3.5G networks, in chapter 10.) The figure shows that, during the 2000s, more EDGE networks were set up around the world than either WCDMA or cdma2000. Although by the end of the decade the total cumulative number of 3G networks (WCDMA plus cdma2000) launched was much larger than the cumulative number of EDGE networks, EDGE systems were a clear alternative for carriers around the world. The figure also shows the rising importance of 3.5G networks, HSDPA especially, from the middle of the decade on.

Figures 7.3, 7.4, 7.5, and 7.6 reveal that the transition to 3G happened at different speeds in different regions of the world—and that, in different areas, different 3G standards were dominant, at least by number of networks launched. Figure 7.3 tracks the cumulative number of networks

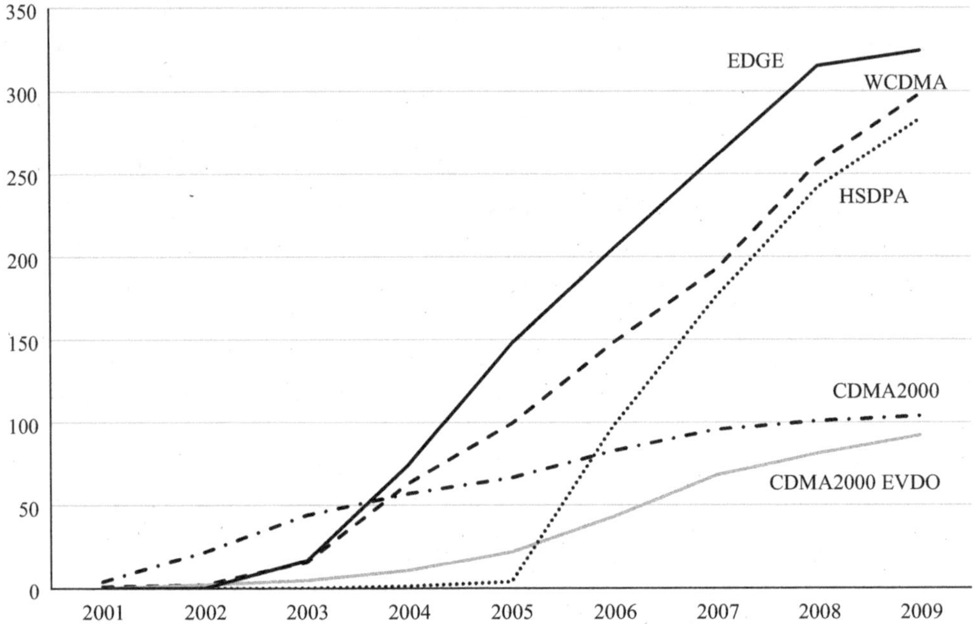

7.2 Cumulative number of networks launched across the world, 2001–2009, by standard. *Source*: Calculated by the authors from P. Curwen and J. Whalley, *Mobile Telecommunications in a High-Speed World: Industry Structure, Strategic Behavior and Socio-economic Impact* (Farnham, UK: Gower, 2010), 68.

launched in Western Europe, figure 7.4 in North America, figure 7.5 in Eastern Europe, and figure 7.6 in Africa.

The figures reveal two important facts. First, different areas of the world had their "preferred" 3G standard: WCDMA was the dominant 3G standard in Western Europe, just as cdma2000 was in North America. Second, in the emerging markets of Eastern Europe and Africa, improvements to 2G networks such as EDGE were more frequent than transitions to either one of the two 3G standards. In Africa, 3G networks did not start taking off until the middle of the decade, and by the end of the decade, more EDGE networks had been launched in Africa than WCDMA and cdma2000 systems combined. Globally, only in 2011 were there more mobile phones shipped for 3G than for 2G networks: this reminds us that the transition to 3G standards was slow and uneven.[16]

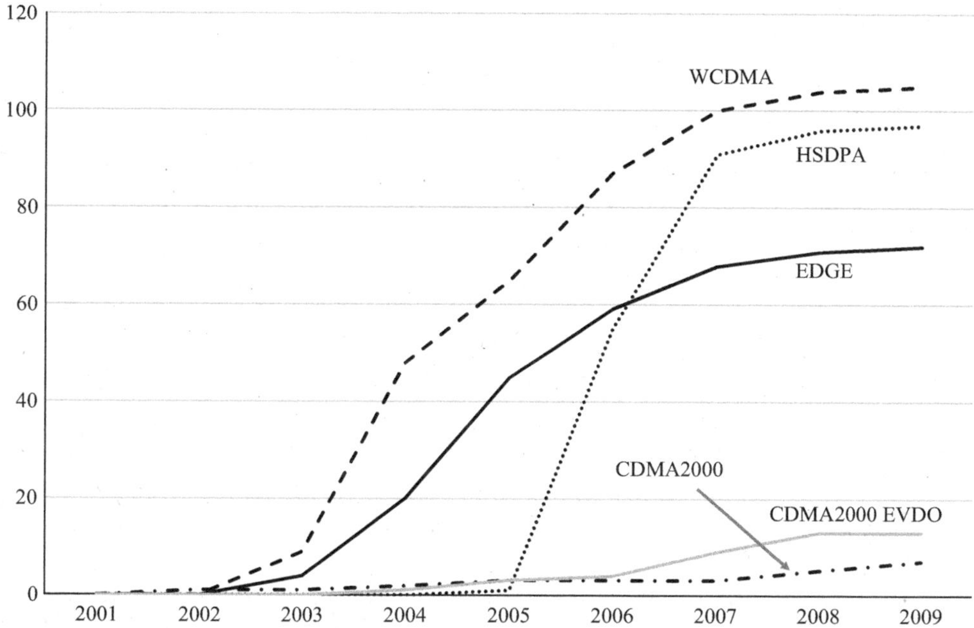

7.3 Cumulative number of networks launched in Western Europe, 2001–2009, by standard. *Source*: Calculated by the authors from P. Curwen and J. Whalley, *Mobile Telecommunications in a High-Speed World: Industry Structure, Strategic Behavior and Socio-economic Impact* (Farnham, UK: Gower, 2010), 68.

THE 3G SPECTRUM AUCTIONS

The new millennium started with 3G spectrum auctions in Europe—they were implemented by the governments of Britain, the Netherlands, Italy, Switzerland, Germany, and Austria in the year 2000, and Belgium, Greece, and Denmark in 2001. Spectrum auctions were not new; they had taken place during the 1990s in New Zealand, the United States, and elsewhere. By the early 2000s, however, governments were more aware of the potential that auctions had to generate revenue for public coffers, and economists were promoting auctions as the best approach to allocate spectrum in an "efficient" manner—that is, in such a way that spectrum would go to those companies that valued it the most.

The first UMTS auction happened in Britain in March and April 2000.[17] The government received the advice of seasoned auction theorists, who considered the virtues and weaknesses of two approaches: the ascending

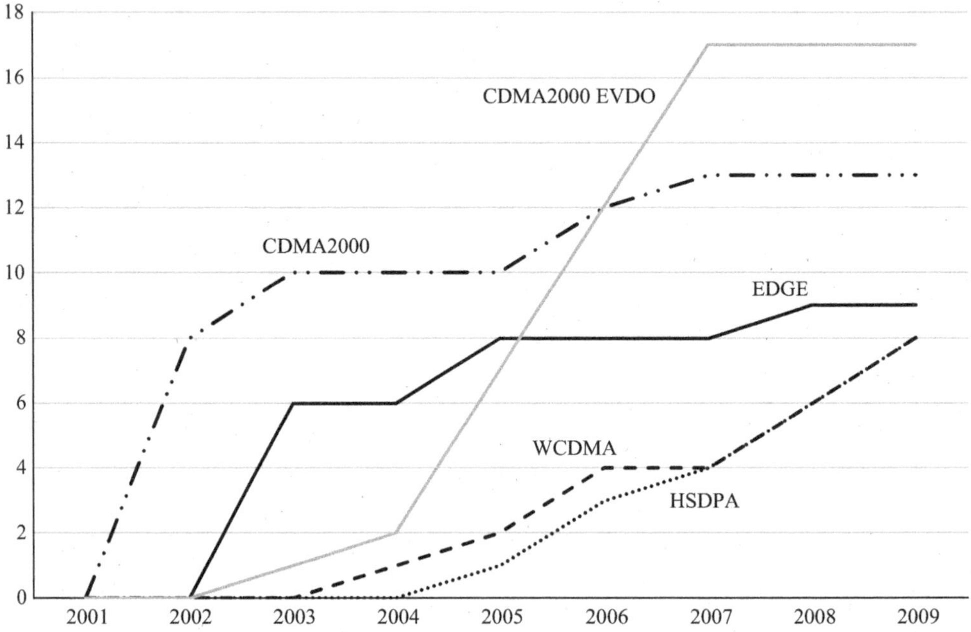

7.4 Cumulative number of networks launched in North America, 2001–2009, by standard. *Source*: Calculated by the authors from P. Curwen and J. Whalley, *Mobile Telecommunications in a High-Speed World: Industry Structure, Strategic Behavior and Socio-economic Impact* (Farnham, UK: Gower, 2010), 68.

(English) auction and the sealed-bid auction. In the English auction, an auctioneer raises the price until only the winner remains, which means that each bidder knows at each stage what the best bid is. (The reverse of the English auction is the Dutch auction, in which an auctioneer starts with an initial high price, and then lowers it until a bidder accepts the current price.) The first-price sealed-bid auction is different from both the English and the Dutch auctions. In a first-price sealed-bid auction, potential buyers submit sealed bids, and the highest bidder gets the item for the price he or she bid. Whereas in the English auction bidders can revise their own bids considering what others are bidding, bidders in a sealed-bid auction submit only one bid.[18]

The core objectives of the spectrum auctions were attracting entry and preventing collusion, and economists believed that the sealed-bid auction was better on both counts. English auctions offered more opportunities for collusive behavior: they allowed bidders in early rounds to signal to one

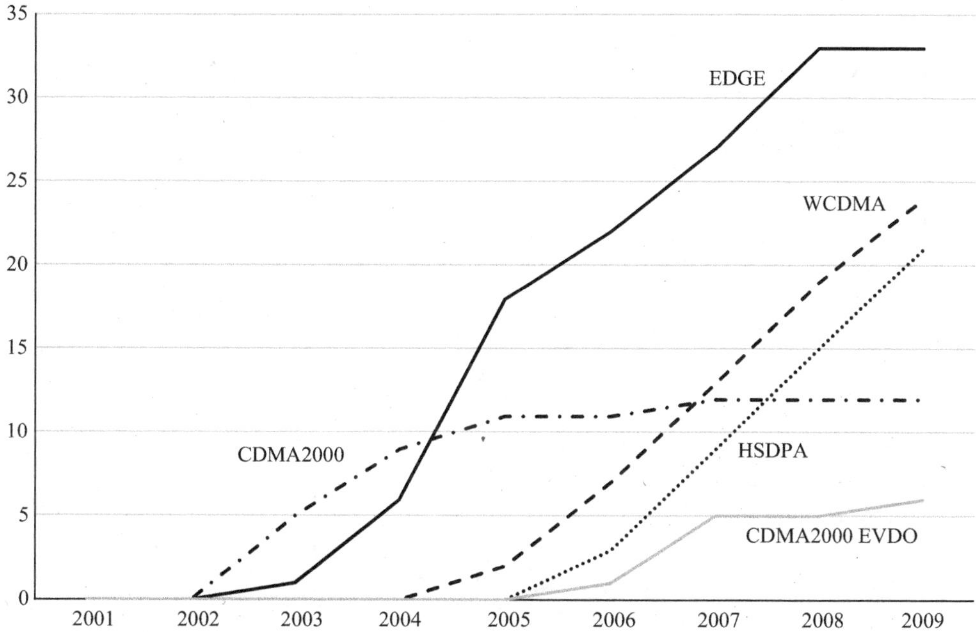

7.5 Cumulative number of networks launched in Eastern Europe, 2001–2009, by standard. *Source*: Calculated by the authors from P. Curwen and J. Whalley, *Mobile Telecommunications in a High-Speed World: Industry Structure, Strategic Behavior and Socio-economic Impact* (Farnham, UK: Gower, 2010), 68.

another how they might divide the spectrum collusively and gave them the chance to punish noncooperating rivals in later rounds. Moreover, ascending auctions tended to deter entry since weaker bidders knew that stronger bidders could always rebid higher in later rounds. By contrast, sealed-bid auctions did not provide opportunities for signaling or punishment to sustain collusion; in addition, they encouraged entry, since weaker bidders knew that they had a higher chance of winning if this format was used.

Since early on it was thought that only four licenses would be available in Britain, a country that had four cellular incumbents, economists were inclined to use a hybrid Anglo-Dutch format to foster entry. Later, however, a fifth license became available, and a simultaneous ascending auction format was chosen. Thirteen bidders participated, and commentators thought that the process, which generated 650 euros per capita, was extremely successful. The auction raised a total of 22.5 billion pounds (USD 34 billion) and was described at the time as "the biggest auction ever."[19]

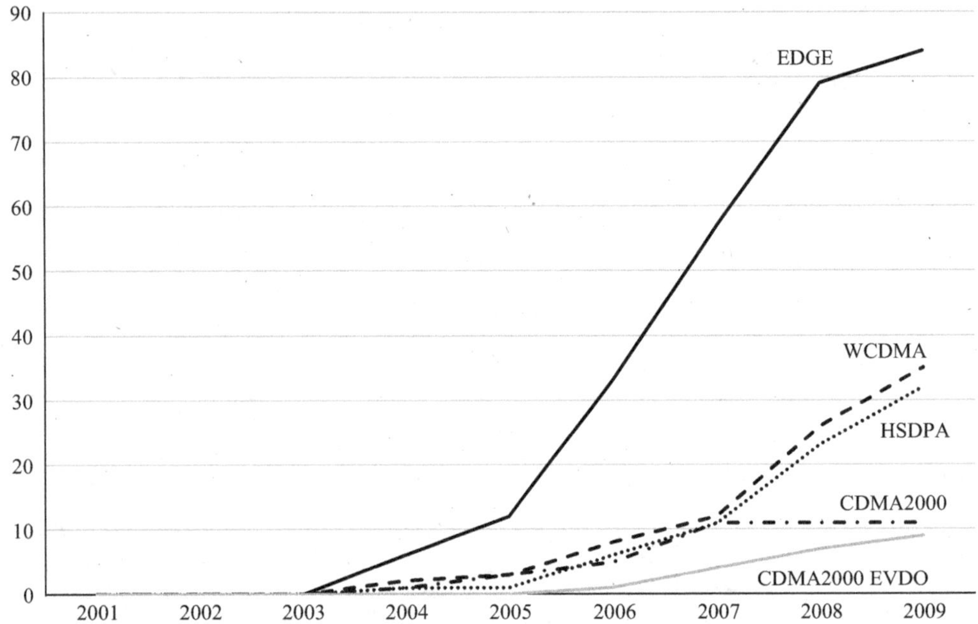

7.6 Cumulative number of networks launched in Africa, 2001–2009, by standard. *Source:* Calculated by the authors from P. Curwen and J. Whalley, *Mobile Telecommunications in a High-Speed World: Industry Structure, Strategic Behavior and Socio-economic Impact* (Farnham, UK: Gower, 2010), 68.

Measured by the revenue they raised for each government, the auctions that followed in the year 2000 were, for the most part, less successful. In July 2000 the government of the Netherlands had its auction. It made the mistake of following the British (ascending) format, even though there were five licenses to be sold and five 2G incumbents. Many of the strongest new entrants partnered with the local incumbents. The one new entrant that competed with the incumbents—Versatel—ended up quitting the process after receiving a threatening letter from Telfort, one of the incumbents.[20] The auction only generated 170 euros per capita; and six months after it was over, the entire process became the subject of an investigation by the Dutch parliament.

The Italian government had its UMTS auction in October 2000. It also followed the British format, and it announced that, if no more serious bidders showed up than licenses were available, the number of licenses could be reduced. Since potential participants had learned at this stage who the strong bidders were likely to be, weak bidders either did not bid or formed joint ventures with incumbents. Only six bidders entered the auction (as opposed to thirteen in Britain) to compete for five licenses, and one quit in an early round. The process generated 240 euros per capita.

The least successful UMTS auction happened in Switzerland, in November and December 2000. Originally, there was considerable interest for the four available licenses, but potential bidders were put off by the (ascending) auction format. To make matters worse, the government allowed for joint bidding agreements right before the auction, which led to a decline in the number of bidders from nine to four. The reserve price was set at a low level, and bidders ended up obtaining the licenses at the reserve price. The government collected 20 euros per capita, just 3 percent of what the process had generated in Britain earlier in the year.

Germany and Austria also had UMTS auctions in the year 2000, although with a modified format. In July and August 2000, Germany auctioned twelve blocks of spectrum from which bidders could create licenses of two or three blocks. Firms could win at most one license each, and the twelve blocks were sold in an ascending auction format. Only seven bidders participated. The two main incumbents, Deutsche Telekom and Vodafone Mannesmann, behaved differently: although Vodafone signaled that its preference was to end the process quickly, Deutsche Telekom kept pushing prices up, so that the government ended up with high revenues and a competitive market. The process generated 615 euros per capita, not far below the British outcome.

The Austrian auction, which took place in November 2000, was similar in format but generated a different outcome. Only six firms competed for the twelve spectrum blocks, and since the government set a low reserve price, participants had incentives to tacitly agree to divide up the market and obtain two bands each. The bidding stopped very soon, with the Austrian government collecting only 100 euros per capita.

The 2001 auctions generated lower per capita revenue than the 2000 auctions for several reasons. By spring 2001, valuations had declined in all markets and, moreover, countries such as Belgium and Greece were less attractive markets than, say, Britain and Germany. Belgium had its UMTS auction in March 2001 and Greece did in July. Each country held auctions for four licenses, and each attracted only three bidders, who obtained the licenses at the reserve price. The government of each country collected 45 euros per capita.

The Danish UMTS auction happened in September 2001 in a context of low spectrum valuations. The government planned to sell four licenses in a country that had four cellular incumbents, so that the revenue expectations were not particularly high. The auction planners, however, changed the format and relied on a sealed-bid auction to give weaker bidders a chance to win. The process attracted a serious bid from a new entrant and the government ended up collecting 95 euros per capita, almost twice as much as analysts expected.

Auctions were used in other countries, in Europe and elsewhere, to allocate 3G licenses. Between early 2001 and early 2002, they were implemented in Australia, Canada, the Czech Republic, New Zealand, Slovakia, Slovenia, and Taiwan. In the early 2000s governments also relied on tenders, beauty contests, and other allocation approaches to license operators in the Czech Republic, Hong Kong, Israel, Japan, Malaysia, Poland, Singapore, South Korea, and the United Arab Emirates.[21]

FROM FEATURE PHONES TO SMARTPHONES VIA PDAS

Although the term *smartphone* was used for the first time in 1995, the first true smartphone was unveiled in 1992 and became available to the public in 1994: it was the Simon Personal Communicator, developed by IBM and marketed by Bell South. The Simon was the first device that combined the capabilities of a phone and a PDA. It had a monochrome LCD screen, came with a stylus, and was priced at USD 899 when it became available. It could be used not only to make voice calls but also for paging and to send and receive emails and faxes. It included an address book, a calendar, a world clock, a way to schedule appointments, and a "notes" feature. If the user managed to free up enough space on the device or was willing to purchase a PC card and insert it in the phone, the Simon was able to run third-party applications.[22]

The early smartphones were built on the technology of PDAs. All the main smartphone operating systems in the pre-2007 period—Palm OS, Windows CE (later called Windows Mobile), Symbian, and BlackBerry OS—came from PDAs.[23]

PALM

Palm Computing was founded in 1992 by Jeff Hawkins, who in the early 1990s, while working for Grid Systems, had created one of the first handheld computers, the GRiDPad.[24] U.S. Robotics, a company that produced computer modems, acquired Palm in 1995, and the next year it introduced the Pilot 1000 PDA with 128k of memory and a monochrome touchscreen display. The Pilot 1000 had four basic functions—calendar, address book, to-do lists, and memos—and sold more than a million units in the first eighteen months on the market. The original Pilot devices were followed by the PalmPilot Personal and the PalmPilot Professional, which used version 2.0 of the Palm OS.

After merging with U.S. Robotics in 1997, 3Com, the new Palm owner, released the third generation of the Palm device, the Palm III, based on Palm OS 3.0. After the Palm III models came the Palm V and the Palm VII lines. The Palm VII, introduced in 1999, was the first Palm product with wireless data access, via a large antenna that connected with the Palm.net service. It supported limited Web browsing and location-based content. The Palm VII was a remarkable device for its time: it allowed users to buy books from Amazon, receive directions from MapQuest, and check scores on ESPN.[25]

In 1998 the Palm founders left 3Com and set up an independent company named Handspring. Until 2001, the company produced Palm OS-based Visor handhelds, which were unique because they incorporated interchangeable modules that enabled the PDA to function as a camera, a GPS device, and a cellular phone. In 2001 Handspring created the first Palm OS-based smartphone, the Treo, introduced the next year. Often described as a "PDA phone," the Treo could be used as a conventional Palm OS PDA, a short messaging service and email communicator, a limited browser, and a mobile phone.[26] Although it had limited Web-browsing capabilities, the first Treo model, the Treo 180, was reviewed as the best device available for surfing the Web at the time.

WINDOWS CE

The success of the Palm PDAs generated a flurry of entry into the PDA market. Microsoft started working on an operating system for mobile computing devices, codenamed Pegasus, in late 1994.[27] By September 1996 it had entered into agreements with six hardware partners—Casio, Compaq, HP, LG Electronics, NEC, and Philips—to bring to market PDAs running on the mobile operating system (renamed Windows CE 1.0). These early Windows-based products were not particularly successful. By late 1997 Palm had about two-thirds of the market for handheld computing devices.

The Microsoft mobile operating system went through various iterations, including Pocket PC 2000, released in April 2000, and Pocket PC 2002, released in October 2001. By this time, Microsoft was already exploring the possibility of bringing to market two lines of PDA-phone combinations: one would be a Pocket PC PDA that also functioned as a phone, and the other a cellular phone with PDA features. Although the company marketed the latter more enthusiastically, the former was the one that reached the market first: named Pocket PC 2002 Smartphone and introduced in Europe in 2002, it was a full-blown Pocket PC device with an integrated GSM phone for voice and data.

SYMBIAN

Symbian had its roots in the Epoc operating system developed by Psion, a British computer company.[28] Psion was founded in 1980 by David Potter, an academic physicist, and in its early years it became a leader in software for microcomputers. In 1994 Psion introduced a small handheld computer, the Series 3, which became a hit in the British market. After it took off, Psion engineers designed an operating system, Epoc, to serve as the foundation for the long-term successor to the Series 3. Psion started exploring the possibility of licensing its (still unreleased) operating system at around the same time that Nokia was getting ready to release the Communicator 9000, a combined mobile phone and PDA. Eventually, Psion and Nokia agreed that Nokia would use Epoc on its future line of Communicator devices.

Between 1996 and 1998 Psion held licensing negotiations with some of the largest mobile handset makers in the world. In 1998 Psion, Nokia, Ericsson, and Motorola announced that they would all co-own a new company, named Symbian Ltd., whose primary focus was to license the Symbian operating system to smartphone manufacturers.

The goal of Symbian was to counteract the rise of Microsoft and its mobile operating systems in the smartphone market. Symbian's owners

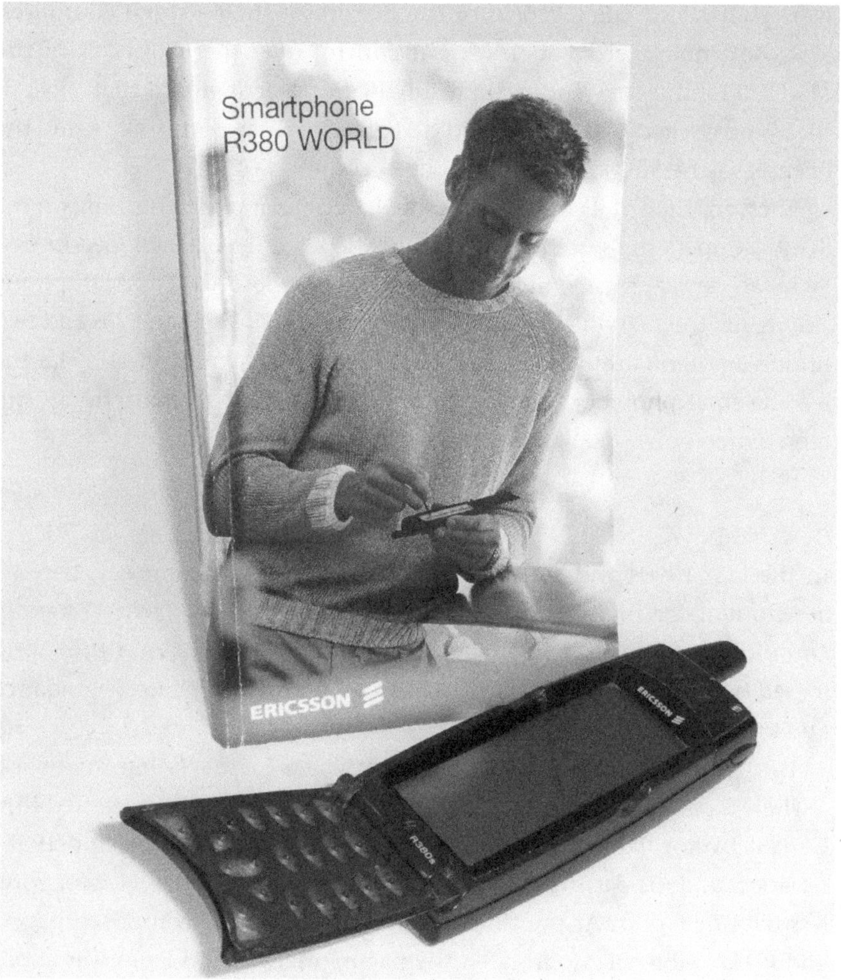

7.7 The Ericsson R380 World smartphone. Developed in Sweden and introduced in late 2000, the smartphone used dual 900/1800 GSM standards. Combining a mobile phone with a PDA, when opened, the whole screen could be used as a PDA using a touch sensitive keyboard or a stylus to enter text. The R380 was the first smartphone to use the Symbian operating system. Courtesy Nigel Linge and Andy Sutton.

were trying to avoid a repeat of the history of the personal computer: Microsoft had exerted such a tight control over the evolution of the MS-DOS (and later Windows) personal-computer standard that most of the value created by the standard had gone to Microsoft itself, while the licensees—the "IBM clones"—had been commoditized.[29]

Ericsson shipped its R380 device, the first cellular phone running on the Symbian operating system, in September 2000, and Nokia shipped its first Symbian device, the Communicator 9210, in June 2001. Symbian phone shipments reached one hundred million units in November 2006 and two hundred million units in March 2008. By this time, Symbian was by far the leading smartphone operating system in the world, but smartphones still represented only a small portion of all cellular phones shipped.

BLACKBERRY

In the late 1990s and early 2000s, RIM, a Canadian company, became a synonym for mobile email. RIM was founded in 1984 in Waterloo, Ontario, by engineers Mike Lazaridis and Doug Fregin. Soon they were joined by Mike Balsillie, a Harvard MBA who provided strategic guidance and business savvy.[30]

Until the mid-1990s, RIM developed products for the Mobitex network, a wireless data network created by Ericsson in Sweden and imported into Canada by Rogers Cantel, a Canadian cellular company. In the mid-1990s Lazaridis and his partners started working on a handheld device for wireless data networks. At the time, the handheld market comprised pagers and PDAs. Motorola dominated the paging market, and Palm was about to become the leader in the PDA market with its Pilot device.

RIM's first handheld, a two-way pager called the Inter@ctive 900, or Bullfrog, and introduced in 1996, failed to take off. Although the Bullfrog improved on Motorola's one-way paging devices, it was too bulky and heavy—and not powerful enough—to undermine Motorola's leadership. The Bullfrog was able to receive email but only after users made a change to their in-boxes on their desktop computers.

The Bullfrog's relative failure did not discourage the RIM partners. Lazaridis's vision of the handheld market had two components. First, he thought that it was a mistake to try to create a device with multiple features; it was better to concentrate on one feature only and to excel at it. Second, he was convinced that that feature should be mobile email.

Lazaridis envisioned a product that would solve a crucial problem faced by all handheld devices offering mobile email services: the two-mailbox problem. Such a device would be able to receive (and send) email automatically, without the user having to log on to his or her desktop and download the messages. Put differently, the handheld had to operate in sync with the desktop. RIM's second handheld, the Inter@ctive 950, or Leapfrog, was introduced in 1998 and made this vision a reality via "push" email.

The RIM 957, introduced in the year 2000, was RIM's first large-screen handheld—it looked less like a Motorola pager and more like a Palm PDA. This product put RIM ahead of both Motorola and Palm in the handheld market. With its on-the-go email service, the 957 was considerably better than the Motorola two-way pager in several ways: Motorola's pagers relayed messages through a network that had limited capacity and reach; in addition, such messages were restricted to five hundred characters and were often dropped or delayed; and finally, active pager users paid heavy monthly fees if they exceeded text limits equivalent to about twelve full-length email messages.

In its mobile email capabilities, the RIM 957 was also better than the Palm VII, the product that dominated the market in 1999–2000. Using email on the Palm VII required users to flip open an antenna and wait for messages to download to the device; and more fundamentally, the Palm VII could not be synched with corporate email boxes. Once users—corporate users especially—started realizing that email service on the RIM 957 was instantaneous, Palm's lead evaporated. In the year 2000 RIM outsold Palm by about 40 percent.

RIM succeeded in the 2000s because it managed to put corporate email on a handheld device. The first RIM handheld that also incorporated a telephone, the BlackBerry 5820, was released in 2001. Described as the "first voice-enabled BlackBerry," it had a browser and could send and receive email and text messages.[31] By the early 2000s, with smartphone offerings from Palm, the Symbian licensees, the Microsoft licensees, and RIM reaching the market, the convergence between voice and data on a handheld device was already in full swing.

THE BEGINNINGS OF THE MOBILE INTERNET

In the early 2000s it was not just electronic mail that migrated from personal computers to handheld devices—Internet access did as well. In June 1997 three of the top handset makers (Ericsson, Motorola, and Nokia) established the WAP Forum in collaboration with Unwired Planet, a Silicon Valley firm that supplied open-standard servers and microbrowsers to wireless companies.

WAP was not an isolated creation. In mid-1998 the three handset companies also created the Bluetooth Special Interest Group and the Symbian joint venture (with British firm Psion). WAP was conceived as an open standard for mobile browsers, Bluetooth as an open standard for wireless communication among devices located reasonably close to one another, and Symbian as an open operating system for mobile phones.

WAP, Symbian, and Bluetooth were developed as alternatives to both Microsoft's Windows CE and Palm Computing's operating systems and browsers.[32] WAP was designed as a carrier-independent and device-independent protocol to access information over a wireless network, and the WAP browser was the browser for mobile devices that used the protocol. Dutch operator Telfort VB created the first WAP site in 1999. Although promotional campaigns generated the expectation that WAP Internet access would be comparable to fixed Internet access in speed and features, the reality was radically different, and WAP failed to take off.

What did take off was i-mode, a service developed in Japan that enabled users to stay permanently connected to the Internet through their cellular phones. I-mode was created by DoCoMo, the cellular subsidiary of Nippon Telegraph and Telephone (NTT), which was spun off from NTT in 1992. At the time of the spinoff, cellular service in Japan was expensive, mobile phones were heavy and inconvenient, and cellular density was low by international standards. In 1994, however, the market was liberalized, competitors flourished, and prices plummeted. DoCoMo responded to competition by developing the world's smallest cellular phones and by setting up branded retail shops all over the country.

DoCoMo executives started considering the possibility of extending the business model beyond voice telephony, and an entrepreneurial trio within the company—Mari Matsunaga, Takeshi Natsuno, and Keiichi

Enoki—identified wireless Internet access as the service DoCoMo should focus on.[33] To make i-mode possible, DoCoMo engineers built a packet-switched network alongside the existing digital cellular network. I-mode, which was officially launched in February 1999, had several ingredients that made it successful. Its interface was easy to use and its fees—the monthly subscription fee, the variable data fees, and the subscribed content-site fees—were low. Early on, the monthly fee was just ¥300 (roughly equivalent to USD 2.78 per month in the year 2000), and it was common for the subscribed sites to set their fees at between ¥100 and ¥300 per month.

The service was designed as a portal site through which users could obtain access to content created by a swath of providers. I-mode content was of four types: entertainment, such as Disney and Pokémon; information, including CNN and Bloomberg; databases, such as Zagat; and transactions, including airline reservations and online banking. DoCoMo classified providers into two groups: first tier and the rest. First-tier providers paid a commission to DoCoMo and, in exchange, they obtained the right to be directly accessible from the menu bar. I-mode operated as a multisided platform with direct and indirect network effects: users were attracted to i-mode by the low monthly fees and the variety and quality of content available on the service, and in turn rapid growth in the number of users attracted content developers, which in turn further accelerated user growth.

The number of i-mode users rose at an impressive pace almost from the beginning. By December 1999 there were already 3.13 million subscribers, and this number had gone up to 6.5 million by April 2000 and to around 11 million by August 2000. It was estimated that, by the end of the year 2000, i-mode accounted for one-third of all Internet users in Japan.[34]

I-mode's growth had a profound impact on the Japanese cellular industry: it enhanced DoCoMo's share of the market, and it created new sources of revenue for the company. It also helped DoCoMo by lowering user churn and, in addition, it had a significant effect on handset manufacturers. Up to that point, users had tended to focus only on how light the mobile device was. Starting in 1999, however, it became important for users that mobile phones be i-mode capable, and DoCoMo responded by developing special phones with browser software and packet-based communications.

By 2003, email accounted for about 14 percent of total i-mode traffic, official Internet providers for 40 percent, and free Internet providers for the remaining 46 percent. With steady subscriber growth from the beginning, by 2004 i-mode had become, by far, the world's most successful approach for accessing the mobile Internet, and an example that many cellular companies in the rest of the world were trying to imitate.

THE NEW GEOGRAPHY OF CELLULAR

The geography of cellular markets also changed radically in the first decade of the twenty-first century. Table 7.4 presents the top cellular markets in the world in 1990, 2000, and 2010—that is, roughly at the beginning of the migration from 1G to 2G, from 2G to 3G, and from 3G to 4G.

The table reveals that by 1990, at the beginning of the transition from 1G to 2G, the cellular revolution had been concentrated in North America, Japan, and Western Europe. Additionally, in 1990 the United States, which at the time was the largest cellular market in the world, had about five times as many subscribers as Britain, the second largest. Around the

Table 7.4 Top 10 cellular markets in 1990, 2000, and 2010

1990		2000		2010	
Country	**Subscriptions**	**Country**	**Subscriptions**	**Country**	**Subscriptions**
United States	5,283,055	United States	109,478,031	China	859,003,000
Britain	1,114,000	China	85,260,000	India	752,190,000
Japan	868,078	Japan	66,784,374	United States	285,118,000
Canada	583,766	Germany	48,202,000	Russia	237,689,224
Sweden	461,200	Britain	43,452,000	Indonesia	211,290,235
France	283,200	Italy	42,246,000	Brazil	196,929,978
Germany	272,609	France	29,052,360	Japan	123,287,125
Italy	266,000	South Korea	26,816,398	Vietnam	111,570,201
Finland	257,872	Spain	24,265,059	Pakistan	99,185,844
Norway	196,828	Brazil	23,188,171	Italy	93,666,088

Source: ITU database.

year 2000, at the beginning of the transition from 2G to 3G, important changes were happening in the geography of cellular systems. In 2000 the United States was still the largest market with over one hundred million subscribers, but China was a close second with about eighty-five million. Further, by the year 2000, a Latin American country (Brazil) and an Asian Tiger (South Korea) were already among the ten largest cellular markets in the world.

By 2010, finally, at the beginning of the switch from 3G to 4G, the geographic core of the world cellular markets had shifted even further. Only three countries that were among the top ten markets in 1990 remained in 2010: the United States, Japan, and Italy. Moreover, by 2010, the United States had been displaced as the top cellular market in the world not only by China but also by India.

Between 2000 and 2010, the number of cellular subscribers grew by a factor of almost 3 in the United States and by a factor of about 10 in China—but it grew by a factor of 73 in Russia, and by a factor of 210 in India. (The caveat to keep in mind in interpreting some of these figures is that in certain countries, such as Russia, there were more cellular subscriptions than people in 2010, which suggests that some subscriptions may have been inactive at the time.)[35] Although China was well ahead of India in cellular subscriptions in the year 2000, a decade later India had almost completely caught up. Further, by 2010, none of the Western European countries other than Italy remained among the top ten cellular markets in the world.

THE 3G ERA: TAKING STOCK

The transition from the 1990s to the 2000s—from 2G to 3G—brought about important changes in the global mobile-phone industry. The 2000s witnessed the diffusion of two new cellular standards: cdma2000 and WCDMA. The former was the successor to the 2G cdmaOne standard, and the latter was the heir to the 2G GSM standard. The migration from 2G to 3G standards, however, did not happen overnight. Several upgrades—including HSCSD, GPRS, and EDGE—were developed for the 2G standards, and these upgrades slowed down the adoption of the 3G standards by offering a less costly, albeit temporary, alternative to operators.

China also developed its own 3G standard, TD-SCDMA. Although the core specifications of the standard were ready by the late 1990s, it was not deployed until the late 2000s, and only by China Mobile, one of the three Chinese cellular carriers. Since it was widely believed to have enabled slower data rates than the other 3G standards, China Mobile's commitment to it was short-lived: the carrier started migrating to 4G technology as soon as it became available.

The late 1990s and early 2000s saw the beginning of the transition from voice-only phones to feature phones first, and from feature phones to smartphones later. Just like the migration from 2G to 3G standards, these shifts did not happen overnight. It took more than a decade—and the arrival of the iPhone and a swath of Android devices in the late 2000s—for smartphones to become pervasive.

The diffusion of the 3G standards facilitated a change—the move from voice to data—that affected mobile-phone users in a fundamental way. In the 1980s cellular phones were used for voice communications, and that continued in the 1990s, even though short messaging services grew in popularity toward the end of the decade. The 2000s were not only about voice but also about data. Two new uses for mobile phones became increasingly popular from the beginning of the 2000s: Internet access and electronic mail. NTT's i-mode became a synonym for mobile Internet access in the early 2000s, and BlackBerry did so for mobile email throughout the decade.

Finally, the 2000s witnessed a shift in the geographic center of gravity of world cellular markets. In the 1980s the diffusion of mobile phones happened mainly in the United States, Japan, and Western Europe. In the 1990s the number of cellular subscribers grew rapidly in countries such as Brazil, South Korea, and China, all of which joined the club of the top ten markets in the world. In the first decade of the twenty-first century, China consolidated its position as the largest cellular market in the world. But the star markets of the 2000s were India and Russia: by 2010 the largest cellular markets in the world were not in the traditional Organisation for Economic Co-operation and Development countries but rather in countries such as Brazil, Russia, Indonesia, India, and China.

8

EMERGENT DIGITAL MARKETS: NEW ZEALAND, JAPAN, SOUTH KOREA, THE ASIA-PACIFIC REGION, ISRAEL, BRAZIL, AND MEXICO

8.1 Picture of mobile-phone shop display, South Korea, 2006. *Source*: Wikimedia.

Chapter 5 covered several case studies of the transition from analog to digital cellular in the 1990s. We examined two cases of enhanced competition (Britain and the United States), three cases of competition for the first time in Europe (Germany, Italy, and Portugal), and two cases in Asia—one of competition for the first time (China) and the other of cellular for the first time (India). We showed that cellular density grew faster in countries that fostered competition *and* established an independent regulator early on. The best examples were Britain and Portugal.

In this chapter, we study the analog-to-digital migration in important geographic markets not covered in chapter 5. We start with New Zealand, usually considered the antithesis of Britain in its approach to regulating telecommunications. Rather than creating an independent regulator, the government of New Zealand decided that conflicts among telecommunications operators would be handled by the courts on the basis of competition-policy principles. We show that this model worked well in New Zealand and explain why.

Then we focus on two large markets in Asia: Japan and South Korea. Japan experimented with limited cellular competition in the late 1980s, and then it enhanced it in the 1990s by licensing additional operators. South Korea had a state-owned cellular monopoly for about a decade starting in the mid-1980s until it liberalized its cellular market in the mid-1990s. Thus, Japan was a case of enhanced competition and South Korea a case of competition for the first time in the 1990s. By the late 1990s, both were among the largest cellular markets in the world.

We also examine a set of countries in the Asia-Pacific region: Thailand, Malaysia, Hong Kong, Singapore, and Taiwan. In all of them, cellular markets were liberalized in the 1990s, but cellular density grew faster in Singapore, Taiwan, and Hong Kong. This happened because these three countries were wealthier and more densely populated than the others, and their economies grew considerably faster. Thus, in the Asia-Pacific region, economics, geography, and demography account for differences across countries in cellular density.

The case of Israel, which we study next, reveals that, even if competitive markets were not the only factor that mattered for fostering growth in cellular density, they did have a powerful impact on it. Israel had a cellular monopoly for several years and liberalized the market in the mid-1990s.

The company that entered the market to compete with the incumbent lowered prices dramatically, and the market exploded. By the late 1990s, Israel was among the top markets in the world in cellular density.

Finally, we study two countries in Latin America, Mexico and Brazil, where cellular systems started operating for the first time in the late 1980s and early 1990s: Mexico introduced cellular phones in 1989, and Brazil did in 1992. Cellular density grew faster in Brazil than in Mexico. The reason is that, although both countries had a period of competitive markets without an independent regulator, in Mexico the state carrier took advantage of the situation to handicap the private competitors, which slowed down cellular growth.

FROM MONOPOLY TO COMPETITION WITHOUT AN INDEPENDENT REGULATOR: NEW ZEALAND

In chapter 5 we showed that, generally, cellular density grew at a faster rate in countries that encouraged cellular competition and established an independent telecommunications regulator early in the process. The two leading examples were Britain and Portugal. At the beginning of the cellular market in Britain, the government fostered competition in cellular services and created an independent regulator, Oftel, to monitor the competitive process. Similarly, the Portuguese government liberalized the telecommunications sector in the late 1980s and early 1990s, while at the same time reactivating a governmental unit whose role was to regulate the sector. In both countries, cellular density grew at a remarkable pace.

New Zealand presents an interesting alternative to the British model of telecommunications reform.[1] During the 1980s, the government of New Zealand introduced reforms in telecommunications and other areas. Although many of these policies were similar to those implemented in Britain, there was a fundamental difference: New Zealand's government chose an approach that has been described as "light touch" regulation, since it relied on general antitrust laws and the courts to deal with disputes among competitors in the cellular (and fixedline) markets.

During most of the twentieth century, New Zealand's Post Office supplied telecommunications, postal, and banking services. By the early 1980s, the economy was worsening, to a good extent because of policies introduced in the prior decade. Faced with rising public deficits and inflation, the Labour government inaugurated in 1984 implemented several market-oriented reforms, not unlike those introduced in Britain at about the same time. The Post Office, widely perceived as inefficient, became an obvious candidate for reform.

A review of the Post Office completed in 1986 recommended that it be reorganized into different units, that regulation be separated from commercial operations, and that the value-added and terminals markets be opened to competition. In 1987 the Post Office was split into three autonomous enterprises owned by the state: Telecommunications Corporation of New Zealand (also known as Telecom), the New Zealand Post, and Postbank, each in charge of one of the sectors—telecommunications, postal services, and banking—that had been previously monopolized by the Post Office. Telecom created a subsidiary, Telecom Mobile, which started operating an analog Advanced Mobile Phone Service (AMPS) network in mid-1987.

The Telecommunications Amendment Act of 1988 mandated that all segments of the telecommunications market be opened to competition starting in April 1989. In 1990 Telecom was sold to a joint venture between local investors and two large US telecommunications companies (Bell Atlantic and Ameritech), and the proceeds were used to retire government debt. A fundamental difference from the British model was that no telecommunications regulator was created. Rather, it was decided that conflicts among telecommunications operators would be dealt with on the basis of general antitrust rules, embodied in the Commerce Act of 1986. The Commerce Commission would monitor the application of antitrust regulations, and the courts would handle disputes among market participants. The High Court would act as a first-instance tribunal, and appeals would be handled first by the Court of Appeals and ultimately by the Privy Council in Britain.

In the early 1990s, the government took steps to dismantle Telecom's cellular monopoly. The first step was to allocate spectrum—some bands for the AMPS standard and others for the Total Access Communications System (TACS) standard—and New Zealand pioneered the use of auctions for this purpose. In 1990 three cellular bands were auctioned: AMPS-A, TACS-A , and TACS-B. Telecom, which owned the fixedline network and was already providing cellular services in the AMPS-B band, won two of them (AMPS-A and TACS-B) while Bell South won the third (TACS-A). The awards resulted in prolonged legal disputes. Telecom Mobile was first deprived of its newly awarded frequencies (AMPS-A and TACS-B), although it later recovered the AMPS-A band. Eventually, the Australian firm Telstra was awarded the TACS-B frequencies. By 1993 there were thus three cellular operators in New Zealand: Telecom Mobile (AMPS-A and AMPS-B), Bell South (TACS-A), and Telstra (TACS-B). Bell South and Telecom had long negotiations on interconnection. The discussions started in early 1992, and an agreement was not reached until mid-1993.

Bell South set up a Global System for Mobile Communications (GSM) digital network that did not attract many subscribers early on, whereas Telecom Mobile's analog AMPS system, later converted to a D-AMPS digital network, thrived. Over the years, however, the Bell South network, acquired by Vodafone in 1998, gained traction and became a serious competitor for Telecom Mobile. By the year 2000, Vodafone New Zealand had about 40 percent of the cellular market (against Telecom's 60 percent), but by 2002, Telecom Mobile and Vodafone New Zealand had roughly equal shares.[2] Cellular subscribers grew at a fast pace in the second half of the 1990s, from 365,000 in 1995 to 1.5 million in 2000, and cellular density rose from about ten to almost forty. The transition to 3G started in January 2001, when five licenses were awarded by auction.[3]

Although it has often been argued that the New Zealand market was harmed by the absence of an independent telecommunications regulator, the reality is more complicated. Figure 8.2 shows cellular density in New Zealand and Britain between 1985 and 2010. The figure shows that New Zealand lagged behind Britain considerably, and thus appears to confirm the theory that New Zealand's "light regulation" approach failed.

Measured from the time of launch, however, cellular density rose faster in New Zealand than in Britain during the first fifteen years of service. Figure 8.3 shows cellular density in New Zealand up to fifteen years after service was launched, with Britain again included for comparison. (With this approach, we "control" for the fact that cellular service started later in New Zealand than in Britain.) The figure shows that New Zealand performed at least as well as Britain for most of the fifteen-year period. This

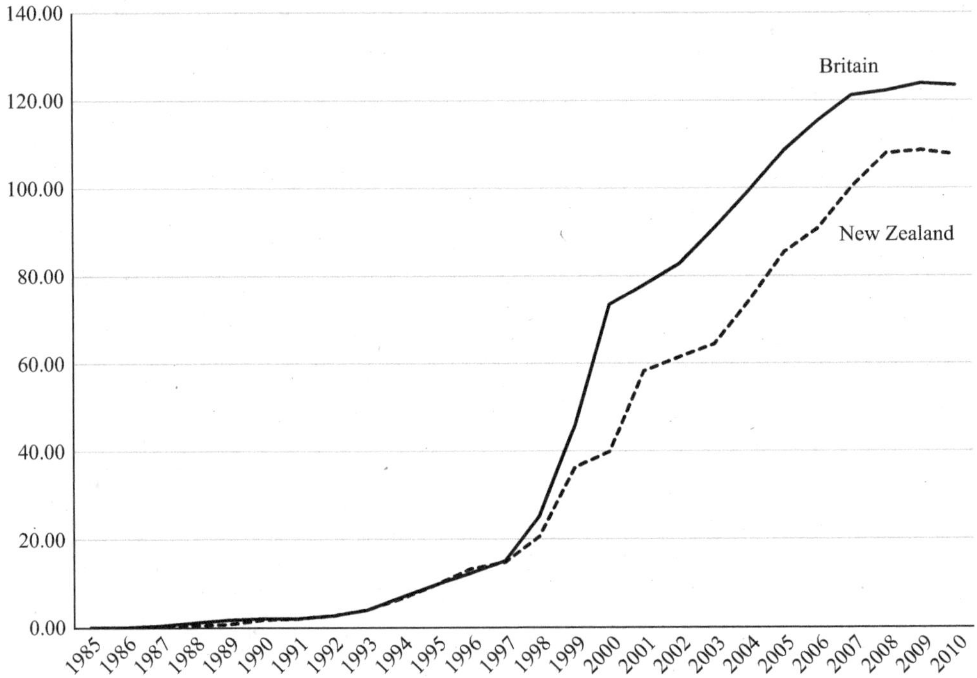

8.2 Cellular density in Britain and New Zealand, 1985–2010. *Source*: International Telecommunications Union (ITU) database.

is remarkable because, as we have shown in earlier chapters, Britain was one of the best-performing countries in the Organisation for Economic Co-operation and Development.

Since, once we control for the lag in the start of service, the cellular market in New Zealand performed at least as well as (or better than) the British market, even without an independent telecommunications regulator, it is worth asking whether an independent regulator was really needed in emerging cellular markets to ensure rapid growth in cellular density. There are at least three reasons why the cellular market in New Zealand grew as fast as it did without a regulator. First, the country's telecommunications reforms were radical. The Post Office was split into three independent entities, the state telecommunications corporation was privatized quickly, and the monopoly period in cellular network services was relatively short. Second, the fact that Telecom was privatized made competition more intense in the cellular market. Finally, and most importantly, the courts intervened actively to restrain behavior they deemed anticompetitive.

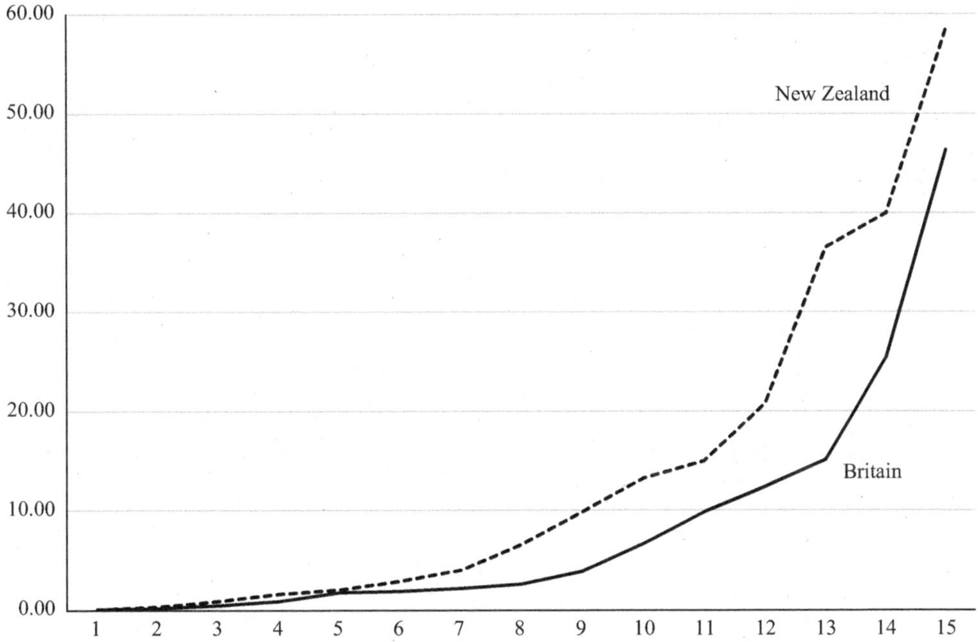

8.3 Cellular density in Britain and New Zealand, one to fifteen years after launch of service. *Source*: Calculated by authors from ITU database.

The main interconnection disputes occurred in the first half of the 1990s and affected both the fixedline and cellular markets. Although the process was complex and involved several appeals, the courts eventually forced the incumbent to improve the interconnection terms it offered to the private competitors. This signaled to both consumers and investors that the reform was irreversible: the courts played the role that a telecommunications regulator would have played had it been set up at the beginning of the reform process. By the end of the decade, the private cellular competitor and the incumbent operator had similar market shares.

DIGITAL COMPETITION IN TWO LARGE ASIAN MARKETS: JAPAN AND SOUTH KOREA

Although the Japanese market became somewhat competitive in the second half of the 1980s, whereas South Korea had a cellular monopoly until the mid-1990s, these countries shared several features. In both countries the government intervened in the domestic cellular market during the

1990s to promote a cellular standard—Pacific Digital Cellular (PDC) in Japan and code division multiple access (CDMA) in South Korea—and to foster the growth of local equipment manufacturers. By the late 1990s both were among the largest cellular markets in the world.

ENHANCED COMPETITION IN JAPAN

The national telecommunications monopoly Nippon Telegraph and Telephone (NTT), operating as NTT DoCoMo, was a monopolistic provider of cellular services in Japan from 1979, the year when the first system was launched, until 1986. In that year, additional licenses were granted to two private consortia, Nippon Ido Tushin (IDO) and Daini Denden Inc.(DDI). After NTT DoCoMo started competing with the consortia, the market changed in two ways: cellular density grew at a faster pace, and the country transitioned from one cellular standard to two.

In 1989 the Ministry of Posts and Telecommunications decided to move the country toward a single standard and created a committee to define the technical requirements of a national cellular system.[4] The standard itself was developed by the Japanese Research and Development Center for Radio Systems, which borrowed heavily from the American digital D-AMPS standard. The Japanese standard was called Pacific Digital Cellular, and foreign firms such as Motorola, Ericsson, and AT&T were invited to join Japanese companies in the development of systems that adhered to the new standard. The PDC standard covered only the air interface—the protocols that deal with communications between mobile devices and base stations—and thus the door was left open for other interfaces to become manufacturer specific.

In 1991 the ministry decided to license two Personal Communications Services (PCS) providers in each region to compete with NTT DoCoMo. (These were the services described in the 1989 "Phones on the Move" document of the British Department of Trade and Industry under the Personal Communications Networks name. They were originally conceived as systems ideally suited for light, handheld portable terminals operating in densely populated areas.) The PCS licenses were granted to two consortia—Tu-Ka, led by Nissan, and Digital Phone, led by Japan Telecom. (Japan Telecom was a private company founded in 1984; it belonged to the

SoftBank conglomerate and should not be confused with NTT.) Both consortia included a variety of Japanese and foreign partners.

After the PCS operators entered the market, as many as five cellular service providers competed for subscribers in the top urban areas: NTT DoCoMo, the two consortia that had entered the market in the 1980s (DDI and IDO), and the two that entered in the 1990s (Tu-Ka and Digital Phone). The ministry gave permission to Tu-Ka and Digital Phone to form joint ventures in less densely populated regions of the country, which were unable to support five operators profitably. In 1993 NTT DoCoMo—created originally as the mobile subsidiary of NTT but at this stage already operating as an independent company—launched its PDC network in Tokyo, and its competitors inaugurated their own digital networks soon thereafter.

Competition among cellular carriers and the liberalization of the terminal market boosted cellular density. The number of cellular subscriptions jumped from about 11.7 million in 1995 to around 66.7 million in the year 2000, and cellular density rose from less than ten to about fifty-three over the same period. By the year 2000, Japan had become the third-largest cellular market in the world.

The number of market players declined at the end of the 1990s. In 1999 Tu-Ka and Digital Phone merged to form J-Phone, and the following year IDO and DDI merged to create KDDI. (The merger also included a telecommunications firm named Kokusai Denshen Dewa [KDD], and that explains the KDDI name.)[5] In 1999 NTT DoCoMo launched its proprietary i-mode Internet access service, which was considerably more successful than the Wireless Application Protocol standard promoted by Ericsson, Motorola, and Nokia. I-mode consolidated DoCoMo's leadership in the Japanese market. The transition to 3G started in June 2000, when licenses were allocated to DoCoMo, J-Phone, and KDDI.[6] By 2004, NTT DoCoMo was the largest cellular carrier, with a 53 percent share, and KDDI and J-Phone followed far behind.[7]

COMPETITION FOR THE FIRST TIME IN SOUTH KOREA

The creation of the Republic of Korea in 1948 resulted directly from decisions made at the Potsdam Conference, which took place in 1945 with representatives from the United States, Britain, and Russia. At the conference, it was determined that the Korean peninsula would be divided at the thirty-eighth parallel, with the North and the South under Soviet and American influence, respectively.[8]

The Ministry of Communications ruled South Korean telecommunications until the early 1980s, monopolizing not only regulation and policymaking but also the supply of services. To foster the development of the phone system in the country, in 1982 the government transferred the operation of telephone services to the Korea Telecommunication Authority, a corporation owned by the government. Supervised by the Ministry of Communications, this corporation was given a monopoly over domestic telephone service and the supply of telephones.

In 1984 the government allowed the Korean Mobile Telecommunication (KMT) company, a subsidiary of the Korea Telecommunication Authority, to begin providing mobile-phone services. KMT set up an analog AMPS system and supplied cellular services as a monopolist for about a decade. The choice of standard reflected the American influence that persisted in the country decades after the Korean War of the early 1950s.

During the 1980s and early 1990s, American telecommunications companies, with the support of the US government, pressured the South Korean government to liberalize telecommunications markets, including cellular services. After several rounds of negotiations, the South Korean government made drastic changes in the market. It first opened the value-added services market to full competition and fostered duopoly competition in long-distance and international calling. Later, it allowed a second mobile carrier to enter and compete with KMT, the government-owned monopolist, while at the same time starting the process of privatizing KMT.

Political infighting broke out between two ministries that had different views on two related issues: when the cellular market would be liberalized and who the second operator would be. The Ministry of Communications wanted a second cellular carrier to start providing services as early as possible and was opposed to giving a mobile license to one of the South Korean chaebols, which were already in the business of manufacturing mobile-phone equipment. (The chaebols are large, family-owned conglomerates that have played a crucial role in the South Korean economy since the 1960s.) The Ministry of Communications was opposed by the Ministry of Trade, Industry, and Energy, whose interests were more aligned with those of the chaebols. The Trade Ministry wanted to delay the introduction of competition in the cellular market and did not view the vertical integration of the domestic conglomerates into cellular services as problematic.

In 1992 the second cellular license was awarded through a beauty contest. Six consortia participated, all of them with large South Korean shareholders and a variety of foreign partners. The Ministry of Communications determined that foreign operators could only bid through joint ventures with domestic companies and that the network would adhere to the American digital D-AMPS standard. The consortium that won the contest had family ties to the South Korean president, which generated allegations of undue influence, and the license was returned voluntarily to the government.

Before a second beauty contest took place, the government decided that the new system would rely on the CDMA standard promoted by the American firm Qualcomm. In 1994 the Shinsegi consortium—which was led by the POSCO iron and steel company and included the American firms AirTouch, SBC (formerly Southwestern Bell), GTE (formerly General Telephone & Electronics Corporation), and Qualcomm—won a new competition for the second license. In 1995 the government awarded a CDMA license to Shinsegi and another one to KMT, which had been partially privatized in 1994. KMT, renamed SK Telecom (SKT), started delivering CDMA services in three cities in January 1996, and Shinsegi launched a more comprehensive CDMA network in April. Later that year, the government encouraged additional carriers to enter the market by awarding three more CDMA licenses.

The South Korean cellular market is particularly interesting as a case study for at least two reasons. For one, South Korea soon became one of the largest cellular markets in the world. Moreover, the disputes between government ministries regarding cellular liberalization also involved a dispute about the choice of cellular standard, and that choice was tied to industrial policy.

In the late 1980s and early 1990s the Ministry of Communications requested that the Korean Electronics and Telecommunications Institute (ETRI), a research organization created in the 1960s, explore the possibility of developing digital cellular technologies domestically. At the time, the ministry was concerned about the presence of foreign phone manufacturers in South Korea, and particularly about Motorola's dominant share in the South Korean market for analog handsets. Since the American firm Qualcomm expressed an interest in transferring CDMA know-how to ETRI, the research institute recommended that the South Korean government adopt CDMA as the official 2G standard.[9] In 1991 ETRI, on behalf of the Ministry of Communications, entered into an agreement with Qualcomm for the development of CDMA technology in South Korea. The ministry expected that, by adopting CDMA as the 2G standard for cellular phones in the country and by fostering technology transfer to South Korean companies, it would help the domestic conglomerates become important players in the international market for cellular equipment.

The industrial-policy implications of the agreement soon took center stage. Four South Korean conglomerates—Samsung, Goldstar (later renamed LG), Hyundai, and Maxon—were invited to participate in the research project. The arrangement stipulated that Qualcomm would share its CDMA knowledge with ETRI and the conglomerates, and also that Qualcomm would receive a royalty for each product sold that used its technology. In turn, the company would donate 20 percent of its royalty income to ETRI to fund further research.

The ETRI-Qualcomm agreement, sponsored by the Ministry of Communications, did not go unchallenged. Other ministries, with the Ministry of Trade, Industry, and Energy at the forefront, were unconvinced that CDMA offered the best path forward for the South Korean cellular market. After all, when the agreement was finalized in 1991, the European GSM standard was much closer to being implemented than the American CDMA. At that stage, CDMA had not even been accepted in the United States as a formal alternative to time division multiple access (TDMA). The trade ministry launched a TDMA development project in South Korea, but the domestic response to this initiative was lukewarm.

The Ministry of Communications and ETRI eventually prevailed, and CDMA was adopted as the official standard for 2G cellular phones in South Korea. The large South Korean cellular market gave the local conglomerates the opportunity to develop CDMA products for the domestic market before competing in international markets, and to break the dominance of foreign equipment manufacturers in the country. By 1998

South Korea had the largest CDMA network in the world; moreover, domestic companies produced about 90 percent of all handsets used and 80 percent of the infrastructure deployed in the domestic market.

The South Korean cellular market grew rapidly in the second half of the 1990s, with subscriptions jumping from around 1.6 million in 1995 to about 26.8 million in the year 2000 and cellular density rising from less than four to about fifty-eight over the same period. By the year 2000, South Korea had become the eighth-largest cellular market in the world.[10] The transition to 3G started in December 2000, when two licenses were awarded by a beauty contest plus a fee. The next year, in August 2001, a third 3G license was allocated to a third carrier.[11]

JAPAN AND SOUTH KOREA: SUMMARY

In the 1990s the liberalization of the cellular markets in Japan and South Korea encouraged subscriber growth. The second half of the 1990s was a period of radical change in both markets, with cellular density jumping by a factor of five in Japan and by a factor of fifteen in South Korea. At the same time, domestic governments intervened actively in the market, allocating licenses to new carriers, influencing the choice of cellular standards, and implementing policies that benefited local producers of cellular equipment.

Figure 8.4 shows cellular density in Japan and South Korea between 1979 and 2010, with Britain included for comparison.

The figure shows that, in both South Korea and Japan, cellular density started growing at a faster pace in the mid-1990s. It also shows that Britain, where cellular service was launched later than in both Asian countries, outperformed them both from the late 1990s.

CELLULAR LIBERALIZATION IN THE ASIA-PACIFIC REGION

Five countries in the Asia-Pacific region—Malaysia, Thailand, Hong Kong, Singapore, and Taiwan—are useful case studies for several reasons. First, three of them—Malaysia, Thailand, and Hong Kong—had competing analog networks in the 1980s, a decade in which cellular competition was uncommon. Second, governments in these countries refrained from imposing a cellular standard for the whole market. They allowed the market to

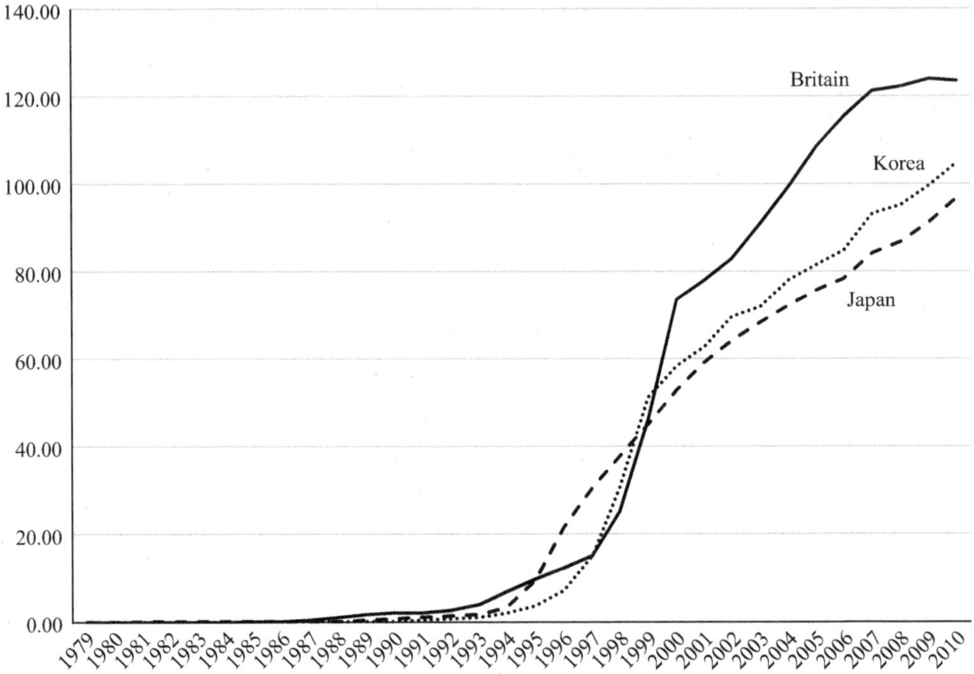

8.4 Cellular density in Britain, Japan, and South Korea, 1979–2010. *Source*: ITU database.

choose a cellular technology, and thus competition among cellular providers often translated into competition among standards. Third, two of them—Singapore and Taiwan—quickly achieved unusually high levels of cellular density. Subscribers grew (relative to population) at a fast pace in Hong Kong as well.

Liberalization in the 1990s had different meanings for different countries: for Malaysia, Thailand, and Hong Kong, it meant enhanced cellular competition, whereas for Singapore and Taiwan, it meant competition for the first time. Nonetheless, during the 1990s Singapore and Taiwan were among the countries with the fastest growth in cellular density in the world, which suggests that factors other than competitive markets mattered for cellular growth.[12]

Malaysia, Thailand, and Hong Kong were pioneers of analog competition in the 1980s. In Malaysia, the domestic postal, telegraph, and telephone (PTT) administration started operating a cellular network in 1985, and a competitor entered the market (with a different standard) in 1989. In

8.5 The Samsung Z500 phone. Launched in 2005 and weighing 95 g, the Samsung Z500 was equipped with a camera, MP3 player, and Wireless Application Protocol browser, and it could conduct video calls. *Source*: Martin Campbell-Kelly.

Thailand, two state entities competed in the analog cellular market in the late 1980s (with networks operating on different standards), and in 1990–1991 they cooperated with commercial organizations to set up additional analog systems. In Hong Kong, the PTT launched two different cellular networks (on two different standards) in the first half of the 1980s, and soon private operators entered the market.

By contrast, Singapore and Taiwan did not have competitive cellular markets in the analog era. In Singapore, the domestic PTT set up three different analog systems between the early 1980s and the early 1990s—that is, there was a monopolistic provider of cellular services that offered subscribers a choice of standard. In Taiwan, one of the state telecommunications entities offered cellular services as a monopolist from 1989 until 1995.

In all of these countries, cellular markets were liberalized in the mid-1990s. In Thailand and Malaysia, private consortia started operating GSM and Digital Cellular System (DCS) 1800 systems in 1994 and 1995. (DCS 1800 was the variant of the GSM standard originally specified for the Personal Communications Networks providers in Britain in the late 1980s.) In Hong Kong, several GSM and DCS 1800 licenses were awarded between 1993 and 1997. In Singapore, the domestic PTT launched GSM and DCS 1800 networks in 1994–1995, and in 1997 a private competitor was allowed to enter the cellular market. In Taiwan, eight different competitors received licenses in 1996–1997. In all of these countries, cellular density started growing at a faster pace in the second half of the 1990s, and in all of them, the transition to 3G started in the early 2000s. The methods for awarding 3G licenses varied considerably, from the case of Thailand, where one license was allocated to a carrier in early 2000, to the case of Taiwan, where five licenses were awarded by auction in February 2002.[13]

Figure 8.6 shows cellular density in all the Asia-Pacific countries discussed here between 1984 and 2010, with Britain included for comparison. The figure shows that, in the long run, the performance of Hong Kong and Singapore was exceptional. Density grew at a remarkable pace in Taiwan until the early 2000s, whereas Malaysia and Thailand only started catching up with Britain in the late 2000s.

Figure 8.7 shows cellular density in all Asia-Pacific countries discussed here up to fifteen years after launch of service in each country. For comparison, Britain is also included.

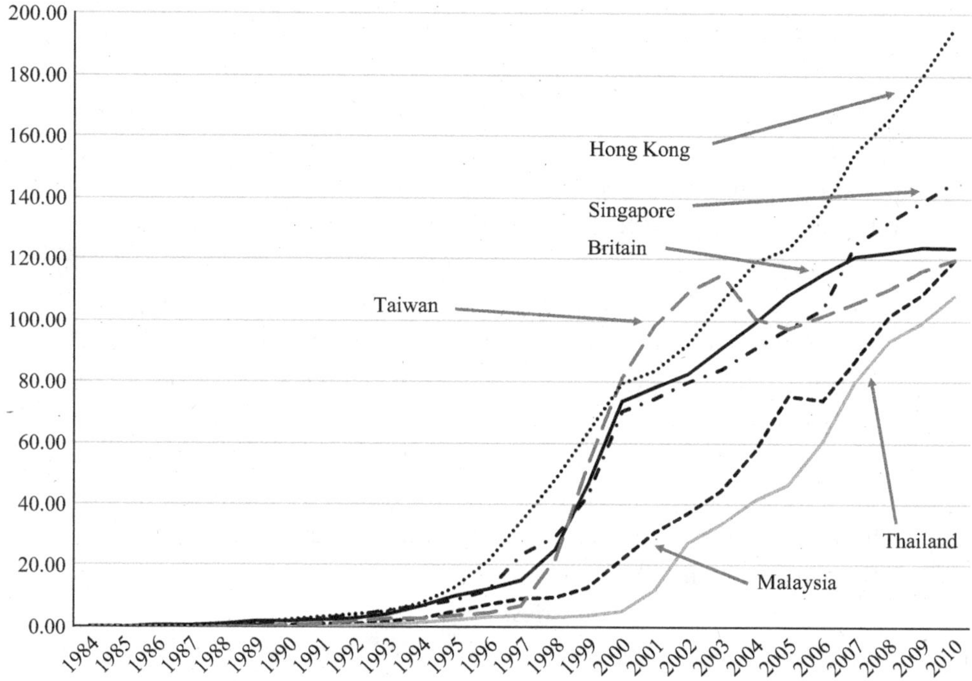

8.6 Cellular density in Britain and the Asia-Pacific region, 1984–2010. *Source*: ITU database.

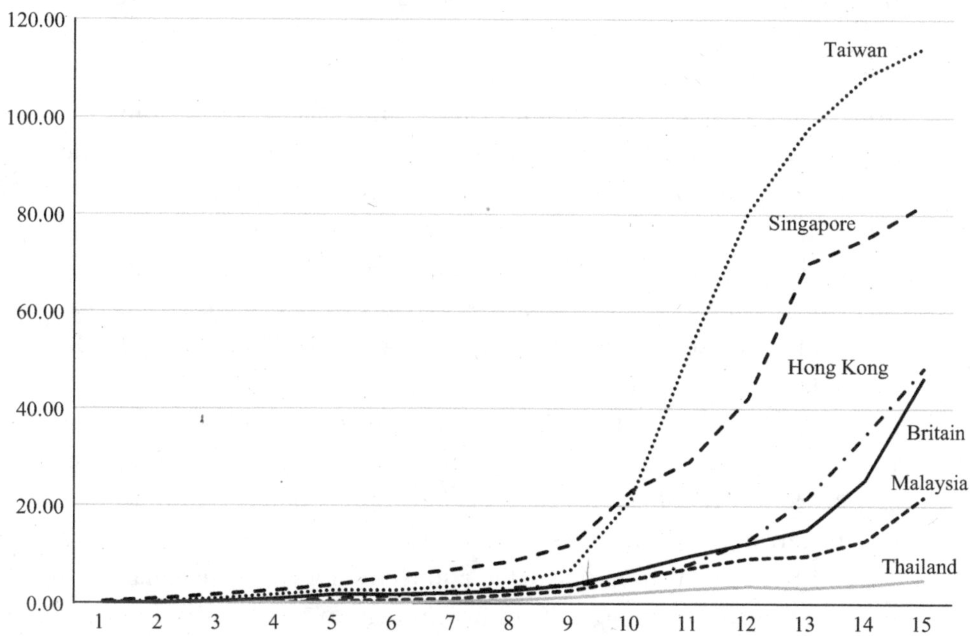

8.7 Cellular density in Britain and the Asia-Pacific region, one to fifteen years after launch of service. *Source*: Calculated by authors from ITU database.

The figure shows that the growth of cellular density accelerated in most countries roughly ten years after launch of service and, further, that density started growing at a faster pace earlier in Taiwan and Singapore. Whereas Thailand and Malaysia performed considerably worse than Britain, Taiwan and Singapore did significantly better (and Hong Kong did at least as well). As noted earlier, Malaysia and Thailand had analog competition in the 1980s, but Taiwan and Singapore did not. This suggests that, although digital competition in the 1990s had a powerful impact on cellular markets in all countries, there were factors other than competition that explain why certain markets grew faster than others. If the pace of growth in cellular density had been driven purely by competition, Thailand and Malaysia, pioneers of competition in analog cellular, should have had an advantage. Economic and demographic factors help explain cellular growth in those countries.

Taiwan, Singapore, Hong Kong, and South Korea have been labeled the Asian Tigers: they are the countries in East Asia that experienced rapid industrialization and economic growth in the second half of the twentieth century. These four countries grew faster than most other countries in Asia between 1960 and 2010. The ratio of gross domestic product (GDP) per capita in 2010 to GDP per capita in 1961 (in 2005 prices and purchasing-power-parity terms) was 16.8 for Taiwan, 15.6 for South Korea, 12.4 for Singapore, 12.4 for China, and 11.5 for Hong Kong. For comparison, it was 8.2 for Thailand and 7.9 for Malaysia. In addition, the Asian Tigers are densely populated, and it is easier for cellular carriers to provide coverage in more densely populated countries. In 1980, when the first cellular systems were being launched around the world, Hong Kong had 4,681 people per square kilometer, Singapore had 3,445, Taiwan had 500, and South Korea had 391. For comparison, Thailand had about 93, Malaysia had 42, and the United States had 25.[14]

Economic growth in the Asian Tigers during the 1960s and 1970s was based on international wage differentials and export-oriented manufacturing. When, in the late 1970s, rising labor costs started to limit further expansion, the governments of all but Hong Kong introduced policies to facilitate the transition to capital- and technology-intensive manufacturing. By fostering foreign direct investment and technology transfer, governmental policies facilitated the development of the electronics and computing industries in these countries. Thus, by the early 1980s, South Korea, Taiwan, and Singapore were technologically advanced, and their populations were savvy in the use of electronics.[15] Hong Kong was somewhat different in that it had been growing at a fast pace for decades and adopted a laissez-faire approach to the economy. By 1960, it had the infrastructure required by a modern economy, and it upgraded and expanded it in subsequent decades, relying mostly on private initiative and funding.[16]

This combination of technology-savvy populations, high per capita incomes, very high population densities, and rapid output growth helps explain why the Asian Tigers reached unusual levels of cellular density by the end of the 1990s. Asia-Pacific countries that were considerably less wealthy and more sparsely populated, such as Malaysia and Thailand, lagged behind, even though they had had competitive cellular markets during the analog era in the 1980s.

FROM MONOPOLY TO COMPETITION IN THE MIDDLE EAST: ISRAEL

The government had a monopoly on telecommunications services in Israel from the time the modern state was created in 1948 through the 1980s. In 1979 the government decided to create a state-owned company responsible for supplying telecommunications services. This was a "corporatization" process by which telecommunications operations were transferred from a state administration to a state-owned corporation, without modifying the monopolistic market structure.[17] The telecommunications law was passed in 1982, and Bezeq, the national telecommunications corporation, was established in 1984.[18]

Two years later, in 1986, the government announced a tender for a cellular license, which attracted limited interest. Motorola Israel, the local representative of Motorola Corporation, joined forces with Bezeq to establish the Pelephone Communication Corporation, which became the first cellular operator in Israel when it inaugurated an analog AMPS system in 1987. The Pelephone system, in which both the calling and the receiving parties paid for the call, was awarded exclusivity until 1994.

In the late 1980s and early 1990s, in a second wave of telecommunications reform, the government decided to privatize Bezeq and liberalize various telecommunications services, including cellular. In 1993 the Ministry of Communications issued a tender for a second cellular license and at the same time changed the Pelephone license into a calling-party-pays regime. Moreover, the ministry established that 50 percent of the score in the beauty contest for the second cellular license would be based on the price would-be operators committed to charge subscribers. Other factors the government would consider when awarding the license were the proposed timetables for geographic coverage and for migrating the system to digital cellular.

The winner of the 1993 license contest was Cellcom, a consortium linking Bell South, Israel Aircraft Industries, the Safra Group from Brazil, and the Discount Investment Group. Cellcom inaugurated an American 2G TDMA network in 1994 and was extremely aggressive with its pricing, setting service rates for the consumer that were about one-tenth of Pelephone's and one-third to one-fourth of the European and American rates at the time. Cellcom's aggressive pricing generated rapid growth in subscriber numbers, from about 133,000 in 1994 to about 1.7 million in 1997.

The business press highlighted the dramatic price changes taking place in the Israeli cellular market. According to commentators, Cellcom entered the market in late 1994 with "the lowest airtime rates anywhere on the planet: two-and-a-half cents for a full minute" of talk any time of the day. Cellular calls suddenly became cheaper than regular fixedline calls. In addition, prices for cellular airtime compared well with prices in any other country. Three hundred minutes of airtime could be bought in Israel for USD 8 versus USD 136 in the US market.[19]

Thus, the transition from monopoly to duopoly in Israel led to drastic price changes, with minimum available per-minute prices dropping by a remarkable 90 percent. This decline in subscription tariffs was driven, to a good extent, by policy design: the rules of the game were that the winner of the tender would be the company that was able to offer the lowest prices to consumers.

In 1997 the Ministry of Communications announced a tender for a third cellular license. This time the contest was for a GSM network, and the score of the competitors would be heavily based on the license fee they were willing to pay. The winner was Partner Communications, an entity in which Hutchison Whampoa, the Hong Kong–based conglomerate, had a large stake. Partner Communications started operating its GSM system in 1999. Its share grew rapidly, to the point that, by 2004, each of the three companies had about one-third of the Israeli cellular market.[20]

The cellular market in Israel changed radically in the second half of the 1990s, with subscriber numbers growing from about 445,000 in 1995 to 4.4 million in the year 2000, and with cellular density rising from about eight to around seventy-three. The transition to 3G started in December 2001, when three licenses were awarded by tender.[21]

Figure 8.8 shows cellular density in Israel and Britain between 1985 and 2010. The figure shows that, for most of the period, Israel performed as well as Britain or better. The figure also shows that cellular density in Israel started growing faster around 1994, precisely when the first competitor (Cellcom) entered the market.

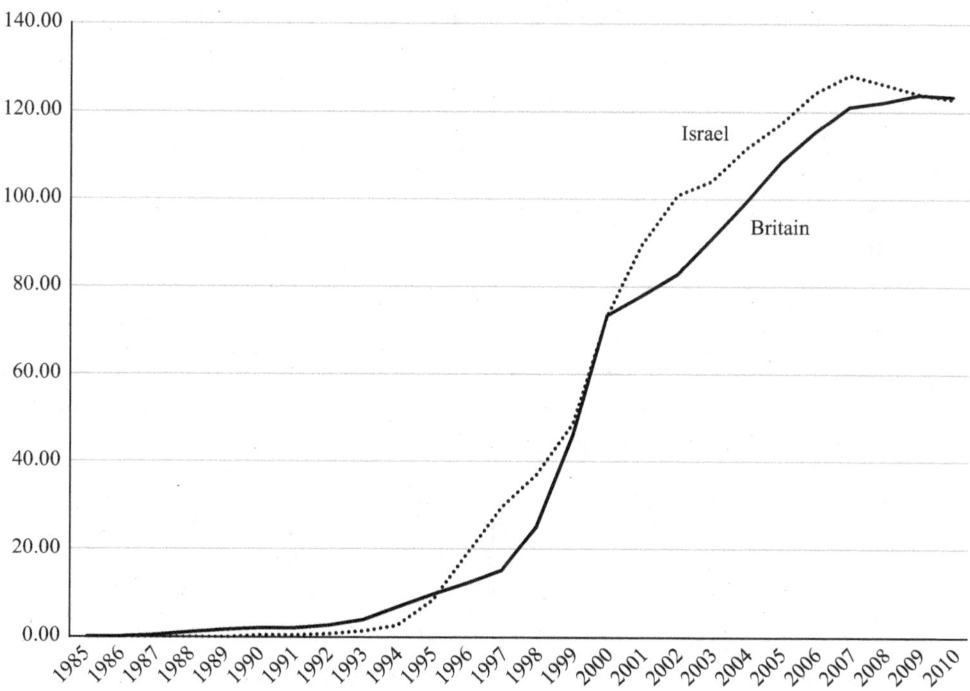

8.8 Cellular density in Britain and Israel, 1985–2010. *Source*: ITU database.

FROM MONOPOLY TO COMPETITION IN LATIN AMERICA: BRAZIL AND MEXICO

Analog cellular service was introduced for the first time in Latin America in the late 1980s. The pioneers were Venezuela in 1988 and Chile, Argentina, and Mexico in 1989. Between 1990 and 1992, systems were also launched in Peru, Bolivia, Uruguay, Paraguay, and Brazil. All of these networks adhered to the American AMPS standard.[22] We study two cases in detail: Mexico and Brazil. Cellular markets in both countries eventually became regional duopolies. Brazil's cellular market had a relatively short monopoly period until liberalization happened in the second half of the 1990s, whereas the Mexican market had competing operators almost from the beginning.

BRAZIL

The modern Brazilian telecommunications system was born between the early 1960s and the early 1970s.[23] In those years, a planning agency was charged with the task of selecting national technical standards, and Embratel, a state operator, was created. The Ministry of Communications was established to act as the sector's regulator, and an organization called Telebrás was founded to plan and coordinate all of the country's telecommunications activities. Telebrás functioned as a holding company that comprised twenty-seven regional operators, which were responsible for domestic local phone service, and Embratel, which was in charge of (national and international) long-distance service. Although Telebrás was at its origins mostly owned by the government, by the mid-1990s it had both private-sector and public-sector stockholders, with the Brazilian government controlling the voting stock but not the total capital of the company.

Up to the mid-1990s, the Brazilian state supplied, as a monopolist, all telecommunications services in Brazil through Embratel and the regional operating companies. In the early 1990s the Telebrás regional subsidiaries started introducing analog cellular services. At the beginning the process was slow, and by 1992 only networks in Rio de Janeiro and Brasília had been inaugurated. From 1993 on, cellular networks were launched in other regions. All of these networks operated on the American AMPS standard. Although plans to introduce competition in cellular services were entertained as early as 1991, the courts ruled that private competitors were permitted in various value-added services, such as paging, but Telebrás had an enforceable monopoly in cellular services.

In the first half of the 1990s, several initiatives were considered to deregulate the telecommunications sector and privatize the key operators. Radical change in the sector did not come until the mid-1990s, however, when a constitutional amendment and two major reform laws were enacted. The new laws encouraged private operators to participate in supplying mobile and satellite services, as well as data communications and value-added services. The laws also laid the foundation for the companies under the Telebrás umbrella to be privatized, and created a telecommunications regulator, the Agēncia Nacional de Telecomunicações (ANATEL), separate from the Ministry of Communications and Telebrás.

Overall, the process of liberalizing the Brazilian cellular market had strong similarities with the approach implemented in the United States in the 1980s, but for the fact that in Brazil international cellular consortia entered regional cellular markets. The regional phone operators were split into two business units, one for fixedline telephony and another for cellular services. In each region, the available spectrum was divided into two bands, A and B, and the mobile unit of the preexisting regional operator received an allocation of spectrum in the A band.

Two important auctions took place in 1997–1998. The first, in 1997, involved the allocation of the B band of spectrum to private cellular competitors. The country was divided into ten regions, and these regions were organized into two groups on the basis of relative regional wealth. It was established that potential buyers could only acquire one license in each group of regions. This auction was viewed as highly successful and generated about USD 7.4 billion for the Brazilian government.

The second auction took place in 1998. In mid-1998 the fixed and mobile Telebrás regional subsidiaries were privatized. The government also imposed restrictions on who would be able to acquire these subsidiaries. Each of the participating consortia would be allowed to buy at most one of the fixed operators and two of the mobile operators in the A band, one in each group of regions. This auction raised a total of USD 19 billion, with the sale of the mobile operators accounting for about 37 percent of the total.

In short, by late 1998 the state regional cellular operators had been privatized and a cellular competitor had been licensed in each region. At this stage, some of the largest cellular operators in the world—including Telefónica from Spain, Telecom Italia, Portugal Telecom, Korea Telecom, Bell South, Bell Canada, and Telia from Sweden—were present in the Brazilian market, owning spectrum in either the A or the B band through consortia that linked them with local companies.

The rationale for awarding regional licenses was that, had national licenses been granted, operators would have tended to roll out their networks only in the most profitable areas of the country. Competition between the A band and the B band operators encouraged investment in cellular infrastructure. Digital upgrades to the original AMPS networks started around 1996 and accelerated after 1998, with operators setting up either CDMA or TDMA systems.[24]

Competition transformed the regional cellular markets. Prices declined quickly, prepaid cellular services became pervasive, the number of subscribers rose from about 1.3 million in 1995 to roughly 23.2 million in the year 2000, and cellular density increased from about one to about thirteen over the same time period. By the year 2000, Brazil had become the tenth-largest cellular market in the world.[25] In 2001 additional licenses were awarded. New operators entered the market in 2002 with GSM networks, and the enhanced competition accelerated cellular growth further. The transition to 3G started in late 2007, when licenses were awarded to four carriers.[26]

MEXICO

The 1917 Mexican constitution determined that telegraphy and other means of communication were monopolies of the state, and a 1938 law placed the national telecommunications industry under the control of the Secretaría de Comunicaciones y Transportes. Telmex (Teléfonos de México), the provider of fixedline telephone and other telecommunications services, was officially established in late 1947 by the merger of the two telephone systems in existence in the country at the time.[27] One of these networks was operated by Ericsson and the other by International Telephone and Telegraph.

During the 1950s, Telmex became majority Mexican owned but remained open to infusions of foreign capital. In 1972 the Mexican government acquired a 51 percent equity holding and gained control of the company's strategy, with the secretary of communications and transportation presiding over the board of directors until the company was reprivatized in 1990.

In 1989 Radiomóvil Dipsa—also known as Telcel, a subsidiary of Telmex—launched an analog AMPS network in Tijuana, not far from the border with California.[28] Telcel was allowed to set up a cellular network in Mexico City in 1990 and later to expand service to the rest of the country. This happened while Telmex was being privatized: in late 1990 the government sold control of the company to a consortium of Grupo Carso (a Mexican conglomerate owned by entrepreneur Carlos Slim), Southwestern Bell (later renamed SBC), and France Télécom.

The Mexican government followed the American model of fostering competition through regional duopolies. The government divided the country into nine regions, with two operators in each. As in the United States, one of the cellular licenses in each region was awarded to the cellular subsidiary (Telcel) of the incumbent fixedline operator, and for this reason it was labeled the "wireline" license. The other was awarded to a competitor that did not have a fixedline network in the region, and for this reason it was labeled the "nonwireline" license.

Bids for the nonwireline licenses were requested in early 1990, and the winning bids were announced in March. The licensees were each required to pay USD 10 million plus a royalty equivalent to 5 to 10 percent of annual revenues. The nonwireline licenses were granted to a variety of consortia, each one linking a Mexican company with an array of foreign partners, including McCaw Cellular, Millicom, Bell South, Bell Canada, and Racal Telecom. Thus, cellular markets in Mexico were duopolies almost from the beginning. All of these early systems adhered to AMPS, the US analog standard.

Over time, the nonwireline cellular licenses became consolidated in the hands of two companies. The first was Iusacell, a conglomerate led by the Peralta family and in which Bell Atlantic took a stake in 1994. Iusacell operated four cellular networks in the central and southern portions of the country, including Mexico City and Guadalajara. The second company was Motorola Network Ventures, which acquired minority stakes in several companies that operated cellular networks in the northern region of the country. In the mid-1990s networks moved to the American TDMA or CDMA digital standard.

By late 1995, Telcel had secured 56 percent of cellular customers, compared with 32 percent for Iusacell and 12 percent for the Motorola networks. Rather than declining as a result of the impact of competition, Telcel's share grew over time, and it was as high as 70 percent in late 2000.[29] This may be explained by the fact that there was no independent telecommunications regulator in Mexico until 1996, when the Comisión Federal de Telecomunicaciones (COFETEL) was established.

Before 1996, Telmex reportedly relied on several tactics to handicap the private cellular operators, including setting high interconnect charges and subsidizing its mobile subsidiary Telcel. Before the foundation of COFETEL, Iusacell complained to the Comisión Federal de Competencia, the Mexican antitrust agency, to which Telmex responded by successfully challenging the commission's authority to rule on telecommunications matters.

In the late 1990s COFETEL implemented policies that led to enhanced competition in the Mexican cellular market. For example, between 1990 and late 1998, the interconnection charges had been asymmetric: mobile operators had to pay a charge to terminate their traffic on the fixedline network, but fixedline operators did not pay to terminate their traffic on the mobile networks. In 1998 COFETEL first lowered the interconnection charges and then introduced a calling-party-pays regime for mobile phones. In the same year, an auction of additional spectrum opened the door for two new players to enter the cellular market: Unefon, which later merged with Iusacell, and Pegaso, which was later acquired by Telefónica.[30]

Cellular growth accelerated in the late 1990s, following the formation of COFETEL. The number of subscriptions rose from fewer than seven hundred thousand in 1995 to about fourteen million in the year 2000, and cellular density increased from less than one to about fourteen over the same period. By the year 2000, Mexico was the thirteenth-largest cellular market in the world. The transition to 3G started in 2005.[31]

MEXICO AND BRAZIL: SUMMARY

Brazil had a period of cellular monopoly—or, more precisely, several regional monopolies—in the first half of the 1990s. Liberalization of the cellular market started around 1997–1998, when cellular competitors were allowed to enter the market and the state-owned carriers were privatized. In Mexico, by contrast, there was cellular competition almost from the beginning. In both countries, an independent regulator was created in the second half of the 1990s.

Figure 8.9 shows cellular density in Brazil and Mexico between 1985 and 2010, with Britain included for comparison. The figure shows that Britain outperformed both Latin American countries in the long run.

Figure 8.9, however, does not "control" for the fact that cellular service started later in Mexico than in Britain, and even later in Brazil than in Mexico. Figure 8.10 shows cellular density in Brazil and Mexico up to fifteen years after the launch of the first cellular system in each country, with Britain again included for comparison. The figure shows the extent to which Brazil outperformed Mexico. From this perspective, Brazil performed even better than Britain for a period of time.

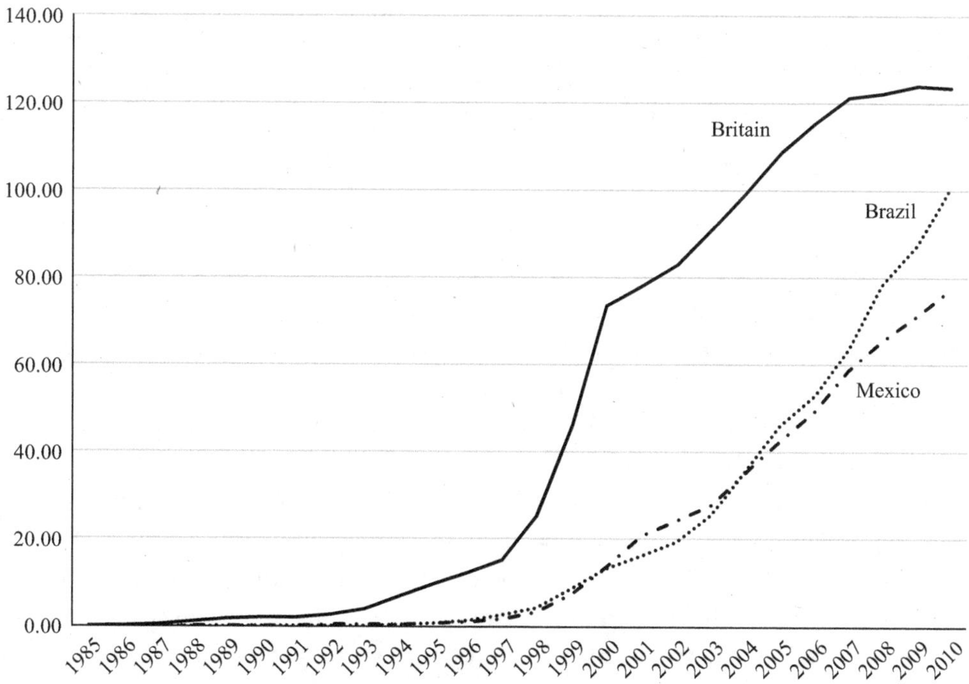

8.9 Cellular density in Britain, Brazil, and Mexico, 1985–2010. *Source*: ITU database.

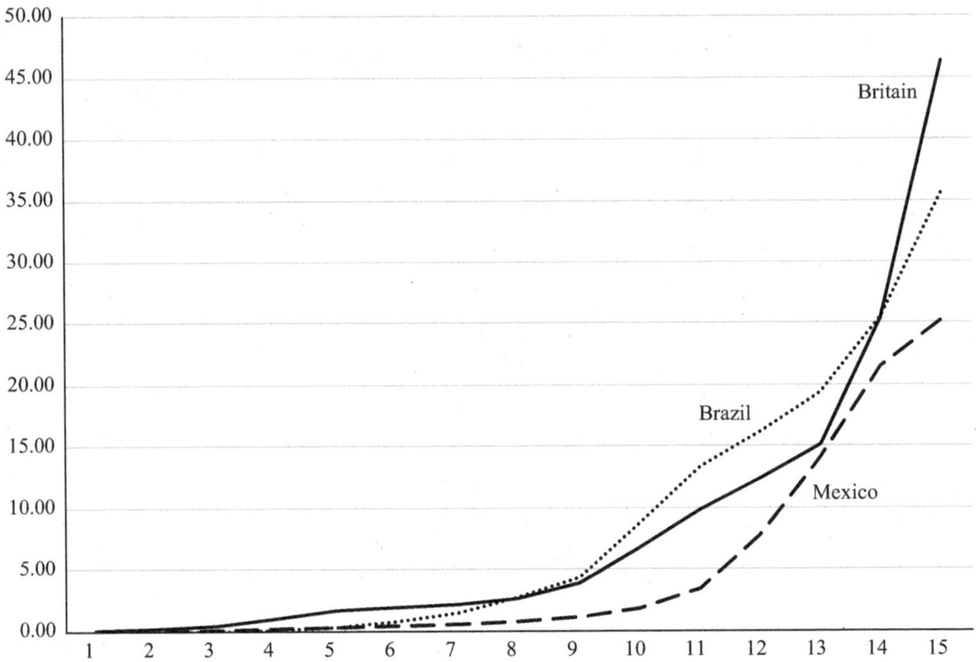

8.10 Cellular density in Britain, Brazil, and Mexico, one to fifteen years after launch of service. *Source*: Calculated by authors from ITU database.

Unlike in the Asia-Pacific region, economic and demographic differences between Mexico and Brazil are poor explanations for the differential performance of their cellular markets. If anything, Mexico had higher per capita income than Brazil during the relevant period and also had considerably higher population density—nonetheless, cellular density grew faster in Brazil than in Mexico.[32]

In a way, Mexico was like New Zealand in that competition authorities were in charge of disciplining the cellular incumbent before an independent regulator was established. There was a substantial difference, however: in New Zealand, operators acknowledged the authority of the competition commission—and the courts—to deal with telecommunications matters. In Mexico, by contrast, the incumbent operator did not. This explains why the lack of a regulator slowed down cellular growth in Mexico but did not have the same effect in New Zealand: the courts were able to discipline the telecommunications incumbent in New Zealand much more effectively than they did in Mexico.[33]

EMERGENT DIGITAL MARKETS: TAKING STOCK

In many countries around the world, the migration from analog to digital cellular in the 1990s also involved a shift toward a more competitive cellular market. In some cases, such as Japan, it was a transition from less competition to more competition, whereas in many others—including New Zealand, South Korea, and Israel—it was a more radical change from monopoly to competition.

The case studies in this chapter reveal different facets of the migration to digital cellular services in different countries. During the 1990s, New Zealand handled competition, and the problems arising from it, without an independent regulator, and did so successfully because the courts played an active role in disciplining the incumbent operator. Japan and South Korea used the transition to digital cellular not only to increase the intensity of competition but also to move each country toward a unified standard and to foster the rise of national champions in the cellular-equipment industry.

The experiences of several countries in the Asia-Pacific region reveal that the economic and demographic characteristics of each country had an impact on how fast cellular density grew. Countries such as Taiwan and Singapore, which were wealthy and densely populated and experienced exceptional rates of per capita income growth, tended also to have brisk growth in cellular density during the 1990s. By the late 1990s, Taiwan, Hong Kong, and Singapore ranked among the top countries in the world in cellular subscribers relative to population. In the Middle East, the case of Israel shows that intense price competition had a substantial impact: a competitor that entered in the mid-1990s with rock-bottom service prices generated explosive growth in the domestic cellular market. By the late 1990s, cellular density in Israel was almost as high as in Taiwan, Hong Kong, and Singapore.

As highlighted by the cases of Brazil and Mexico, the presence of an entity that acted as an effective arbiter of the competitive process mattered for cellular growth. In Mexico, the incumbent challenged the authority of the competition authority to deal with disputes in the cellular market, and this slowed down growth in cellular density. It was not until the very late 1990s that the Mexican independent telecommunications regulator was able to intervene in the market to lower interconnection rates and enable the entry of additional operators. Brazil started later than Mexico and had a longer monopoly period, but after introducing a strong telecommunications regulator and increasing the intensity of competition, it experienced faster cellular growth.

9

COMPETING IN 3G CELLULAR MARKETS

9.1 The Siemens ME45 phone. Introduced in 2001 and weighing just 99 g, the Siemens ME phone was designed for durability—it was water resistant and dust and shock proof. The phone included a Wireless Application Protocol browser for Internet access. The phone was used by Jason Bourne in *The Bourne Supremacy* (2004). *Source*: Martin Campbell-Kelly.

Until the late 1990s, there were fundamentally three cellular markets: network services, handsets, and infrastructure. In the market for network services, cellular carriers in a country—such as the postal, telegraph, and telephone administrations and their rivals in Europe—competed with one another to acquire subscribers for their cellular networks. In the handsets market, companies such as Nokia, Ericsson, and Motorola supplied mobile devices to subscribers (often indirectly through carriers). In the infrastructure market, firms such as Nokia and Ericsson supplied network operators with base stations, switches, and other pieces of cellular equipment.

In the late 1990s and early 2000s, new product and service markets developed. The main one was the market for mobile operating systems. This market grew with the rise of smartphones, which replaced voice-only phones and feature phones throughout the 2000s, slowly at first and more rapidly toward the end of the decade. Palm, Microsoft, Symbian, and Research in Motion were the main players in the operating-system market before 2007—that is, before Apple (with the iOS) and Google (with Android) became the dominant players.

Another market that barely existed in the late 1990s and grew during the 2000s was the market for baseband processors, the semiconductor devices that allow cellular handsets to communicate with a network's base stations. The little information that is publicly available about baseband processors in the 1990s suggests that, during this period, they were mostly designed and manufactured in-house by the handset makers themselves. During the 2000s, a "merchant" market for baseband processors developed in which companies such as Qualcomm and Ericsson designed and manufactured mobile semiconductors for third-party customers rather than for self-supply.

In this chapter, we analyze the evolution of the network-services market from two different perspectives. First, we focus on market structure and competition. We identify patterns of market structure that developed during the 2000s and explore the extent to which these patterns varied across countries and regions. Later, we examine network-services markets from the perspective of the cellular carriers. We study the rise of international operators—that is, carriers that had a presence in multiple countries and often on multiple continents.

After that, we present two case studies—company histories—that give us a better understanding of the strategies and business processes cellular carriers implemented. We track the rise of two of the most important network operators of the twenty-first century: VimpelCom in Eastern Europe and Vivo in Latin America. The focus on these two companies allows us to examine the peculiarities of two of the largest cellular markets in the world: Russia and Brazil. We also use the stories of VimpelCom and Vivo to highlight trends, such as consolidation, that characterized network-services markets more generally in the first decade of the twenty-first century.

Next, we study competition in the market for cellular handsets, highlighting the consolidation of Nokia, Ericsson, and Motorola as the top vendors in the worldwide market. Finally, we explore the dynamics of competition in mobile-semiconductor markets.

COMPETITION AND MARKET STRUCTURE IN NETWORK SERVICES

During the 3G era, certain patterns of competition and market structure became evident in cellular network-services markets around the world. We focus on Western European countries first to highlight these trends, and then cover other parts of the world.

Between 2000 and 2005, more 3G licenses were awarded in Europe than in any other region of the world, with a significant portion of them granted in 2000 and 2001.[1] Governments had the option to increase competitive intensity relative to the 1990s, and twenty-three countries chose to do so by issuing more 3G licenses than there were 2G incumbents. During the 2000s, about thirty-seven companies (or consortia) entered cellular network-services markets in Europe for the first time. There were new entrants, for example, in Austria, Britain, Denmark, Finland, France, Germany, Ireland, Italy, Norway, Poland, Portugal, Spain, Sweden, and Switzerland. Only some of the entities that were awarded 3G licenses launched their networks, however: as of late 2009, only twenty-one of the thirty-seven new entrants had networks in operation.

Figure 9.2 presents the key features of market structure in many Western European markets as of the second quarter of 2004. For each country, the figure shows the combined share of the largest two cellular operators.

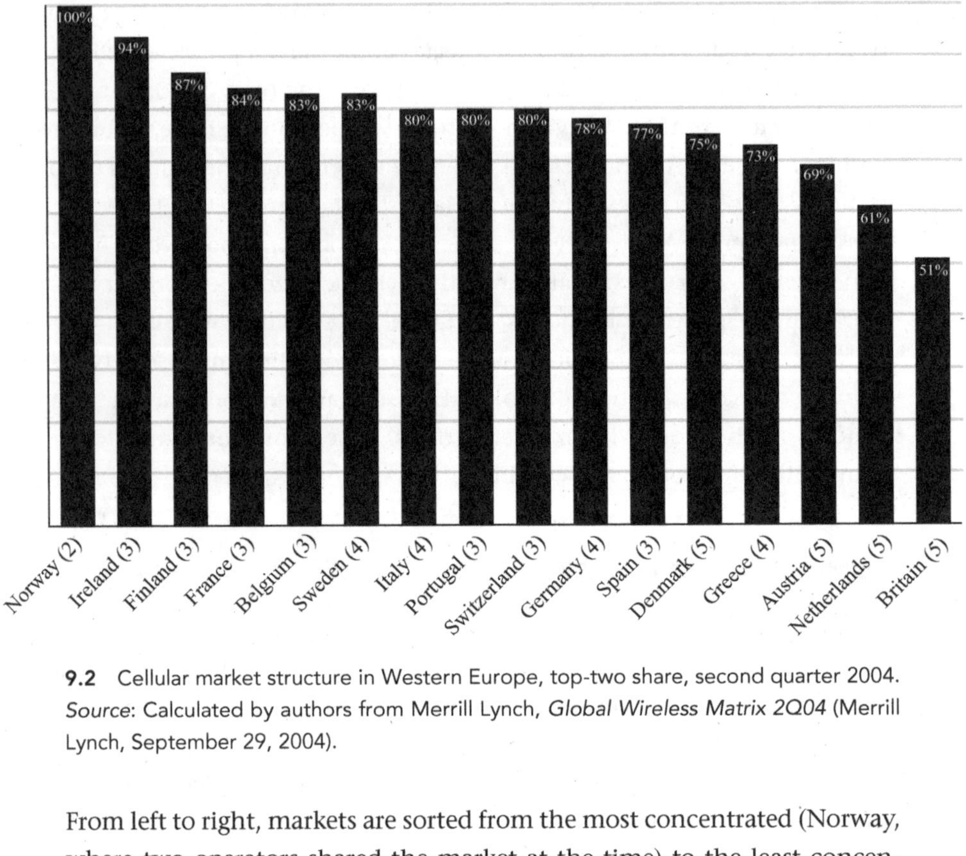

9.2 Cellular market structure in Western Europe, top-two share, second quarter 2004. *Source*: Calculated by authors from Merrill Lynch, *Global Wireless Matrix 2Q04* (Merrill Lynch, September 29, 2004).

From left to right, markets are sorted from the most concentrated (Norway, where two operators shared the market at the time) to the least concentrated (Britain, where there were five operators and the largest two combined only had a 51 percent share). At the bottom, between parentheses next to each country's name, is the total number of cellular operators in that country.

Certain features of the data stand out. First, as of 2004, most markets had between two and four operators. Second, in most countries, the top two operators combined usually had roughly between 70 and 90 percent of the market. Third, similar patterns of market structure tended to prevail across European countries, regardless of the size of each national market. Germany had close to seventy million cellular subscribers at the time, it had four operators, and the largest two combined had a 78 percent market share. Greece was a much smaller market—it had fewer than eleven million subscribers. Nonetheless, it also had four operators, and the largest two combined had a 73 percent market share.

Markets with just a few operators became prevalent in most regions of the world during the 3G era. Figure 9.3 shows the situation in Central and Eastern Europe as of 2004.

Markets with three cellular carriers were present in Hungary and Poland, even though the latter had more than twice as many subscribers as the former. Even Russia, which had almost fifty million subscribers at the time, had just three large operators—MTS, VimpelCom, and Megaphon. As in Western Europe, the top two operators usually had between 70 and 90 percent of the market.

During the 3G era, markets with just a few carriers became common outside the European continent as well. Figure 9.4 presents the core

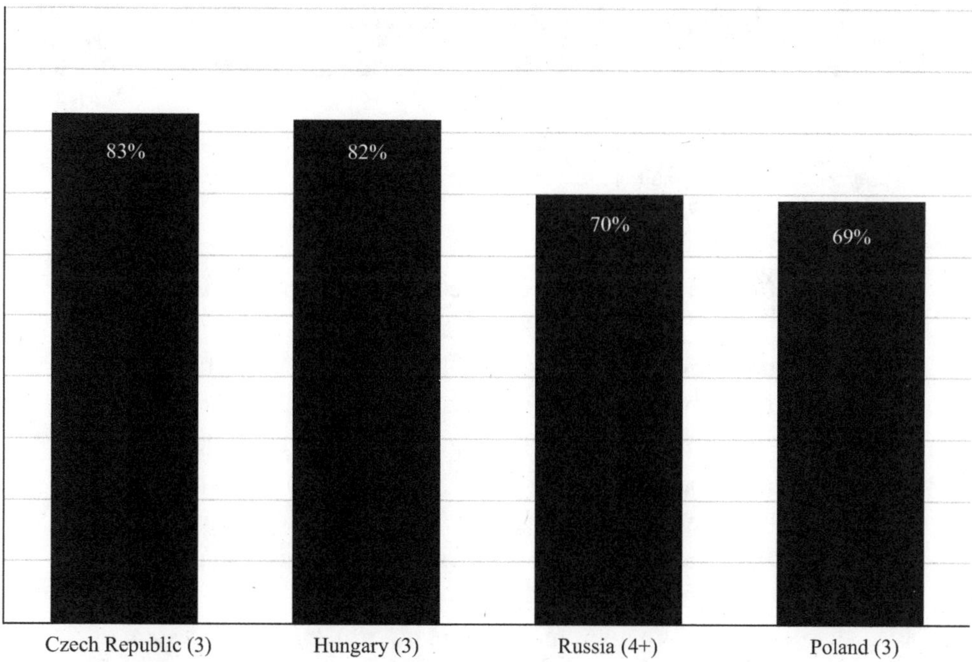

9.3 Cellular market structure in Central and Eastern Europe, top-two share, second quarter 2004. *Source*: Calculated by authors from Merrill Lynch, *Global Wireless Matrix 2Q04* (Merrill Lynch, September 29, 2004).

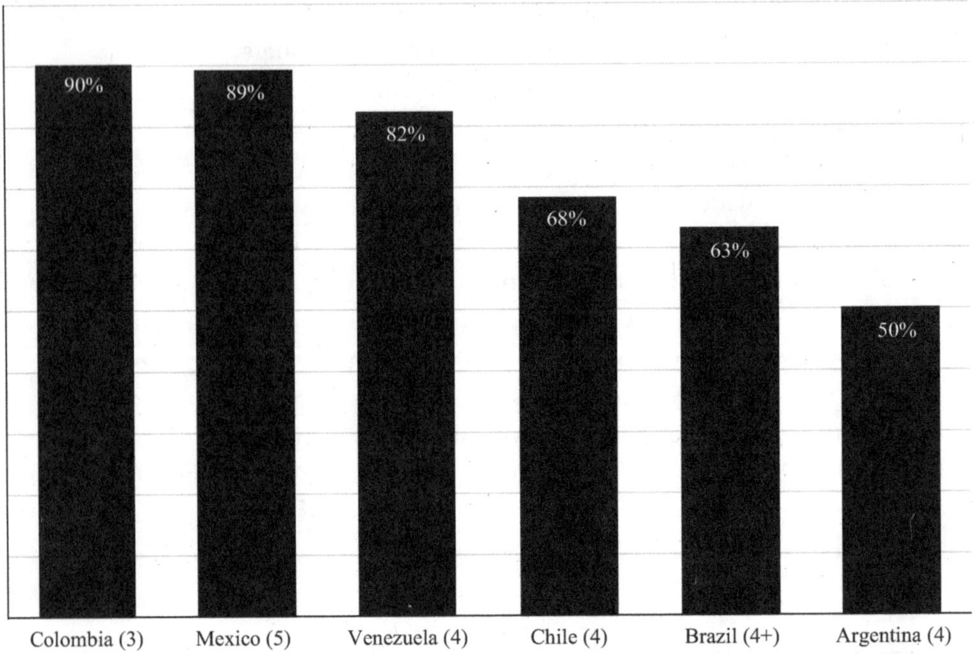

9.4 Cellular market structure in Latin America, top-two share, second quarter 2004. *Source*: Calculated by authors from Merrill Lynch, *Global Wireless Matrix 2Q04* (Merrill Lynch, September 29, 2004).

features of the market structure in several Latin American countries as of 2004.

The figure shows that many large Latin America markets had four or five operators. There was, nonetheless, some degree of variation across countries: Argentina and Chile had four operators and markets that were reasonably competitive, whereas Mexico, at the other extreme, had a total of five carriers but the largest two, combined, accounted for 90 percent of the market. The largest Mexican carrier—Telcel—had 77 percent of the market in 2004.[2]

Figure 9.5, finally, shows market structures in Asia and Oceania.

In this region, there were several countries with at least four carriers. It was still the case that, in most countries, the largest two carriers combined had between 70 and 90 percent of the market.

In the United States, the trend that started in the late 1990s toward the formation of networks with national coverage continued in the 2000s. By

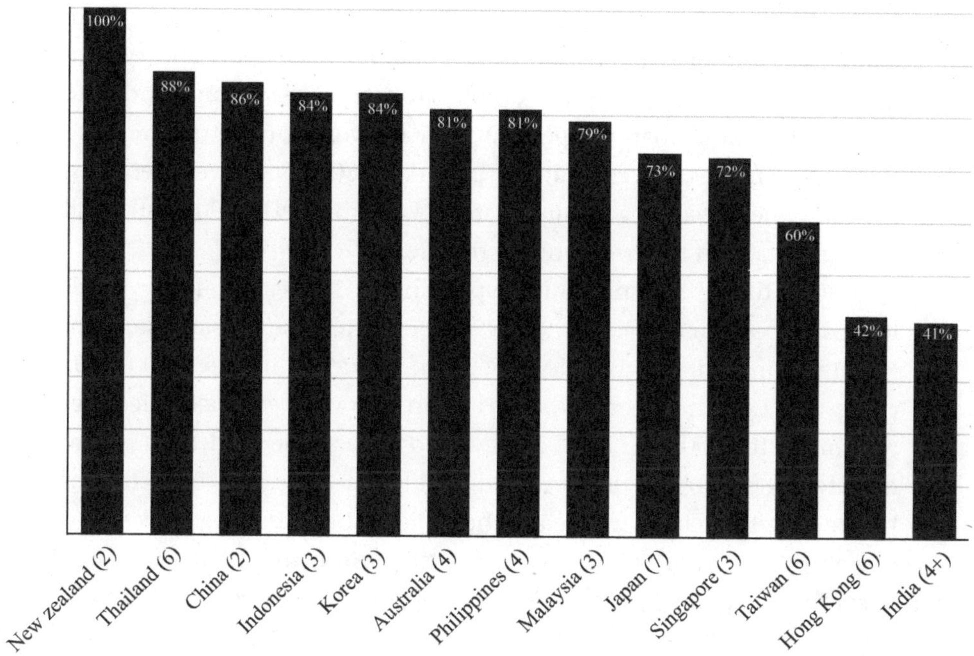

9.5 Cellular market structure in Asia and Oceania, top-two share, second quarter 2004. *Source*: Calculated by authors from Merrill Lynch, *Global Wireless Matrix 2Q04* (Merrill Lynch, September 29, 2004).

late 2003, there were six companies that could be described as national operators: AT&T Wireless, Cingular Wireless, Nextel Communications, Sprint PCS, Verizon Wireless, and T-Mobile. From then on, these six carriers used mergers and acquisition to consolidate into four: AT&T, Verizon, T-Mobile, and Sprint, of which the first two were by far the largest.[3]

It is worth posing the question why the network-services market had the oligopolistic nature just described. (Oligopolies are markets with just a few suppliers.) The first answer—namely, that governments awarded a limited number of cellular licenses—is not an answer at all: government agencies in charge of spectrum allocation knew that it was unreasonable to offer dozens of licenses when only a few operators would be able to operate profitably. Thus, the next question is why the network-services market was a "natural oligopoly," if indeed it was so.[4]

One answer to this question is that setting up a mobile telecommunications network was expensive, although costs declined over time.[5]

Network setup costs varied with geography, the distribution of the population, and the operator's strategy, but radio transmission and switching equipment was usually the dominant cost: base stations tended to account for more than 50 percent of a network's infrastructure cost. Industry estimates suggest that, in the early 1980s, total investment cost per subscriber for analog cellular systems was around USD 900, although it fell gradually in the course of the decade.

Among digital systems, those operating in higher frequency bands tended to be more expensive: in 1990, for example, a Global System for Mobile Communications (GSM) network operating in the 1,800 MHz range was 1.8 times more expensive than one with equivalent features running in the 900 MHz band. This was so because the GSM 1800 system operated in a higher frequency band, and for this reason its base stations had to be set up closer to one another. The cost disadvantage of systems operating in higher frequency bands fell during the 1990s, however. According to some estimates, by the year 2000, infrastructure investment costs per subscriber had stabilized at around 350 euros (roughly USD 382), although there was considerable variability across firms and countries.

The declining setup costs should have facilitated entry. But they did not, mainly because, as investment costs declined, license fees went up. During the 1980s and, to some extent, the 1990s, it was common for governments to assign radio frequencies to operators in exchange for nominal fees. During the 1990s, several countries started to charge substantial upfront license fees, and in the early 2000s, several countries in Europe and beyond resorted to spectrum auctions to allocate cellular frequencies. Many of those auctions resulted in operators paying high prices for each potential subscriber.

After the auctions, the view spread that auctions that had raised abundant revenues for governments had also generated prices that were likely too high in light of the potential flow of income from 3G services.[6] Many auction winners developed the perception that they may have fallen prey to the "winner's curse"—that is, that they may have paid too much for the right to deliver 3G cellular services. To make matters worse, technical analyses in the early 2000s suggested that the infrastructure investments required to set up a 3G network would be substantially higher than those required for a 2G network.[7]

Firms responded with strategies that tended to limit the number of market participants. The most obvious one was mergers and acquisitions, which happened frequently in the American market during the 2000s. In other regions, firms that had won a 3G license chose to delay the building of 3G infrastructure and the supply of services. In some cases, firms decided to hand back their license to the regulator and forgo the license fee paid. In others, companies started building 3G networks shared with their competitors, and regulators ended up accepting such arrangements to accelerate the introduction of 3G services. All of these trends led to a world in which markets with only a few operators became pervasive.

NETWORK SERVICES: TWO CASE STUDIES

During the 3G era, many operators continued expanding in countries other than their home base to become multicountry, and often multicontinent, carriers. Table 9.1 shows the number of countries where each international operator was present in each region of the world in late 2009. In the table, "MTN" stands for Mobile Telephone Networks, an important African carrier.

The table reveals several patterns. First, as expected, operators were heavily represented on the continent where they had their own domestic market. For example, France Télécom, Vodafone, Telefónica, and Deutsche Telekom—all international operators with a home base in a European country—were often present in European countries other than their home market. Second, all of the European operators also had a strong presence in at least one other region of the world. Usually, this second region was dictated by historical, cultural, and linguistic factors. For example, France Télécom and Vodafone were strong in Africa, whereas Telefónica was prominent in Latin America. Third, a few operators were present almost exclusively on one continent. MTN was fundamentally an African carrier, Zain and Etisalat had their networks in Africa and the Middle East, and América Móvil was a Latin American operator. In any case, all of the large carriers operated in multiple countries, and often on more than one continent.

We develop two case studies that convey the details of how network markets functioned in different countries, and how some operators became important players in those markets. The first case study focuses

Table 9.1 Presence of international operators in countries around the world, late 2009, number of countries by region

Company	Western Europe	Eastern Europe	Middle East	Asia-Pacific	Latin America	North America	Africa	Total
France Télécom	12	1	2	1	6	1	17	40
Vodafone	14	1	4	7	0	1	10	37
Telefónica	9	0	0	4	16	0	5	34
MTN	1	0	3	1	0	0	17	22
TeliaSonera	8	4	4	6	0	0	0	22
Zain	0	0	5	0	0	0	17	22
Telenor	4	5	2	10	0	0	0	21
Deutsche Telekom	11	6	0	0	0	3	0	20
Etisalat	0	0	4	5	0	0	10	19
América Móvil	0	0	0	0	16	2	0	18
Millicom	0	0	0	1	6	0	7	14
Hutchison	6	0	0	7	0	0	0	13

Source: P. Curwen and J. Whalley, *Mobile Telecommunications in a High-Speed World: Industry Structure, Strategic Behavior and Socio-economic Impact* (Farnham, UK: Gower, 2010), 195.

on the rise of Vivo in Brazil, and the second on the evolution of VimpelCom in Russia.

BRAZIL: CONSOLIDATION AFTER LIBERALIZATION

In the second half of the 1990s, the Brazilian privatization and liberalization process was extremely effective at generating cellular competition. It was so effective that, by the early 2000s, the Brazilian cellular market was heavily fragmented.[8]

In each Brazilian region, the available spectrum was split into an A band and a B band, not unlike what happened in the United States in the 1980s, and the mobile unit of the preexisting regional phone operator received an allocation of spectrum in the A band. (The regional phone operators, and their cellular subsidiaries, were known as the Telebrás companies.) In 1997–1998 two auctions changed the nature of cellular markets in Brazil. In 1997 the auction of the B band opened the door for several private consortia to enter the market. Then, in 1998, the fixedline and mobile Telebrás companies were privatized, which raised the intensity of competition in Brazilian telecommunications generally, and especially in cellular markets. After the Telebrás auction, the owners of the A spectrum band—that is, the consortia that acquired the newly privatized cellular companies—quickly implemented digital upgrades to code division multiple access (CDMA) and time division multiple access and launched new services.

By 2001 the government had decided to further liberalize the cellular market and auctioned three additional bands of spectrum. Strong competition among cellular operators, plus new regulations, generated a rapid increase in the number of subscribers. By 2005 Brazil had become the fifth-largest cellular market in the world, after China, the United States, Russia, and Japan.

The three auctions—1997, 1998, and 2001—accomplished the goals of liberalizing the market and privatizing the state operators, but at the same time they generated a heavily fragmented market. The 1997–1998 auctions created fifteen different groups of mobile companies controlling about forty mobile operators, and the 2001 auctions added ten operators. In 2001 the largest operator, Telesp Cellular, controlled less than 18 percent of the Brazilian market. In that year, a process of cellular consolidation began.

Vivo, which at the time of writing is still one of the largest Brazilian operators, was born when Telefónica and Portugal Telecom combined their cellular assets in Brazil. Telefónica entered Brazil in 1997 through its participation in the management of the Companhia Riograndense de Telecomunicações, a telecommunications company based in Rio Grande do Sul. In 1998, during the privatization of the Telebrás entities, Telefónica participated in a consortium that acquired Telesp, the largest fixedline operator, and several carriers that operated Advanced Mobile Phone Service (AMPS) networks. After that, Telefónica made two important decisions—it started developing a digital network using CDMA technology, and it acquired a municipal company in the São Paulo area that provided both fixedline and mobile-phone services.

Portugal Telecom entered the Brazilian market in 1998, when it acquired the largest mobile operator in the country, Telesp Cellular, and decided to develop a digital network based on the CDMA standard. This choice gave it common ground with Telefónica, the only other operator in Brazil that relied on CDMA.

In 2002 Telefónica and Portugal Telecom created a joint venture named Brasilcell. Several factors underpinned this agreement. First, both partners were operating CDMA networks and, starting in 2002, both transitioned to the 2.5G CDMA evolution. Second, they had contiguous operating areas. Finally, they had already had a successful experience of collaboration in Morocco. At launch, Brasilcell had 13.7 million subscribers, was the largest cellular carrier in Latin America, and operated in an area that generated more than 70 percent of Brazil's gross domestic product.

From then on, Brasilcell expanded through acquisitions. In 2003 it rebranded itself as Vivo with the goal of developing a strong national brand. In 2004 it launched the first 3G service in Brazil. As of 2005, Vivo had 29.8 million subscribers and 34.5 percent of the Brazilian cellular market.

Vivo's main competitors in Brazil were Telecom Italia Mobile and Claro (América Móvil). Telecom Italia Mobile entered Brazil in 1997. After that, it expanded in Brazil by acquiring cellular licenses in various regions of the country, to the point that, by 2005, it had 20.2 million subscribers and a 23.4 percent share of the cellular market.

The origins of Claro go back to September 2000, when América Móvil was created as a spinoff from Telmex with the goal of making forays in Latin American cellular markets. América Móvil established a joint venture with Bell Canada and SBC (formerly Southwestern Bell) called Telecom Americas, which entered the Brazilian market by acquiring stakes in three cellular networks. From then on, Telecom Americas expanded in Brazil by buying shares in existing operators and by participating in auctions for new licenses. In 2003 it rebranded itself as Claro. As of late 2005, it had 18.7 million subscribers in Brazil and a 21.6 percent share of the cellular market.

RUSSIA: THE RISE OF PRIVATE OPERATORS AND THE TREND TOWARD CONSOLIDATION

Cellular-phone systems were launched in Russia immediately after the reforms introduced by Mikhail Gorbachev between 1986 and 1991, a period in Russian history usually associated with the terms *glasnost* (transparency) and *perestroika* (restructuring). The origins of VimpelCom, one of the most important cellular carriers in Russia, go back to the collaboration between a young American entrepreneur, Augie Fabela II, and a Russian scientist, Dmitry Zimin.[9] Augie Fabela Sr. was a Mexican businessman who had immigrated to the United States. His son, Augie Fabela II, was a Stanford graduate who shared his father's entrepreneurial flair.

In the early 1990s the Fabelas acquired Plexsys, an Illinois-based manufacturer of cellular equipment with excellent products but no international experience. They took a trip to Russia to explore opportunities to sell the Plexsys products there, but once in Russia, they realized that the operation of cellular networks offered the most promising opportunities. In 1991 the Fabelas had several meetings with representatives of MAK-Vimpel, a military-industrial conglomerate. In these meetings, they identified a scientist, Zimin, who they thought would be the ideal business partner for their Russian ventures.

Zimin had worked at Russia's Applied Radio Research Institute (RTI). In 1988, when Gorbachev authorized private enterprise through the Soviet Law on Cooperatives, Zimin started searching for business opportunities. Of the firms he helped start, VimpelCom proved to be the most consequential in the long run. It was founded by Zimin and a team of RTI colleagues with the goal of finding practical applications for the radio technology they had become experts in. The firm's principal shareholders were Zimin, RTI, and the Vimpel conglomerate.

The meetings the Fabelas had with Zimin and others, first in Russia and later in the United States, led to a partnership that redefined VimpelCom: the company would focus on becoming a player in the Russian cellular market. The Americans would provide financing, marketing, and managerial skills, and the Russians would contribute their technical expertise and political connections.

Mobile telephony as a service available to the public started in Russia in 1991, when a network was launched in Saint Petersburg. Earlier, the Ministry of Communications had decided that the country would have two cellular standards, Nordic Mobile Telephone (NMT) 450 and GSM. In 1993 the first GSM license was awarded in Moscow to Mobile TeleSystems (MTS), a firm that would become VimpelCom's main rival in the Russian market.

By the time VimpelCom was ready to roll out its networks, Moscow had no spectrum available in either the 450 MHz or the 900 MHz band, so neither NMT nor GSM could be implemented. Since Plexsys was a manufacturer of AMPS equipment, which operated in the 800 MHz band, Zimin and Fabela decided to try launching an AMPS network in Russia, even though the Ministry of Communications had expressed a clear preference for NMT and GSM.

Zimin was forced to rely on his political connections and lobbying skills. After long negotiations, the Ministry of Communications announced that AMPS would be a regional standard and issued an operational license to VimpelCom for the city of Moscow. The firm started its test operations in Moscow in 1994. The next year, it used Ericsson equipment to roll out a full-scale D-AMPS network in Moscow and its surroundings.

At its inception, VimpelCom was more focused on technology than on consumer needs. In order to address this issue, Fabela hired a marketing expert who radically changed VimpelCom's culture. The company invested substantial resources in a rebranding effort that paid off: in the second half of the 1990s, VimpelCom's BeeLine cellular brand became one of the most recognizable trademarks in Russia. In 1996 VimpelCom became the first Russian company to be listed on the New York Stock Exchange.

The year 1998 was crucial for VimpelCom in several ways. First, in the midst of a severe economic crisis that Russia underwent that year, the company struggled with a large amount of dollar-denominated debt and lost subscribers to MTS, its main rival. Second, VimpelCom started exploring the possibility of cooperating with global telecommunications operators. Telenor, a then state-owned telecommunications firm based in Norway, had minority stakes in three Russian regional cellular operators and was interested in establishing a presence in the Moscow market. VimpelCom and Telenor reached an agreement that gave the latter a stake in the Russian cellular operator.

The most fundamental decision that VimpelCom made in 1998 was its commitment to the GSM standard. At the beginning of the year, VimpelCom had two networks. Its D-AMPS service was marketed under the BeeLine name, covered most of the Moscow region, and generated most of the company's revenues, whereas its GSM 1800 network offered limited coverage in the city of Moscow. The dilemma was whether the company should keep on investing in expanding its D-AMPS service or should rather build up its GSM infrastructure. In the midst of the economic crisis, VimpelCom committed to GSM.

In 1999 VimpelCom launched a marketing campaign to promote its GSM services in the Moscow region. It offered, for the first time in Russia, handsets at reduced prices and prepaid cards for service. This put the company in direct confrontation with MTS, which operated a GSM service in the area. With its marketing push, VimpelCom made cellular telephony available to the mass market in Moscow: with aggressive price cuts, it was able to reach new segments, including young adults and small businesses. By 2002 GSM represented 95 percent of VimpelCom's annual revenues.

MTS was VimpelCom's main rival in Russia. In 1993 MTS was founded as a closed joint-stock company by a consortium led by the Moscow City Telephone Network, Deutsche Telekom (through its T-Mobile subsidiary), and Siemens. At the time, the Moscow City Telephone Network and its Russian partners held a 53 percent ownership stake.[10]

MTS launched GSM services in Moscow in 1994. In 1996 Sistema, a Russian conglomerate, acquired a majority stake in MTS, and Deutsche Telekom bought out Siemens. The next year, MTS started expanding its cellular network outside Moscow. Through license acquisitions, agreements with regional companies, and a merger, MTS managed to become one of the largest cellular operators in the Russian market. In June 2000 MTS had its initial public offering on the New York Stock Exchange, the second ever by a Russian company. In the early 2000s, it continued its expansion in Russia and abroad.

Competition between VimpelCom and MTS in the late 1990s and early 2000s fueled rapid cellular growth in Russia. Between 1995 and 2000, the number of cellular subscribers in the country grew from about 88,000 to about 3.26 million, and cellular density rose from 0.06 to 2.22. Growth accelerated in the early 2000s. Between 2000 and 2005, the number of subscribers grew to a remarkable 120 million, and cellular density skyrocketed to around 83. By mid-2004, MTS had 37 percent of the market, VimpelCom had 33 percent, and Megaphone, the third-largest carrier, had 18 percent.[11]

CELLULAR OPERATORS IN BRAZIL AND RUSSIA: TAKING STOCK

There are interesting similarities between patterns of competition and market structure in three large markets such as Brazil, Russia, and the United States. In all of them, governments awarded regional rather than national licenses. In all of them, however, a consolidation trend emerged that led to the rise of supraregional carriers first and of national operators later.

Although the reasons why governments awarded regional licenses in the first place may have differed across countries, the rationale for consolidation was the same: cellular-phone users valued the possibility of traveling around the country and being able to make and receive calls anywhere without incurring roaming charges. An operator that was able to offer national coverage had an advantage over another that supplied comparable network quality but only local or regional coverage—operators were able to differentiate their service offerings by supplying better geographic coverage than their competitors.[12]

In the 1990s national versus regional coverage was not an issue in Western Europe, since licenses were usually national, and operators offered national coverage by default. In larger countries such as Russia, Brazil, and the United States, however, licenses were regional and national coverage could only be achieved through mergers, acquisitions, and the purchase of new regional licenses when they became available. These were the mechanisms that companies such as VimpelCom in Russia and Vivo in Brazil relied on to become national operators, and similar mechanisms were used by large cellular carriers in the United States as well.

COMPETITION IN HANDSETS

By the late 1990s and early 2000s, there were many types of handheld devices. There were, first, pen-based personal digital assistants (PDAs), including some based on the Palm OS and others on Microsoft's Pocket PC OS. Second, there were keypad-based PDAs, such as those marketed by Research in Motion and Psion. Third, there were voice-only phones, the traditional mobile phones that allowed subscribers to make voice calls and send and receive text messages. Fourth, there were a variety of feature phones, data-enabled phones with rudimentary operating systems that often incorporated a camera and a music player. And finally, there were the early smartphones, devices with sophisticated operating systems developed first for PDAs. There was competition among PDAs, on the one hand, and among mobile phones, on the other. Here we disregard the former and focus on the latter.

We have already highlighted at least two technological discontinuities in the history of cellular phones. The first happened at the beginning of the 1990s, when digital cellular replaced analog cellular—this was the transition from 1G to 2G. The second happened at the beginning of the twenty-first century, when applications such as mobile email and mobile Internet access started playing an important role in the cellular experience—this was the transition from 2G to 3G. Each of these discontinuities created an opportunity for new companies to take the lead in the markets for cellular equipment (handsets and infrastructure). In each of these transitions, leadership in the market for cellular devices changed.[13]

Motorola was dominant in the world market for cellular handsets during the 1980s, mainly because it was the leader in the American market, which was the largest in the world. In the early 1990s the rise of digital cellular brought about a first technological discontinuity. The transition from analog to digital cellular happened throughout the 1990s, with digital handset sales surpassing analog sales in 1998.

Motorola made two strategic decisions in the 1990s that undermined its leadership in the world market for mobile handsets. First, it continued investing heavily in analog cellular technologies despite the steady rise of digital handsets. The company was convinced that its new handset model, the StarTAC phone, would outperform the new digital phones that were reaching the market. It even attempted to dictate terms to network operators: it offered its new device only to operators that were willing to buy three-quarters of their handsets from Motorola, which alienated important carriers such as Bell South. Second, it made huge investments in developing a new cellular infrastructure—the Iridium project, an array of low-orbiting satellites conceived to enable subscribers to make calls from anywhere in the world. Motorola focused so heavily on Iridium in the 1990s that it missed the (more traditional) digital cellular revolution and, in addition, Iridium declared bankruptcy in 1999. By the time Motorola recognized the need to develop digital handsets, Nokia and Ericsson were already marketing their lines of digital handsets with some success.

A spectacular failure, Iridium deserves some discussion because of what it reveals about the nature of cellular services at the time. The assumption behind Iridium was that it would be appealing to both carriers and subscribers. It would appeal to carriers because, by setting up base stations in the sky, it would address the most expensive component of a network's infrastructure—the cellular towers. It would appeal to subscribers because it would enable them to make calls from anywhere in the world and it would do away with the panoply of analog and digital standards that made life complicated for cellular users. The problem was that, given its cost structure, Iridium required about a million customers in the first year just to break even, and it could only achieve that figure with reasonable pricing. Given that the Iridium venture had promised to pay its parent, Motorola, USD 140 million per quarter in perpetuity for operations and maintenance, Iridium's prices were stratospheric: handsets cost about USD 3,000 and subscribers paid about USD 3 per minute to use the system. In short, the economics of Iridium meant that it was destined to die well before it was even launched.[14]

Nokia emerged as the leading handset company in the late 1990s. It made several strategic moves that positioned it ahead of its rivals. First, it committed to digital cellular earlier than its competitors. It was the first firm to introduce a GSM portable phone, and it also became a leader in

Lightweight, portable subscriber units for the Iridium system communicate directly with satellites, using a small antenna.

Motorola's Iridium system is based on a costellation of 77 satellites in low earth orbit that use cellular technology to communicate with users on earth.

9.6 Motorola publicity images from 1996 promoting Iridium, a satellite-based cellular infrastructure and handset that promised to eliminate costly cell towers. Iridium was a spectacular failure and went bankrupt in 1999. Courtesy The National Museum of Computing, UK.

GSM infrastructure in Europe. It played a crucial role in the development of several important innovations on the digital standard, including text messaging, games, and email clients. Second, it understood that mobile phones were becoming commodities and developed a highly diversified product line: toward the late 1990s and early 2000s, Nokia was the company that introduced the largest variety of new phone models every year. Its handsets had a user-friendly interface and excellent design. By the early 2000s, Nokia had displaced Motorola as the leading firm in the world market for cellular terminals.

The three leading handset companies—Nokia, Ericsson, and Motorola—collaborated in the development of the Symbian operating system to fend off Microsoft's forays into cellular phones. Figure 9.7 shows the evolution of handsets sold by the original Symbian founders (individually and for the three of them in combination). Figure 9.8 shows the share of non-Symbian and Symbian phones among all handsets sold by Nokia, Ericsson, and Motorola.

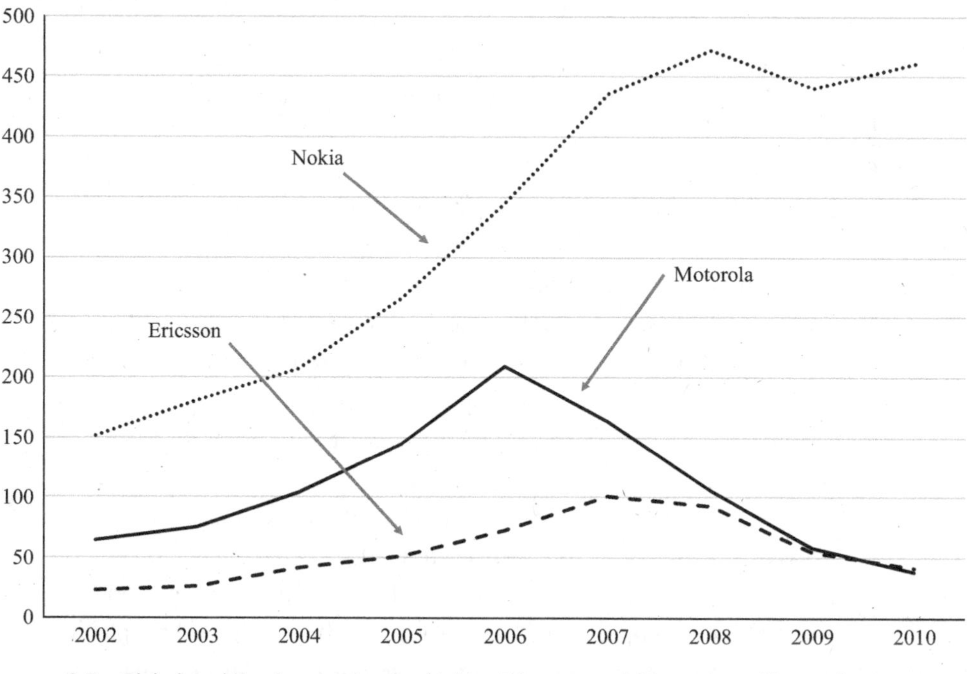

9.7 Global mobile-phone sales for Nokia, Ericsson, and Motorola, millions of units, 2002–2010. *Source*: Gartner press releases.

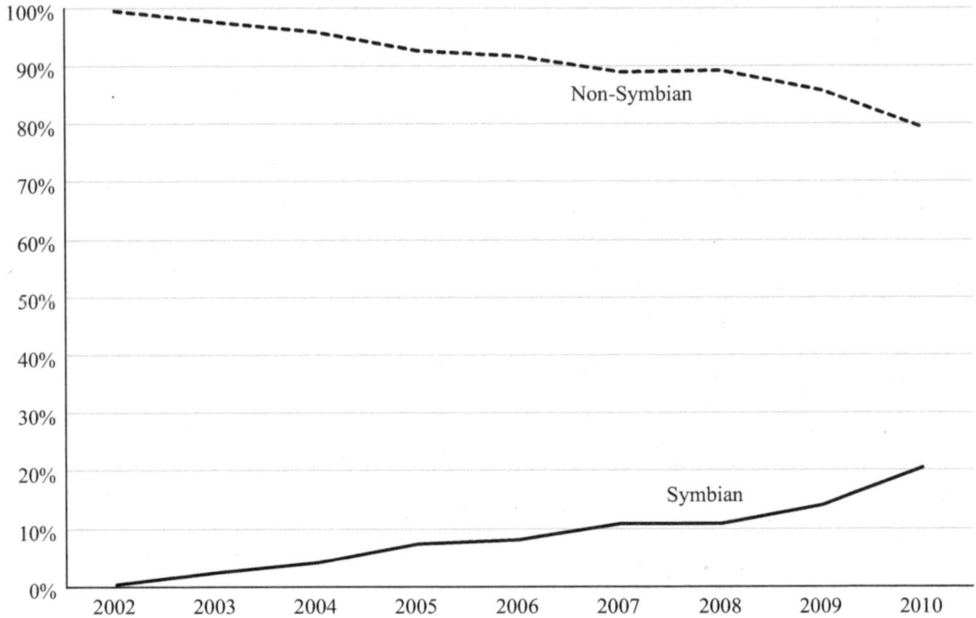

9.8 Symbian and non-Symbian phones, percentage of Nokia, Ericsson, and Motorola sales, 2002–2010. *Source*: J. West and D. Wood, "Evolving an Open Ecosystem: The Rise and Fall of the Symbian Platform," in *Advances in Strategic Management*, vol. 30, *Collaboration and Competition in Business Ecosystems*, ed. R. Adner et al. (Bingley, UK: Emerald Group, 2013), 27–67.

Figure 9.7 shows that, during the 2000s, Nokia outsold the other two handset vendors combined. Figure 9.8 shows that the share of Symbian phones sold by the founders rose over time but always represented a small minority of the handsets they sold. Most of the devices that Nokia, Motorola, and Ericsson marketed during this period were not smartphones but rather voice-only phones or feature phones. Symbian was an important project for the founders, but none of them bet their corporate existence on its success.

In the early years of the twenty-first century, companies such as Nokia, Motorola, and Ericsson thus competed in three areas: voice-only phones, feature phones, and smartphones. In voice-only phones and feature phones, they competed on the basis of attributes—including phone design—and prices. In smartphones, however, one of the key features—the operating system—was common to all of them because their smartphones relied on the Symbian OS.

In order to give licensees some room to differentiate their products, Symbian was designed so that the user interface was easy to change. Five interfaces were developed for Symbian. These custom user interfaces allowed handset makers to offer distinctive phones—that is, vendors differentiated their products by modifying the look and feel of the operating system.

The three handset leaders had their golden age in the 2000s. Their sales grew from the beginning of the decade until 2006–2008. Motorola's sales peaked in 2006 at about 209 million units, Ericsson's peaked in 2007 at about 101 million units, and Nokia's peaked in 2008 at about 472 million units. It is no coincidence that the mobile-phone sales of these companies declined after 2006–2008: the iPhone was introduced in 2007, and the first Android phone was launched in 2008.

The irruption of iOS and Android transformed mobile-handset markets around the world. First, iOS and Android accelerated the transition from a world dominated by feature phones to one dominated by smartphones. Second, they undermined Symbian's leadership in the world smartphone market and BlackBerry's in the American smartphone market. Finally, iOS opened the door for Apple to become one of the leading handset makers in the world, and the Android OS played a similar role for Samsung and other Asian handset manufacturers.

THE GROWING MARKETS FOR WIRELESS SEMICONDUCTORS

During the 2000s, the transition from voice-only phones to feature phones first and to smartphones later fostered the growth of wireless semiconductor markets. Mobile-phone semiconductors became increasingly important during the 2000s because the silicon content of smartphones was considerably larger than that of voice-only phones.[15] Since the demand for wireless semiconductors was driven by the demand for smartphones, the explosive growth of smartphones that started in the late 2000s fueled the demand for wireless semiconductors.

The growth of smartphones accelerated in the second half of the 2000s. Figure 9.9 presents trends in sales of more traditional, voice-centered phones and smartphones between 2006 and 2012.

The figure reveals that, although volumes shipped of both traditional phones and smartphones grew at a fast pace, smartphones grew considerably

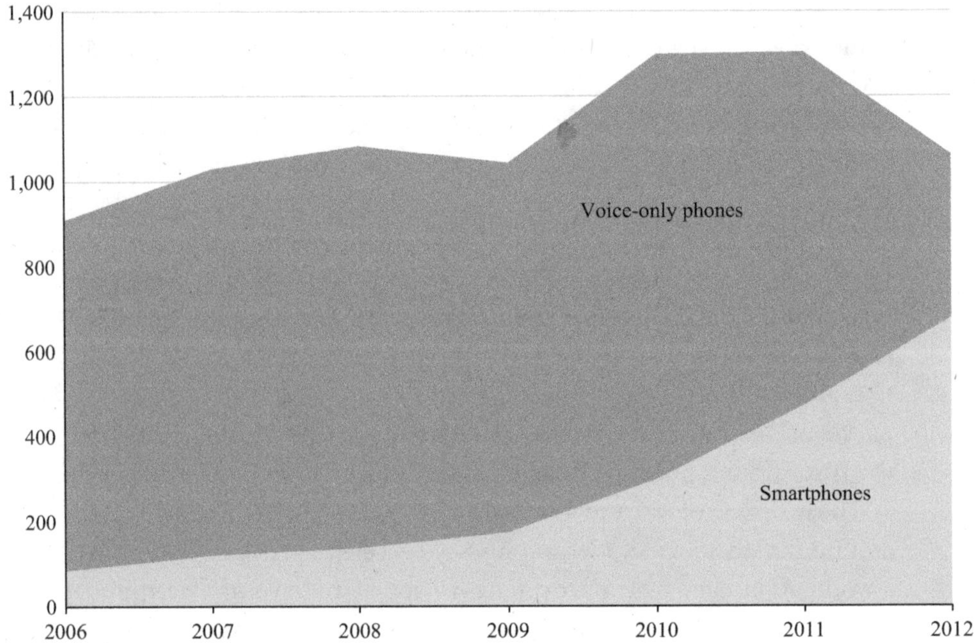

9.9 Sales of voice-only phones and smartphones, millions of units, 2006–2012. *Source*: Nomura Equity Research, *U.S. Semiconductor Primer* (Nomura Equity Research, December 11, 2013), 22.

faster. In 2006 smartphones represented about 8 percent of total units shipped, but by 2012 they already accounted for about 39 percent.

Figure 9.10 presents total sales (in USD billions) of voice-only phones and smartphones.

Smartphones were considerably more expensive: the average selling price of one was USD 378 in 2006, versus USD 148 for a traditional mobile phone. In addition, the price of traditional phones declined faster than the price of smartphones. Thus, by 2006 smartphones accounted for about 19 percent of total handset revenue, but by 2012 they already accounted for about 81 percent. The increasing importance of smartphones in the context of the cellular industry drove up the demand for wireless semiconductors.

Table 9.2 presents a breakdown of the key components of a voice-only phone and a smartphone with their costs. "RF" stands for radio frequency, "NAND" is a type of memory that takes its name from the namesake logic gate, and "DRAM" stands for dynamic random access memory.

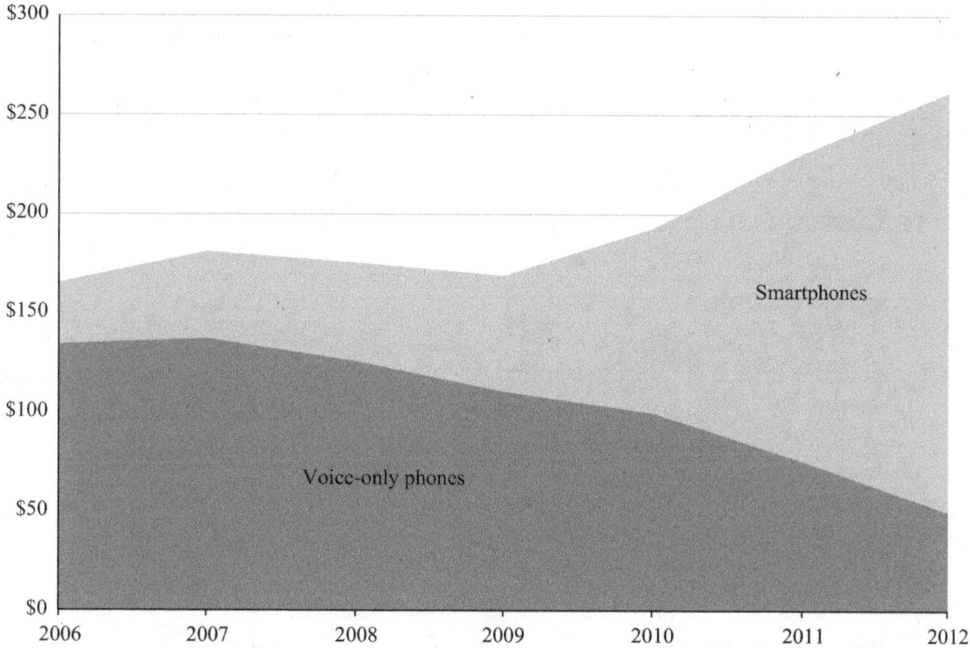

9.10 Sales of voice-only phones and smartphones, USD billions, 2006–2012. *Source*: Nomura Equity Research, *U.S. Semiconductor Primer* (Nomura Equity Research, December 11, 2013), 22.

The table compares the bill of materials for the Nokia 105 and the iPhone 5S. Although the Nokia 105 was announced in 2013, it was designed following the Nokia 1100, which was a simple, voice-only phone introduced in 2003. The Nokia 1100 was the best-selling mobile phone between 2003 and 2009, when it was retired. Thus, in the table the Nokia 105 represents the typical voice-only phone with its components and their costs, and the iPhone 5S, introduced in 2013, is a proxy for the typical smartphone of the early to mid-2010s with its components and their costs.

Given how complex present-day smartphones are, the simplicity of the Nokia 105 is striking. A voice-only phone, such as the Nokia 105, had some cheap semiconductor components (baseband, radio frequency [RF], and memory), plus enclosures and connectors, a charger and a battery, and a display. This was representative of the typical mobile phone of the 1990s and the early 2000s, and its total content cost, as of 2013, was around USD 13.50. In the iPhone 5S, by contrast, everything was more sophisticated and expensive: the semiconductor components (baseband,

Table 9.2 Bill of materials for voice-only phone and smartphone

	Voice-only	Smartphone
Baseband, RF, and memory	$5.25	
Baseband		$32.00
Processor		$19.00
Connectivity or RF		$4.20
NAND/DRAM		$20.40
Enclosures and connectors	$3.50	$28.00
Charger and battery	$2.50	$10.60
Display	$2.25	$41.00
Camera		$13.00
Sensors		$15.00
Power management		$7.50
Total content cost	$13.50	$190.70

Source: Nomura Equity Research, *U.S. Semiconductor Primer* (Nomura Equity Research, December 11, 2013), 133.

processor, connectivity, and memory), the enclosures and connectors, the charger and battery, and the display. In addition, the iPhone 5S had components (camera, sensor, and power management) that the simple voice-only phone did not have. The total content cost of the iPhone 5S as of 2013 was USD 190.70, higher than that of the Nokia 105 by a factor of 14. Additionally, in the Nokia 105 the combined semiconductor components cost USD 5.25, whereas in the iPhone 5S they cost USD 75.60.

The table shows that smartphones have at least four crucial semiconductor components: baseband, application processor, connectivity, and memory.[16] The baseband processor—also referred to as baseband, baseband chip, or cellular modem—manages the radio control functions and the transmission of signals: it allows the mobile phone to communicate with the network's base stations. The baseband has always been among the most important semiconductor component in mobile phones, smartphones included.

Another semiconductor component in smartphones is the application processor, a central processing unit similar to the one installed in a personal computer. One of the key roles of the application processor is to manage the software programs loaded on the device. The connectivity-related semiconductor components are the means of connecting wirelessly to the Internet and to other devices. Smartphone connectivity includes protocols such as Bluetooth, wireless local area network, and Global Positioning System but excludes the protocols related to cellular standards, which are handled by the baseband processor. Finally, the memory-related semiconductor component is captured in the table in the NAND/DRAM row.

Although all of these semiconductor markets grew in importance in the second half of the 2000s with the rise of smartphones, the baseband-processor market has always played a crucial role. Because it manages a phone's antennas and decodes voice and data signals, the baseband processor is the heart of the mobile phone. In 2007 basebands accounted for about USD 14 billion in worldwide revenue, compared with about USD 3 billion for application processors, and less than USD 2 billion for connectivity.

During the 2000s, the presence of multiple cellular standards forced basebands to become more sophisticated. The rise of the 3G standards—wideband code division multiple access (WCDMA) and cdma2000—did not imply that 2G networks vanished overnight. Third-generation networks were rolled out slowly and unevenly even in the wealthiest countries in the world. Additionally, in emerging markets, 2G networks persisted for many years after 3G technology became available. This meant that, during the 2000s, a mobile phone had to be able to operate on 2G and 3G networks, since a subscriber could easily move from a 2G area of a country to a 3G area and vice versa (or from a 2G country to a 3G country and vice versa). In the late 2000s the situation became even more complicated with the advent of the 4G Long-Term Evolution (LTE) standard. A smartphone's baseband chip had to be able to seamlessly switch among all existing networks and minimize latency—the time that goes by between the moment a signal is sent by a transmitter and the moment it is received by the handheld device. As the smartphone market matured and became segmented on the basis of prices and features, having a powerful baseband became one of the defining features of the most sophisticated and expensive smartphones.

In the course of the 2000s, companies tended to specialize in designing and manufacturing baseband processors for specific standards.[17] By 2008, the main players in the worldwide baseband market were Qualcomm, with almost 37 percent of global revenues; Texas Instruments, with almost 23 percent; and Ericsson, with almost 13 percent.

The role of the various semiconductor companies varied across standards. In 2008 Qualcomm, for example, had a 98 percent share in the CDMA baseband market and a 39 percent share in the WCDMA market. Texas Instruments was not active in the CDMA baseband market but had a 31 percent share in the WCDMA market. Similarly, Ericsson was not a supplier of CDMA baseband chips but had a 15 percent share in the WCDMA market. Other players in the worldwide baseband market included Infineon, a German semiconductor firm later acquired by Intel, and several Asian companies such as MediaTek, Spreadtrum, and Samsung.

Although only 19 percent of all 2G handsets were cdmaOne handsets, and only 13 percent of 3G handsets were cdma2000 devices, Qualcomm played a crucial role in the baseband-processor market.[18] Even though CDMA-only mobile phones were a minority in 2G and 3G, many devices relied on (multimode) baseband chips that enabled the subscriber to continue to operate the phone if the subscriber moved into a CDMA-based region of the country (or into a CDMA-based country). In addition, Qualcomm had a reasonably large share of the WCDMA standard-essential patents and played an important role in the WCDMA baseband-processor market during the 2000s.

Texas Instruments, which was still a large player in the baseband market as of 2008, decided to progressively eliminate its baseband business starting that very same year, and by 2012 it was no longer a significant player in this market.[19] Samsung, which entered the business in 1999, developed basebands for the GSM and WCDMA standards but tended to develop and manufacture these products for its own handsets only, rather than for the "merchant" market.[20] Moreover, in the first half of the 2010s, several companies—including Broadcom, Nvidia, and ST-Ericsson—exited the baseband market. Apart from Qualcomm, significant players that remained in this market once the 4G LTE standard started to gain traction were MediaTek and Spreadtrum, two Asian companies that originally focused on China and other Asian markets, and Intel, which acquired Infineon in 2010.

High, and increasing, research and development (R&D) costs drove the exodus from the baseband market. With baseband technology becoming increasingly complex during the 2000s, and especially toward the end of the decade, companies were forced to ramp up R&D spending, which in turn made it necessary for them to achieve a certain scale to operate profitably. Many vendors were not able to reach such scale and decided to leave the market.[21]

COMPETITION IN 3G MARKETS: TAKING STOCK

Competition in cellular markets intensified as the industry transitioned from the 1990s into the 2000s. Until the late 1990s, there were three product and service markets in the cellular world: network services, handsets, and infrastructure. In the 2000s the market for network services settled into an oligopolistic pattern all over the world: in most countries, there were two to four operators, with the top two having a combined share of 70 to 90 percent. In many countries, the top three operators had a combined share of almost 100 percent of the market. In places such as Brazil and Russia, carriers tended to consolidate into large operators with national scope, such as VimpelCom in Russia and Vivo in Brazil. For these companies, being able to offer a national network without roaming charges was a way to differentiate their services from those offered by regional carriers.

In the market for handsets, the 2000s witnessed the transition from voice-only phones to feature phones first and to smartphones later. Further, in the late 1990s and throughout the 2000s, Nokia became the leader in the world market for cellular handsets, displacing Motorola. Nokia's rise was based on sound strategic decisions made in the 1990s, including an early commitment to digital cellular, an emphasis on phone design, and a systematic attempt to supply devices for a wide variety of market segments. Motorola's decline was driven by its late embrace of digital cellular and its commitment to ventures such as the Iridium satellite project.

In the 2000s there was competition both within and across smartphone operating systems. Nokia, Ericsson, and Motorola competed with one another within the Symbian platform: they adopted different Symbian interfaces with the goal of making their products attractive to consumers. At the same time, their Symbian smartphones competed with devices that ran on operating systems other than Symbian, including the product offerings of Research in Motion, Palm, and Microsoft's hardware partners. The Symbian OS dominated the world market for smartphone operating systems until the arrival of iOS in 2007 and Android in 2008.

Finally, the 2000s witnessed the steady development of markets for wireless semiconductors. In the second half of the decade, smartphones grew at a fast pace, and the rising demand for smartphones increased the demand for semiconductor components. Among such components, four were important: baseband chips, application processors, connectivity, and memory. And of these four, baseband processors became the crucial one during the 2000s, especially because of the coexistence of several cellular standards. In a market with high R&D costs, many baseband vendors tried to achieve a scale large enough to survive profitably, but not many succeeded. As the 4G LTE standard ramped up in the early 2010s, several baseband suppliers decided that the time had come to exit the market, while the ones that stayed engaged in fierce competition.

IV

THE FOURTH GENERATION, THE 2010s

10

NEW STANDARDS, OPERATING SYSTEMS, DEVICES, AND MARKETS IN THE WORLD OF 4G

10.1 A 4G cell tower located in rural Scotland, 2018. Courtesy EE.

In the transition from the 2000s to the 2010s, the cellular industry underwent important transformations in standards, operating systems, and devices. During the 2000s, two new cellular standards were developed—Worldwide Interoperability for Microwave Access (WiMAX) and Long-Term Evolution (LTE)—that made it possible for cellular networks to achieve significantly faster data rates. In the late 2000s it was unclear which one would eventually prevail. In the early 2010s LTE became *the* 4G cellular standard.

The introduction of the iPhone in 2007 and the Android phones starting in 2008 changed the world market for smartphone operating systems. Steve Jobs, Apple's CEO, presented the iPhone as a "revolutionary mobile phone" that combined the capabilities of a phone, a music player, and what he called an "Internet communicator."[1] In the second quarter of 2007, when the iPhone became commercially available, Symbian was the leader in the world market for smartphone operating systems with 66 percent of all units sold. The share of the iPhone grew steadily to about 16 percent in late 2009, while the shares of Symbian and Windows CE declined steeply. (The Microsoft smartphone operating system was originally called Windows CE and was rebaptized Windows Mobile in 2003.) The share of the BlackBerry OS kept on growing through late 2009, mainly because BlackBerry was less of a consumer item and was embedded in many organizations, including many governmental ones.

By late 2008 the iPhone was well on its way to becoming an influential player in the world smartphone market—and the iPhone OS, as the iPhone's operating system was known before 2010, was quickly gaining share in the global smartphone operating system market. Its ascent, however, was slowed down by the rise of a challenger: in September 2008 T-Mobile released the first smartphone running on Android, the operating system acquired and further developed by Google. From then on, the global smartphone operating system market became a contest between Apple's iOS and Google's Android.

In the transition from the 2000s to the 2010s, cellular product markets changed as well, especially because Chinese companies became important vendors of cellular infrastructure, handsets, and semiconductors. In cellular infrastructure, Huawei and ZTE displaced some of the traditional leaders. In cellular handsets, Lenovo, ZTE, Huawei, and Xiaomi frequently showed up among the world's top vendors. And MediaTek and Spreadtrum competed with the likes of Qualcomm in the world market for baseband processors. The transition to a market economy in China, which started in 1978, facilitated the rise of all these companies.

Finally, as many developed countries reached cellular density of one hundred and higher, other areas of the world drew the attention of the cellular industry. By 2010 China and India had become the largest cellular markets in the world. In the transition from the 2000s to the 2010s, the African continent became the new frontier for cellular carriers and equipment manufacturers.

NEW STANDARDS: WIMAX AND LTE

During the 2000s, the dominant 3G standards—wideband code division multiple access (WCDMA) and cdma2000—were adopted all over the world, more so the former than the latter. As new 3G networks were set up, the Third Generation Partnership Project (3GPP) continued making progress in the definition of cellular standards, keeping the industry up to date via periodic "releases." (3GPP was the standard-setting organization established in 1998 that initially focused on developing the WCDMA 3G standard.) Table 10.1 presents the evolution of 3GPP releases starting with Release 99 in the year 2000.

3GPP defined the 3G standard WCDMA in a release that was targeted for 1999—and so it was called Release 99—but was made public in the year 2000.[2] From then on, releases were identified by a release number rather than a year.

UPDATING THE 3G STANDARDS

In the first half of the 2000s, 3GPP published releases that defined "evolutions" of WCDMA as opposed to new, 4G standards. The crucial WCDMA evolution was called High-Speed Packet Access (HSPA) and comprised two different enhancements to the 3G standard: High-Speed Downlink Packet Access (HSDPA) and High-Speed Uplink Packet Access(HSUPA). (*Downlink* refers to the process by which data travels from the base station to the cellular device, and *uplink* to data going in the opposite direction.)

Table 10.1 3GPP releases, 2000–2011

Year	Release	Definition
2000	99	3G Universal Mobile Telecommunications System with WCDMA
2001		
2002	5	High-Speed Downlink Packet Access (HSDPA)
2003		
2004	6	High-Speed Uplink Packet Access (HSUPA)
2005		
2006		
2007	7	High-Speed Packet Access Plus (HSPA+)
2008	8	HSPA+; first version of LTE standard
2009	9	WiMAX / LTE / Universal Mobile Telecommunications System interoperability
2010		
2011	10	LTE-Advanced

Source: A. Ghosh et al., *Fundamentals of LTE* (Upper Saddle River, NJ: Prentice-Hall, 2011).

Since Internet use in the late 1990s and early 2000s showed that applications required higher throughput on the download than on the upload, 3GPP focused first on improving the download. HSDPA was specified in Release 5 in 2002 with a downlink data rate of up to 1.8 Mbps; typical user data rates were in the 500 Kbps to 2 Mbps range. HSUPA was specified in 2004 in Release 6 with an uplink data rate of up to 2 Mbps, with practical network deployments offering rates between 500 Kbps and 1 Mbps. Usually deployed as a software upgrade to existing WCDMA networks, HSPA was first installed in 2005 by AT&T and quickly gained acceptance among WCDMA operators all over the world.

The CDMA 3G standard, cdma2000, also had its evolution. It was called EVDO, for "Evolution, Data Only." Originally developed by Qualcomm as a solution for fixed applications, it was later upgraded for fully mobile systems and deployed for the first time in 2002, three years before HSPA.

Figure 10.2 shows the cumulative number of WCDMA, HSDPA, cdma2000, and cdma2000 EVDO networks launched across the world during the 2000s. (EDGE, a 2G evolution, is included for comparison.)

The figure shows that the cumulative number of cdma2000 and WCDMA networks launched grew rapidly from the early 2000s, with the latter growing faster than the former. It also shows that the number of HSDPA networks launched rose at an even faster rate from the moment they were introduced in the mid-2000s. By the end of the 2000s, about three hundred WCDMA networks—and almost as many HSDPA systems—had been set up all over the world. By that time, about one hundred cdma2000 networks had been set up and almost as many cdma2000 EVDO systems.

In Release 7 (2007) and Release 8 (2009), 3GPP specified what was called HSPA+, a collection of enhancements to HSPA that improved system capacity, latency, and data rates. (Latency is the time that goes by from the

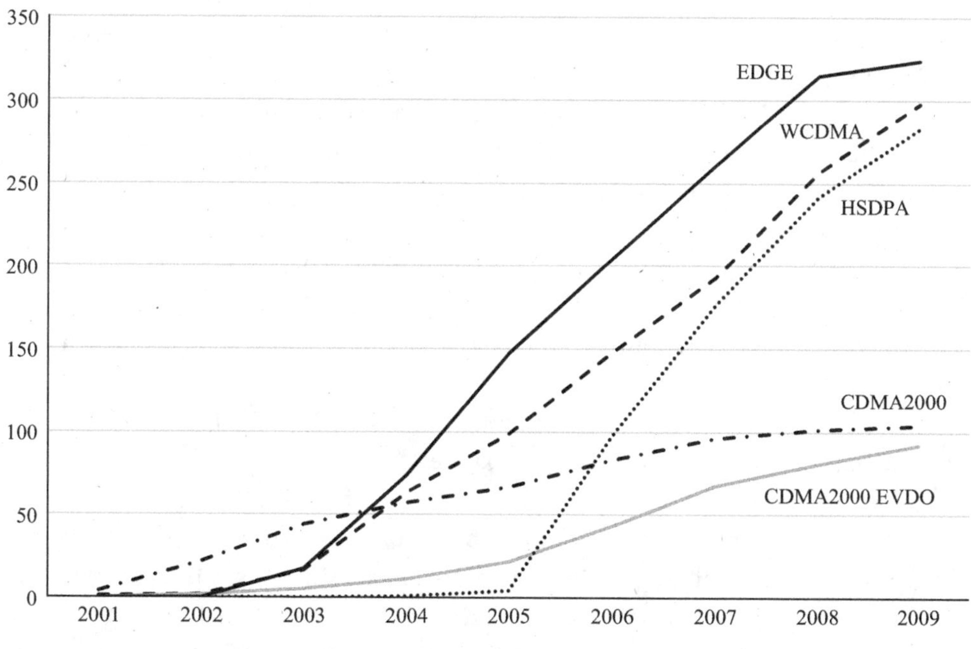

10.2 Cumulative number of networks launched around the world, 2001–2009, by standard. *Source*: Calculated by authors from P. Curwen and J. Whalley, *Mobile Telecommunications in a High-Speed World: Industry Structure, Strategic Behavior and Socio-economic Impact* (Farnham, UK: Gower, 2010), 68.

moment a signal is sent by the base station to the time it is received by the mobile device.) One of the key features of HSPA+ was a modulation approach that increased the peak downlink rate to 21.2 Mbps and the peak uplink rate to 11.5 Mbps. In addition, HSPA+ defined the use of up to two antennas for transmission in the base station and two for reception in the mobile device for what is technically called MIMO (multiple input, multiple output) transmission. The use of MIMO increased the peak downlink rate to 28 Mbps. The simultaneous use of the special modulation approach and MIMO, defined in Release 8, raised the peak rate to 42 Mbps.

DEFINING THE 4G STANDARDS

While 3GPP was publishing its releases during the 2000s, another institution, the Institute of Electrical and Electronics Engineers (IEEE), was also making progress in defining standards. In 1998 IEEE formed the 802.16 group with the task of developing a standard for what was called a wireless metropolitan area network. The group produced a specification for fixed wireless applications in 2001 and later enhanced it to allow for mobility. Completed in 2005, the revised standard was called IEEE 802.16e and is often referred to as Mobile WiMAX. In 2001, while the standard was in development, an industry consortium was formed under the name the WiMAX Forum to promote and make contributions to the standard.

The 3GPP Release 8, published in 2008, specified not only portions of the HSPA+ evolution but also the first version of the LTE standard, which eventually became the (single) 4G cellular standard. Several trends during the 2000s led to the development and adoption of LTE. During the 2000s, mobile applications evolved from text messaging, ringtones, and rudimentary Web access to music downloads and video sharing. Sites with embedded video content proliferated on the Web, video-sharing sites such as YouTube gained in popularity, and people increasingly used their mobile devices to access and share videos.

Additionally, in the first decade of the twenty-first century, smartphone shipments grew faster than shipments of voice-only and feature phones. By 2009 smartphones accounted for more than 30 percent of all mobile-phone sales in the United States. The proliferation of smartphones transformed users into heavy consumers of wireless data services. Finally, with many countries reaching a mobile-phone density of one hundred and higher during the 2000s, competition among operators intensified. With flat-rate pricing, mobile-phone usage grew faster than revenues, and operators were faced with the need to develop a network infrastructure that would lower the cost per megabyte of data. All these trends, combined, led to the rise of LTE.[3]

Purists argue that LTE, as defined in the 3GPP Release 8, was not truly 4G, but rather generation 3.9 or "super 3G."[4] True 4G arrived only with LTE-Advanced, which was defined in Release 10, published in 2011. LTE-Advanced offered faster data rates, softer hand-offs when users moved across cells, and higher spectrum efficiency than "regular" LTE.

Strictly speaking, two different versions of the LTE standard have been deployed around the world: FDD-LTE and TDD-LTE (also known as FD-LTE and TD-LTE).[5] FDD stands for *frequency division duplexing* and TDD for *time division duplexing*. Duplex communication happens when the phone transmits and receives at the same time, as opposed to simplex communication, which happens when the device either transmits or receives (but not both) at a given time. In TDD the device transmits and receives on the same frequency, with transmission and reception alternating in tiny time intervals. In FDD, by contrast, the device transmits and receives on different frequencies.

In one key area—the approach adopted for allocating scarce spectrum to multiple users—both WiMAX and LTE represented a break with WCDMA. While HSPA+ relied on CDMA, just as WCDMA did, both LTE and WiMAX used a technique called orthogonal frequency division multiple access. More precisely, LTE and WiMAX relied on orthogonal frequency division multiplexing as their modulation technology and on the closely related orthogonal frequency division multiple access technology as their multiple-access approach.[6]

BATTLES AMONG STANDARDS

In the late 2000s WCDMA operators could either upgrade their networks to HSPA+ or transition to a new, 4G standard. If they decided to adopt a 4G standard, they could choose either WiMAX or LTE. While HSPA+ and

LTE were both developed by 3GPP as evolutions of the Global System for Mobile Communications (GSM) and WCDMA networks that were in use at the time, WiMAX was specified independently by IEEE and the WiMAX Forum, and it had no backward-compatibility constraints.

While the standard-setting entities were specifying the standards, companies took sides and started promoting one standard or the other. This was not new: as we have shown in earlier chapters, stakeholders took sides every time new cellular standards were defined and introduced. In the second half of the 2000s, Intel became a key WiMAX promoter. As early as 2006, it introduced a system on a chip called Intel WiMAX Connection 2250, designed to support both fixed and mobile WiMAX networks.[7]

By September 2008, Intel and Sprint were ready to launch the first WiMAX network in Baltimore. The service would be available in Baltimore first and would then be extended to Portland, Oregon; Washington, DC; and Chicago. Initially, the target market for this network was not cellular phones but rather laptops and other small mobile devices, for which Intel would supply the WiMAX chipset. Several manufacturers of tablets and laptops—including Samsung, Dell, Sony, Acer, Asus, Lenovo, Panasonic, and Toshiba—had already indicated that they would introduce WiMAX products.[8] Sprint launched its service in Baltimore and, in December 2008, formed a joint venture with Clearwire, another carrier, to offer WiMAX services under the Clearwire name. Comcast, Intel, and Google made investments in the joint venture.[9]

During the 2000s, as IEEE developed WiMAX and 3GPP worked on LTE, Qualcomm and the Third Generation Partnership Project 2 (3GPP2) developed Ultra Mobile Broadband, which they thought would become a 4G cellular standard. In 2008, however, as many of the major carriers started committing to LTE, Qualcomm decided to terminate the Ultra Mobile Broadband project.[10] After that, Qualcomm started promoting LTE.

Industry observers interpreted WiMAX as an attempt by Intel and others to generate a low-cost alternative to existing cellular networks. WiMAX would be less costly because WiMAX implementers would not have to pay royalties to Qualcomm and others for their intellectual property. In 2008 six large WiMAX supporters—Cisco, Intel, Samsung, Spring Nextel, Alcatel-Lucent, and Clearwire—started planning for the creation of a patent pool that would keep royalty rates low and would facilitate the adoption of the standard.[11] This explains why Qualcomm not only supported LTE but also actively tried to prevent WiMAX from expanding. Reportedly, Qualcomm even acquired spectrum in India to prevent WiMAX from gaining a foothold in that country, one of the largest cellular markets in the world.[12] In the late 2000s and early 2010s, carriers committed to LTE. Sprint shut down its WiMAX network in the United States in early 2016.[13]

LTE ADOPTION

The first LTE networks were launched in countries that had a long history with cellular phones, including some of the first-generation pioneers. Sweden and Norway led the way in December 2009. The next year, several European countries inaugurated their LTE systems, including Poland, Germany, Finland, Austria, Estonia, and Denmark. In the second half of 2010, a few countries in Asia (Uzbekistan, Hong Kong, and Japan) joined the LTE club as well. In December 2010 Verizon Wireless inaugurated its LTE network in the United States, and AT&T followed suit about six months later.[14]

As had happened in prior cellular generations, other regions of the world adopted 4G with a lag. Australia had its first LTE inauguration in September 2011, and Uruguay became the first South American country to launch LTE a few months later, in December 2011. In Africa, several countries started LTE networks between April and December 2012, including Angola, Namibia, Mauritius, and South Africa.

NEW SMARTPHONES: THE IPHONE AND THE ANDROID PHONES

In the late 2000s two new smartphone operating systems radically altered the cellular world: iOS, designed by Apple and embedded in the iPhone, and Android, supplied by Google and embedded in many smartphones manufactured by Google's hardware partners.

THE IPHONE AND ITS OPERATING SYSTEM

The smartphone market changed with the arrival of Apple's iPhone in June 2007.[15] Apple had started exploring the development of an Apple phone soon after introducing the iPod music player in late 2001.[16] One of the problems consumers were facing in the early 2000s, and especially after the introduction of the iPod, was that mobile devices were proliferating and people were carrying too many of them—a PDA, a phone, a music player, and sometimes a digital camera. Some attractive "convergent" devices had started appearing on the market. In 2003, for example, the Palm Treo 600 had become popular by offering a solid combination of phone, PDA, and email capabilities.[17]

By 2004 the iPod business, which already accounted for 16 percent of Apple's revenues, looked promising but also vulnerable—mobile phones were becoming increasingly sophisticated, and the price of storage was plunging. Moreover, music stores rivaling Apple's iTunes, which had opened for business in April 2003, were proliferating. Apple responded with the first "music phone," the ROKR, jointly developed with Motorola and introduced in 2005. The ROKR—which had an unfriendly interface, could hold no more than one hundred songs, and had to be synced with a personal computer to complete a purchase from the iTunes store—sold well below expectations despite benefiting from an expensive marketing campaign.

10.3 The Apple iPhone. Launched in 2007, the iPhone revolutionized the market for handsets. The phone was *Time* magazine's Invention of the Year in 2007. Courtesy Nigel Linge and Andy Sutton.

As the ROKR episode was unfolding, Apple established a business relationship with Cingular, the wireless carrier. After protracted negotiations, Apple and Cingular—which was acquired by AT&T in December 2006—struck a deal centered on the introduction of an Apple phone, internally known as Purple 2. Apple's iPhone, developed in an atmosphere of complete secrecy during 2005–2006, was announced by Steve Jobs in January 2007. It became commercially available a few months later. The entire front surface of the iPhone was a touch screen, and all of its functions were activated by touch. It ran the Mac OS X operating system (later rebranded iOS when used on the iPhone), offered a new and improved Internet-access capability, and was able to wirelessly download music and movies from the iTunes store.

GOOGLE'S ANDROID

In September 2008 T-Mobile released the first smartphone running on Android, the operating system supplied by Google.[18] The history of Google's Android goes back to Andy Rubin, a computer and electronics hobbyist who had worked for Apple and WebTV.[19] In the late 1990s Rubin and a group of engineer friends founded the Danger Inc. start-up, which in the early 2000s developed the Sidekick, one of the first smartphones to effectively combine Web access, email, instant messaging, and other applications. In 2002 Rubin met Larry Page and Sergey Brin, the Google founders, at Stanford University. Page and Brin, who were Stanford alumni, were pleasantly surprised to find out that Google was the default search engine on the Sidekick.

The Sidekick developed a cult following among hip teenagers but never became a contender in the broader smartphone market. Rubin left Danger Inc. in 2004 and launched a new startup, Android, whose goal was to design a free and open-source mobile operating system.[20] The company would generate revenues by selling support services for the system. In 2005 Rubin approached Page to obtain Google's endorsement for Android. Rather than just endorsing the start-up, Google acquired it in July 2005, which fueled speculations that Google would become a smartphone vendor.

When it acquired Android, Google was reportedly concerned about two facts: first, Web surfing had started to migrate from personal computers to mobile phones, and second, the Internet capabilities of existing smartphones were limited. Since mobile phones were not particularly effective platforms for searching the Internet and viewing mobile ads, consumers' migration toward the mobile Web had the potential to undermine the foundations of Google's success.

Android's acquisition by Google did not turn Google into a smartphone vendor, at least not immediately. Unlike Apple, which sold the iPhone as an integrated and proprietary bundle of hardware and software, Google followed a strategy similar to the one Microsoft had employed in the 1980s in its relationships with PC makers: Google supplied an operating system and licensed it to any and all handset makers. The key difference between Google's Android and Microsoft's DOS and Windows was that Android was open source, and handset manufacturers did not pay a licensing fee for it.

In November 2007, only a few months after the iPhone reached the market, a group of technology companies that included Google, T-Mobile, HTC, Qualcomm, and Motorola announced that they would form the Open Handset Alliance to develop the Android operating system.[21] The first Android smartphone, introduced in September 2008 by T-Mobile, was reviewed poorly and did not sell well. The first successful Android smartphone was Motorola's DROID, which reached the market in October 2009.[22]

THE DECLINE OF THE INCUMBENTS

The rise of the iPhone OS (known as iOS after 2010) and Android and the corresponding decline of the pioneers (Symbian and the BlackBerry OS) transformed the mobile-phone industry. Apple introduced the iPhone 3G in June 2008. Soon thereafter Nokia, which owned 48 percent of Symbian, announced that it would buy the remaining 52 percent from the other stakeholders and would then set up the Symbian Foundation to "unify" and "open source" the Symbian operating system for smartphones. The key objective of the foundation was to "drive the development of web applications for use by consumers on cell phones."[23] The Symbian Foundation—whose founding members included Nokia, Sony Ericsson, Motorola, DoCoMo, AT&T, LG Electronics, Samsung Electronics, Texas Instruments STMicroelectronics, and Vodafone—was an (unsuccessful) attempt by some of the key industry players to stop the decline of what had been the leading mobile operating system in the world.

Although, in theory, Nokia continued supporting the development of the Symbian operating system, in practice it searched for alternatives to remain relevant in the changing smartphone market, and in February 2011 it announced a mobile-phone partnership with Microsoft. The agreement called for Nokia to use Microsoft's Windows 7 as the "primary" operating system on its mobile phones; but it went beyond that by explicitly announcing the partners' intent to establish a "third ecosystem," apart from Apple's and Google's, encompassing app developers and network operators.[24] Nokia made an alliance with Microsoft, as opposed to adopting Android as its mobile operating system, because it realized that it would find it difficult to differentiate its brand in the Android ecosystem.[25] Research in Motion, the other incumbent in decline, tried a different strategy to stay relevant in the market. It developed a new operating system, BlackBerry 10, and unveiled it in early 2013 while changing the company name to BlackBerry.[26]

Figure 10.4 tracks worldwide market shares (in units shipped) for smartphone operating systems between the first quarter of 2007 and the

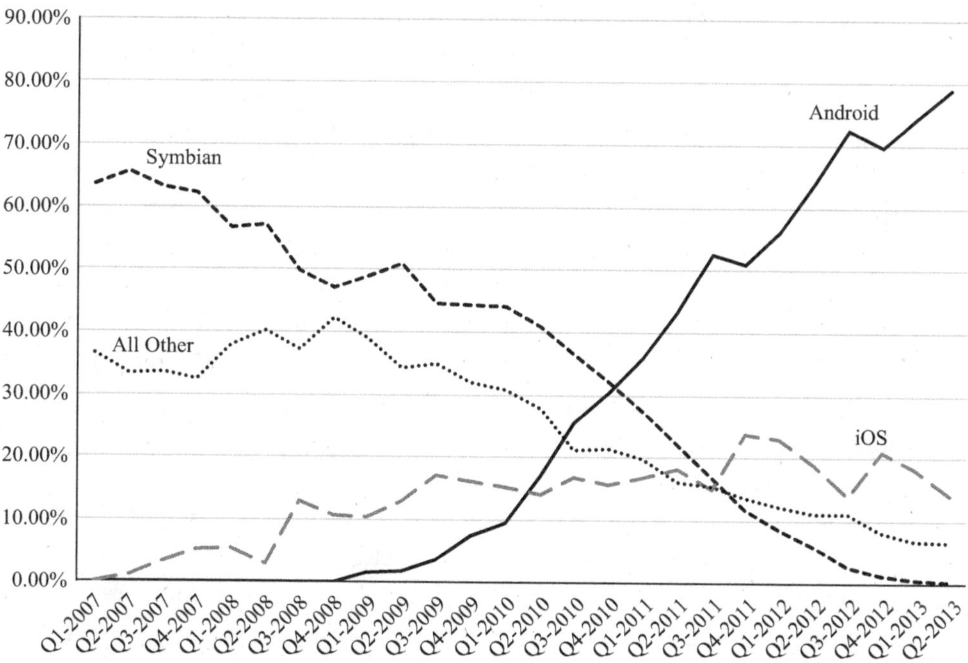

10.4 Global shares of smartphone operating systems, in units sold, 2007–2013. *Source*: Gartner press releases.

second quarter of 2013. The figure clearly shows the collapse of Symbian and the rise of iOS first and Android later.

In 2007 Symbian had between 60 and 70 percent of the global smartphone operating-system market (in devices sold). By the end of the period, in 2013, it barely had any sales at all. Although the decline in the installed base of Symbian users was surely much less dramatic, Symbian's collapse is striking. Other operating systems that saw substantial declines in sales, captured in the "All other" category, included BlackBerry, Windows Mobile, and a few smaller players.

The figure also shows the rise of the newcomers. The share of Apple's iOS rose to about 25 percent in late 2011, declining somewhat thereafter. By the end of the period covered in the figure, the dominant smartphone operating system was Android, whose share grew to almost 80 percent of devices sold in the second quarter of 2013. The Android surge was as impressive as the Symbian plunge.

NEW VENDORS OF CELLULAR TECHNOLOGY: THE RISE OF CHINA

The economic reforms introduced in China starting in 1978 set the stage for the rapid rise of Chinese suppliers of cellular infrastructure, handsets, and semiconductors. In 1978 the Chinese government started implementing a set of reforms that would eventually transform the Chinese economy and society.[27] Rather than trying to eliminate the planned economy in a single brush, the government introduced what has been described as a dual-track system in which there were two prices for most goods: a relatively low price set by the state and a usually higher price determined by the market. Steel plants, for example, received an output allocation from both the central government and the local government of the region in which they operated, and everything they produced above and beyond the allocations was left to the control of the enterprises to sell on the market at the prevailing market price. Over time, the government allocations declined and market sales increased, so that the dual-track system became an instrument for transitioning into a market economy.

In 1979 the Chinese government started liberalizing the foreign-investment regime by setting up four "special investment zones," of which Shenzhen was the largest.[28] These original zones were similar to areas that had been set up in other Asian countries starting in the 1970s: they were regions in which foreign investment was fostered by means of lower taxes, simplified administrative and customs procedures, and duty-free import of components. In China, a country that had started the transition toward a market economy from a planned economy, the zones also acquired additional meaning—they became ideal laboratories for market-economy experiments. Shenzhen, for example, hosted some of the early experiments involving labor, land, and equity markets. Shenzhen played a dual role: it became China's first export powerhouse region and also one of the core areas through which "parallel imports" entered China—branded goods from all over the world brought into the country (without the permission of the intellectual-property owners) by small importers that avoided custom duties and value-added taxes.

Shenzhen also became the cradle for many high-technology companies. Among them were Huawei and ZTE, two important producers of cellular infrastructure and phones, and the *shanzhai* vendors of cellular phones, which thrived in China until mobile users in that country transitioned from 2G to 3G in the early 2010s. (In Chinese, "shanzhai" originally means mountain fortress, but over time came to mean counterfeit, imitation, and copycat.)

NEW MARKETS: THE RISE OF AFRICA

In the transition from the 2000s to the 2010s, Africa became the new frontier for cellular carriers and cellular-equipment manufacturers.[29] Figure 10.5 shows the rise of world cellular subscriptions from around 2.2 billion in 2005 to around 8.3 billion in 2019. (In this figure, and in the next two, the asterisk associated with 2019 indicates that the figure for that year is estimated.) Over this time period, global subscriptions rose by a remarkable 277 percent.

Figure 10.6 shows the growth of global cellular subscriptions broken down by region. (The Commonwealth of Independent States, or CIS, encompasses several post-Soviet republics in Eurasia.) The figure shows that cellular subscriptions rose in all world regions but at different speeds. In Europe and the Americas, for example, subscriptions grew at a slower pace than in Asia and Africa.

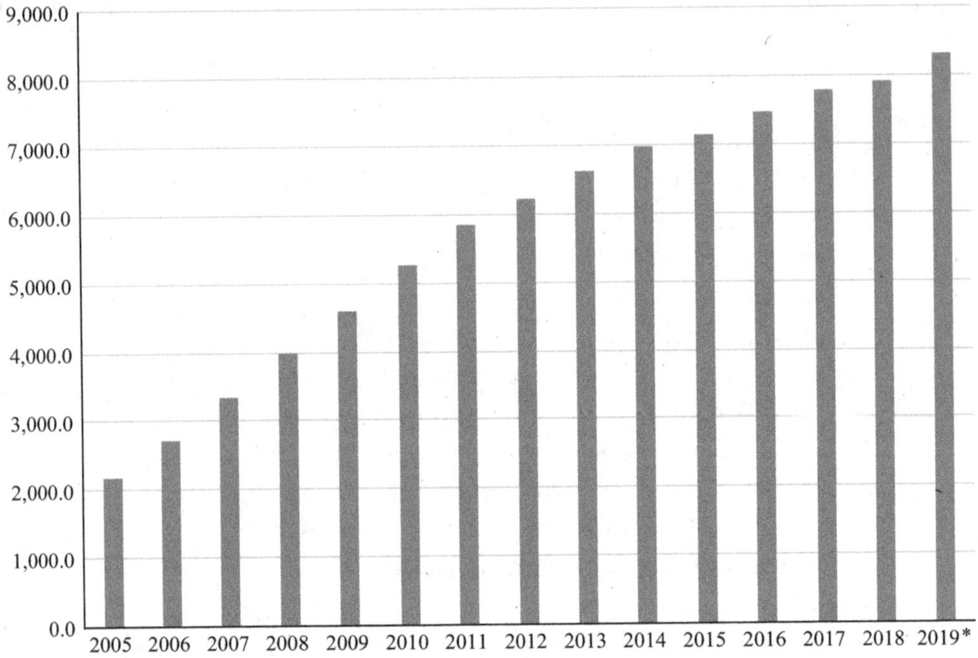

10.5 World cellular subscriptions, in millions, 2005–2019. *Source*: International Telecommunications Union database.

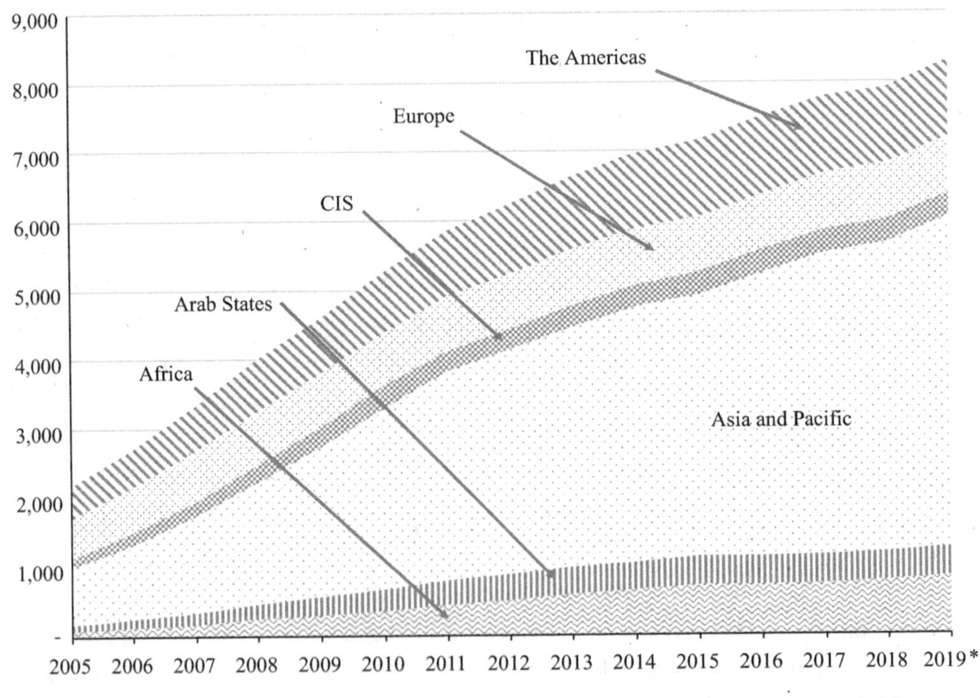

10.6 Cellular subscriptions, 2005–2019, in millions, by region. *Source*: International Telecommunications Union database.

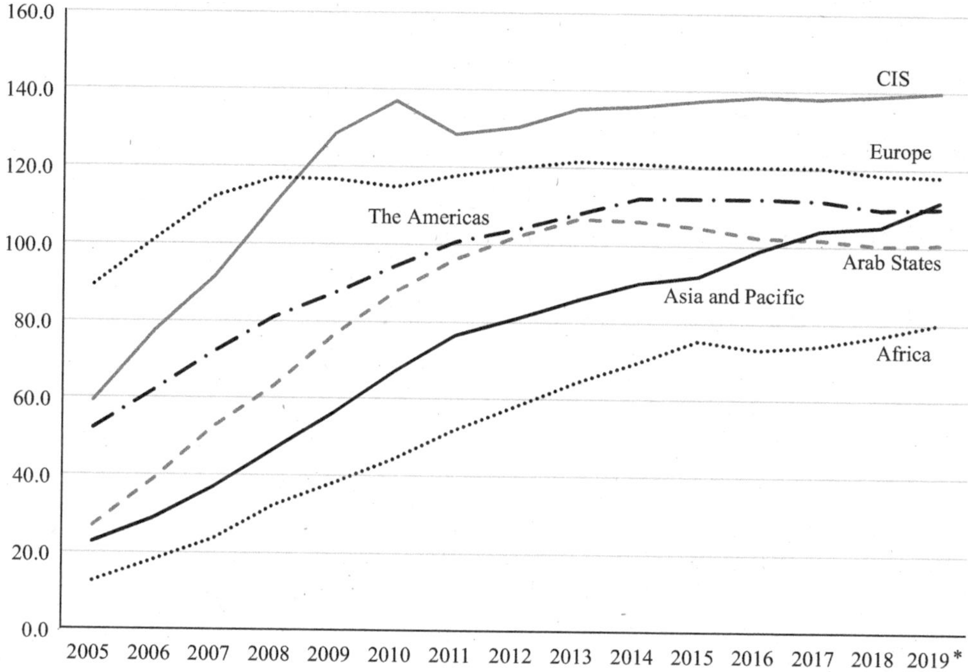

10.7 Cellular density, 2005–2019, by region. *Source*: International Telecommunications Union database.

Figure 10.7, finally, shows the evolution of cellular density—cellular subscriptions per one hundred inhabitants—for the same regions of the world over the same period.

The figure shows that, although Africa was among the fastest-growing regions for cellular subscriptions, at the end of the period the African continent was the region with the lowest density in the world—about eighty. In 2019 it was the only region that had cellular density under one hundred. In addition, the population of Africa is expected to double by 2050, at which point Africa will have more than a quarter of the world's population.[30] This combination of facts explains why Africa has become the new frontier for cellular carriers and equipment manufacturers.

11

CELLULAR SYSTEMS IN AFRICA

11.1 Picture of an M-PESA mobile-money agent, Nairobi, Kenya, 2016. Courtesy Charles-Axel Pauwels.

Cellular phones arrived late to the African continent. Figure 11.2 shows, by year, the number and identity of the African countries that introduced cellular systems between 1985 and 1996. Of the twenty-seven countries that launched cellular networks during this period, only six did so between 1985 and 1990.

As the figure shows, of the six countries that launched cellular systems between 1985 and 1990, four were in North Africa (Tunisia in 1985, Egypt in 1987, Morocco in 1988, and Algeria in 1990). The other two, the sub-Saharan systems, became operational in South Africa in 1986 and Zaire in 1988. (Zaire is known today as the Democratic Republic of the Congo. Sub-Saharan Africa is the area of the continent located south of the Sahara Desert.) Except for Zaire, these countries had one feature in common: they were among the wealthiest countries on the African continent in the decades before the arrival of mobile phones.[1]

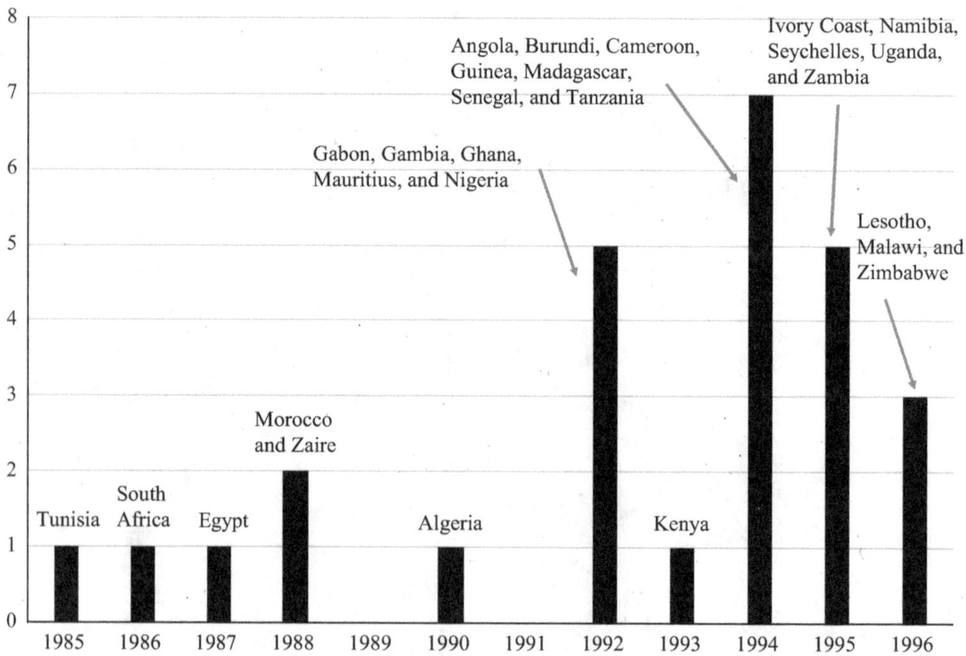

11.2 Cellular systems launched in Africa between 1985 and 1996, number of countries by year. *Source*: G. Garrard, *Cellular Communications: Worldwide Market Development* (Boston: Artech House, 1998), 404, 406.

In thirteen of the twenty-seven countries that launched cellular systems before 1997, the domestic postal, telegraph, and telephone (PTT) administration was initially the only service provider. In an additional three countries, the local PTT formed a joint venture with a private company and the venture was initially the monopolistic cellular carrier. In the remaining eleven countries, a private company or a consortium of private firms was responsible for introducing cellular services.[2] Table 11.1 presents the breakdown of African countries according to the type of organization that introduced cellular services. In the table, NMT stands for Nordic Mobile Telephone, TACS for Total Access Communications System, AMPS for Advanced Mobile Phone Service, and GSM for Global System for Mobile Communications.

In this chapter we explore the African experience with mobile phones. We analyze the status of telecommunications systems when mobile phones first appeared on the continent, the peculiarities of the market for cellular service in Africa, the impact of telecommunications reform, the main mobile operators and their strategies, and the market for cellular infrastructure. We highlight the role played by Africa-based cellular carriers such as Telecel, Celtel, and Mobile Telephone Networks (MTN) and explore the many ways cellular phones have transformed the everyday lives of people in Africa.[3] We examine the role of mobile money and the impact of mobile phones on movements of political protest such as the Arab Spring. Toward the end of the chapter, we highlight the increasingly important role played by China in the telecommunications sector of many African countries, especially through the financing of investments in mobile infrastructure.

FIXEDLINE AND CELLULAR PHONES IN AFRICA

With a few exceptions, telecommunications systems on the African continent were set up by the colonial powers—Belgium, Britain, France, Germany, Italy, Portugal, and Spain—as mechanisms for control and governance. By the time African countries achieved independence in the 1960s, telecommunications infrastructure on most of the African continent was obsolete and insufficient, as obsolete as the PTT system that

Table 11.1 First cellular systems and carriers in African countries, 1985–1996

Group 1: PTT introduced cellular			Group 2: Joint venture that included the PTT introduced cellular			Group 3: Private firm or consortium introduced cellular			
Country	**Year**	**Standard**	**Country**	**Year**	**Standard**	**Country**	**Year**	**Standard**	**Carrier**
Tunisia	1985	NMT450	Nigeria	1992	TACS	Zaire	1988	AMPS	Telecel
South Africa	1986	C-Netz	Tanzania	1994	TACS	Ghana	1992	TACS	Millicom
Egypt	1987	n/a	Namibia	1995	GSM	Mauritius	1992	TACS	Millicom
Morocco	1988	NMT450				Burundi	1994	AMPS	Telecel
Algeria	1990	NMT900				Guinea	1994	AMPS	Telecel
Gabon	1992	AMPS				Madagascar	1994	AMPS	Telecel
Gambia	1992	TACS				Ivory Coast	1995	GSM	Comstar
Kenya	1993	TACS				Seychelles	1995	GSM	Cable & Wireless
Angola	1994	AMPS				Uganda	1995	GSM	Celtel (Vodafone)
Cameroon	1994	GSM				Lesotho	1996	GSM	Vodacom
Senegal	1994	R2000				Malawi	1996	GSM	Telekom Malaysia
Zambia	1995	AMPS							
Zimbabwe	1996	GSM							

Sources: G. Garrard, *Cellular Communications: Worldwide Market Development* (Boston: Artech House, 1998), 404; J. Funk, *Competition between and within Standards: The Case of Mobile Phones* (Houndmills, UK: Palgrave, 2002), 43–45; and S. Mulavu, *2007 Zambia Telecommunications Sector Performance Review,* Lukasa, Zambia: The University of Zambia, 2007

African nations inherited from the colonial powers.[4] The newly independent African nations, however, did not dismantle their PTT systems since such systems were generally aligned with the political and economic orientation of the new governments, which emphasized state control, centralized planning, and national sovereignty.[5]

The African governments formed after independence intervened often in their countries' telecommunications systems, and they did so in ways that undermined the financial viability of such systems. Governments used taxes to appropriate large portions of the revenues earned by incumbent fixedline operators and forced operators to provide subsidized services to the state and its officials. Government personnel often paid their phone bills weeks or months after they were due, if they paid them at all.[6] As a consequence, fixedline incumbents frequently faced severe budget constraints and lacked the resources to invest in new telecommunications technology.

When, during the 1980s, cellular systems started taking off in Organisation for Economic Co-operation and Development (OECD) countries, they encountered well-developed fixedline telephone networks. In the early days of the cellular industry, cellular phones were viewed in OECD countries as complements to traditional phones, and interconnection problems were always understood as affecting the relations between cellular systems and the fixedline incumbent in each country.

Africa was different in that, when cellular systems first appeared there, fixedline density was trivially low. By 1994, around the time when many African countries launched cellular systems for the first time, fixedline density was 4.19 in the countries of North Africa (Algeria, Egypt, Libya, Morocco, and Tunisia), 9.00 in South Africa, and 0.45 in the rest of sub-Saharan Africa. Fixedline density on the African continent as a whole was 1.52, compared with around 57.88 in the United States, 48.90 in Japan, and 67.89 in Sweden.[7]

Scant fixedline density in Africa had several consequences. The most fundamental was that collecting and transmitting information without the benefit of a telephone was very costly. Setting up a business meeting, scheduling a home repair, arranging a family gathering—all of these activities involved costly trips that often failed to achieve their goals. Rural farmers who wanted to know at what price their crops were selling had to spend a day or two traveling to the marketplace. People were employed as runners to take messages from one side of a city to another or from one region to another.[8]

Cellular phones in Africa quickly became substitutes, rather than complements, for traditional phones. By 2009, 90 percent of Africans living in cities and 48 percent of Africans living in the countryside were within reach of a mobile network. By contrast, fixedline networks grew much more slowly after 1990, with fixedline density in 2009 averaging slightly over one.[9] Most Africans never had access to a fixedline phone, so that the mobile phone became the one and only telecommunications device they knew and used. In addition, since fixedline customers were so few, interconnection issues in Africa often involved not only relationships between each of the cellular systems and the fixedline incumbent but also relationships among two or more cellular networks.

In Africa the incumbent fixedline operator often continued to dominate the fixedline voice and Internet (broadband) markets. The importance of these markets in the overall scheme of telecommunications systems in African countries, however, shrank considerably after the early 1990s. By 2016 there were about 770 million cellular subscribers in Africa but only 11 million fixedline subscribers; in that year, there were also 280 million mobile broadband Internet subscriptions but only 6 million fixed broadband subscriptions.[10]

Two types of reform facilitated the growth of cellular markets in Africa. First, the 1990s witnessed political reform. After most African countries became independent in the 1960s, many of them came under the rule of authoritarian governments. By the end of the 1970s, only three countries in sub-Saharan Africa—Botswana, the Gambia, and Mauritius—were multiparty democracies, and more than two-thirds had had periods of military rule. This changed during the 1990s, when countries often held elections to choose new leaders: by 1994, thirty-five sub-Saharan countries had held elections, and most other countries did in the years that followed. For this reason, the 1990s have been described as the second wave of democratization in Africa.[11]

In addition, the 1990s witnessed telecommunications reform. For some governments, reform meant simply licensing additional cellular operators. Others went further: they partially privatized the fixedline operator, licensed a second fixedline network, and allowed several cellular carriers to enter the market. Many governments created at least one telecommunications regulator, which was more or less independent of the government depending on country. Before exploring telecommunications reform in Africa, we discuss certain peculiarities of mobile-phone markets—supply and demand—on the continent.

CELLULAR MARKETS IN AFRICA: SUPPLY, DEMAND, AND USES OF MOBILE PHONES

Because of social and economic conditions, cellular markets in Africa had several peculiar features, on both the supply and the demand sides. In addition, mobile phones were used to host mobile-money platforms such as M-PESA and to fuel movements of political unrest. (*Pesa* means "money" in Swahili, a language widely used in East Africa, and thus M-PESA stands for "mobile money.")

SUPPLY

The rise of the GSM standard in Europe in the early 1990s facilitated the diffusion of cellular phones not only on that continent but also in the rest of the world, including Africa, because it made it feasible for equipment and handset vendors such as Ericsson and Nokia to benefit from economies of scale in manufacturing.[12] When GSM networks were first set up in Europe in the early 1990s, the equipment cost per subscriber was about USD 1,000; a few years later, however, when GSM networks were launched in Africa, costs had declined to around USD 150.[13]

In Africa, however, launching and maintaining a cellular network involved costs that were absent in OECD countries. Because most African countries had a geographically restricted electrical-power grid, cellular operators often used portable generators to power base stations—two generators for each base station, one for regular operations and another for backup. Generators worked on diesel, and diesel was the one input that needed to be protected from theft. Thus, operators usually paid for a security guard who lived close to the base station and cultivated the surrounding land.

Companies attempting to set up cellular networks in African countries often found unstable political situations. In the late 1990s and early

2000s, Celtel, one of the pioneering African operators, started providing service in three countries that were in the middle of, or just coming out of, a civil war: Congo Brazzaville, Sierra Leone, and the Democratic Republic of the Congo. When Celtel tried to expand its network beyond Kinshasa, the capital of the Democratic Republic of the Congo, it had to ask permission from the rebel leaders that controlled the eastern portion of the country. Further, the equipment had to be flown there because of the lack of roads.[14]

DEMAND: NETWORK SERVICES

In addition, the cellular business model had to be adapted to the nature of demand on the African continent. In the early days of cellular in Africa, people called first and paid later, which created problems—only wealthy people had access to mobile phones and, in spite of that, bills were rarely paid on time. In the late 1990s prepaid cellular systems started appearing in Europe, and they were quickly imported into Africa. By 2001 most cellular customers in Africa were on a prepay system: they paid upfront for their mobile service. This facilitated the rapid growth of cellular density on the continent during the 2000s.[15]

It became common for Africans to buy minutes of cellular service on "scratch cards." (Each card had a secret code that the user revealed by scratching, and once this code was entered on the mobile phone, credit was added to the user's account.) Distributors bought such cards from carriers and then sold them to cellular users, often through street sellers. Operators soon discovered that the lowest-value cards sold the fastest: people often bought one-dollar or two-dollar cards several times during the month. Later, carriers developed the capacity to transfer additional minutes electronically to subscribers, which lowered transaction costs considerably, but scratch cards continued to be used.[16]

Thus, in Africa, cellular minutes were bought and sold like fast-moving consumer goods: they were purchased several times per month, and often more than once per week. Further, people who could not afford a cellular phone by themselves shared a handset with others: a survey of some of the poorest neighborhoods in Nairobi, Kenya, revealed that cellular phones were shared, on average, by four people.[17]

DEMAND: HANDSETS

By 2017 most people in sub-Saharan Africa owned mobile phones. Among mobile-phone owners, however, few owned smartphones. A Pew Research Center survey conducted in 2017 explored mobile-phone ownership in six countries with relatively high cellular density: South Africa, Ghana, Senegal, Nigeria, Kenya, and Tanzania.[18] South Africa was the country with the highest proportion of mobile-phone owners—91 percent. In the other countries, the proportion of mobile-phone owners among the surveyed individuals varied between 75 and 80 percent.

South Africa was unique because there were more smartphone owners than basic-phone owners. (In the survey, "basic phones" included voice-only phones and feature phones.) Fifty-one percent of surveyed individuals in South Africa owned smartphones, and 40 percent owned basic phones.

In all other countries, smartphone owners never represented more than 35 percent of all surveyed individuals. Smartphone ownership varied between 13 percent of all surveyed individuals in Tanzania and 35 percent in Ghana. The survey also found that smartphone ownership increased between 2013 and 2017. It went up from 33 percent to 51 percent in South Africa, and from 8 percent to 13 percent in Tanzania, for example.

In 2017 individuals who owned mobile phones—basic phones or smartphones—used them for sending text messages, among other things. Seventy-eight percent of mobile-phone owners in the six countries covered by the survey had used them for texting in the year before the survey. (The survey reported the median proportion across the six countries.) The next most popular uses were taking photos or videos (59 percent) and making or receiving payments (59 percent). In Kenya, the country where the mobile-money platform M-PESA originated, 82 percent of mobile-phone owners used the phone for making or receiving payments in the year before the survey. Making and receiving payments via a mobile phone was feasible on all device types: 79 percent of all basic-phone owners and 88 percent of all smartphone owners in Kenya reported having used their mobile phones to make and receive payments in the year before the survey.

MOBILE MONEY IN AFRICA

Just as the limited number of roads in Africa made it difficult for carriers to set up their cellular networks, the limited number of banks prompted people to start using their airtime as money, which led to the rise of mobile money. Vodafone's M-PESA, the leading mobile-money network in Africa, probably would not have grown as it did if Africa had had a mature banking infrastructure.[19]

In the early 2000s cellular carriers in Africa realized that people were using airtime as money. Often, cellular subscribers used airtime as a store of value, buying minutes today and reselling them in the future. Since bank branches were scarce, especially in rural areas, but airtime resellers were relatively plentiful, it was easier and safer to use cellular minutes than to use cash as money. Over the years, mobile money evolved beyond the buying and selling of airtime.

The original idea for what became M-PESA was conceived in 2003 by Nick Hughes, who was then head of social enterprises at Vodafone.[20] When the pilot was implemented in Kenya in the mid-2000s, it was expected that people would use the text-based system to repay microfinance loans. However, the team in charge soon realized that pilot participants were using the system for a variety of purposes—businesses made payments to one another, and they also used M-PESA as an overnight safe because banks closed too early for businesses to make deposits; for safety reasons, travelers would sometimes deposit mobile money at the beginning of a journey to collect it at the end of the trip a few hours later; people bought airtime with M-PESA and sent it to their relatives in rural areas; and they also made payments to people outside the pilot areas.[21]

The system was launched in 2007 in Kenya and in 2008 in Tanzania, and as of this writing it is available in ten countries.[22] To use the system, Vodafone customers first opened an account with an M-PESA agent. When they did so, they received a four-digit PIN that allowed them to access their mobile-money account. They then exchanged traditional cash for digital cash: they gave traditional cash to the M-PESA agent, who in turn sent mobile money to the customer's account in exchange for the cash received.

M-PESA agents fueled the growth of the mobile-money platform. In fact, one of the reasons the system grew as it did in Kenya is that airtime resellers were pervasive in the country and many of them became M-PESA agents. In countries where agents were scarce, the system grew slowly, if at all. M-PESA agents were recruited by Vodafone and were regularly monitored and trained. Agents bought mobile money in advance so that they had it available when customers wanted to "cash in"—that is, when they wanted to give the agent traditional cash and receive digital cash in return. Agents also had traditional cash available for situations when customers wanted to "cash out"—that is, give the agent digital money and receive traditional cash in return. M-PESA took off because it managed to develop a critical mass of both users and agents.

Once customers cashed in, they could use their digital money to, among other things, make remittances to other M-PESA customers, pay utility bills, add airtime to their cellular phones, pay for goods and services at merchants, withdraw money from an ATM, receive money from other customers (including from abroad), and pay and receive salaries. The balances of traditional cash and digital money that M-PESA agents stored and managed belonged to them rather than to Vodafone.

Studies by economists have shown that M-PESA has helped families in Kenya avoid extreme poverty. A study found that poor families that used M-PESA had more stable incomes, since they had a tool available that helped them save and they were able to rely on transfers from relatives and friends when they were facing financial hardships.[23] Another study found that families that used mobile money accessed health care services more frequently when they needed them (in comparison with families that did not rely on mobile money): families were able to use mobile money for informal borrowing from friends and relatives and thus were able to visit a clinic, consult a doctor, and purchase the required medicines.[24]

MOBILE PHONES AND THE ARAB SPRING

The Arab Spring was a wave of protest and civil unrest that affected several countries in North Africa and the Arabian Peninsula starting in late 2010.[25] The movement of antigovernment protest started in Tunisia in December 2010 and in a short period of time led to the demise of authoritarian governments in Tunisia and Egypt. Popular revolt spread to about seventeen other Arab countries, including Yemen, Libya, Bahrain, Syria, Jordan, and Morocco.[26] Commentators have pointed out that one of the reasons the movement spread rapidly was the availability of social media and mobile phones.[27]

The immediate origin of the Arab Spring was the self-immolation of a street vendor, Mohamed Bouazizi, in a small town in Tunisia on December 17, 2010. The story of Bouazizi, who died protesting against mistreatment by the police and the local bureaucracy, spread through social media, including YouTube, Twitter, and Facebook, and inspired people to organize protests and criticize their governments. Historians of Africa have pointed out that the Arab Spring was fueled by years of accumulated grievances over poverty, rising prices, unemployment, and authoritarian governments.[28] On January 14, 2011, Ben Ali, Tunisia's ruler, fled into exile. Inspired by the events in Tunisia, crowds in Egypt took to the streets demanding the end of Hosni Mubarak's rule. On February 11, 2011, Mubarak was forced to resign. After that, the movement spread to other countries.

Anecdotal evidence suggests that social media and mobile phones played an important role in mobilizing protests both before and after the death of Bouazizi. In addition, several studies by political scientists and economists have attempted to explore more precisely the impact of social media and mobile phones on the evolution of the Arab Spring.

One such study by political scientists focused on the role of social media.[29] In Africa, however, many more people have access to social media on mobile phones than they do on desktops or laptops, and thus this study indirectly explored the role of mobile phones in fueling political revolt. The study drew three fundamental conclusions. First, social media played a crucial role in shaping political debate both before and during the Arab Spring. In Tunisia, and well before the Arab Spring, advocates of democracy used social media to embarrass Ben Ali by, for example, streaming videos of his wife using a government jet to make expensive shopping trips to Europe. Bloggers in Tunisia and Egypt used social media to publish information critical of the government.

Second, a spike in online revolutionary conversations usually preceded major events on the streets and squares. Online discussions, the study argued, played an important role in driving street protests in Tunisia and Egypt. Third, social media helped spread democratic ideals across borders. Advocates of democracy in Tunisia and Egypt used social media to connect with sympathizers in other countries. By facilitating these connections, social media may have contributed to fuel revolt outside the countries where the movement originated.

Another study by economists focused more specifically on mobile phones and their impact.[30] This study used detailed geocoded information on mobile-phone coverage in Africa between 1998 and 2012 in combination with aggregate data on the incidence of political revolt and information on individual participation in protests. The study found that mobile phones played a crucial role in fostering mass mobilization in Africa during periods of economic downturn. According to the study, mobile phones encouraged revolt because they made individuals more responsive both to deteriorating economic conditions and to the participation of their neighbors. The authors labeled the first effect "enhanced information" and the second "enhanced coordination." The study found that mobile phones fostered political mobilization not only during the Arab Spring but also during the food riots that happened between 2007 and 2012.

TELECOMMUNICATIONS REFORM IN AFRICA

In Africa, cellular phones would not have taken off as they did without telecommunications reform. We have collected information on the main features of the telecommunications-reform process for forty-eight African nations at two points in time: 1998 and 2009. For space reasons, we present such information only for twelve of those countries in table 11.2.[31]

The sample of countries in the table is not random. It is designed to show a variety of outcomes and to highlight factors that likely contributed to such outcomes. The table includes several of the African cellular pioneers: Egypt, Morocco, South Africa, and Tunisia. These were countries that introduced cellular systems before 1990. It also includes a sample of countries that have been identified as early reformers: the Ivory Coast, Ghana, Mauritius, Senegal, South Africa, and Uganda. (See later in this section for the definition of early reformers; South Africa was both a cellular pioneer and an early reformer.) The table presents information for one country, Malawi, that undertook a limited telecommunications reform in the 1990s. Finally, it includes information on two countries, Eritrea and Ethiopia, that introduced their first cellular systems late and that, as of 2009, had not yet allowed for cellular competition.

The table presents information on whether reform legislation had been passed by 1998, whether a separate regulator was created and when,

Table 11.2 Telecommunications reform, cellular competition, and cellular density in Africa, selected countries

	Reform law passed as of 1998	Separate regulator as of 1998	Cellular launch	Cellular competition as of 1998	Number of mobile operators in 2009	Per capita income in 1992 (PPP)	Cellular density in 2000	Cellular density in 2009
Egypt		Y (1998)	1987	Y		2,365	2.1	72.1
Eritrea	Y (1998)	Y (1996)	2005	N	1		0.0	2.8
Ethiopia	Y (1996)	Y (1996)	1999	N	1	531	0.0	4.9
Ghana	Y (1996)	Y (1997)	1992	Y	5	1,770	0.7	63.4
Ivory Coast	Y (1995)	Y (1996)	1995	Y	5	1,980	2.7	63.3
Malawi	Y (1994)	Y (1998)	1996	N	2	646	0.4	15.7
Mauritius	Y (1988)	Y (1998)	1992	Y	3	9,357	15.2	85.2
Morocco	Y (1997)	Y (1997)	1988	N	3	3,755	8.2	80.9
Senegal	Y (1996)	N (2001)	1994	Y	3	1,789	2.5	55.1
South Africa	Y (1991)	Y (1997)	1986	Y	3	6,840	19.0	94.2
Tunisia	Y (1995)		1985	N	2	5,452	1.2	93.2
Uganda	Y (1996)	Y (1997)	1995	N	5	672	0.5	28.7

Source: Prepared by authors from information in International Telecommunications Union, *General Trends in Telecommunications Reform, volume 2: Africa* (Geneva: ITU, 1998); International Telecommunications Union, *General Trends in Telecommunications Reform, volume 4: Arab States* (Geneva: ITU, 1998); International Telecommunications Union, *Measuring the Information Society 2018, Volume 2* (Geneva: ITU, 2018); G. Garrard, *Cellular Communications: Worldwide Market Development* (Boston: Artech House, 1998); M. Williams et al, *Africa's ICT Infrastructure* (Washington DC: World Bank, 2011); and ITU database

whether the cellular market was competitive by 1998, the number of cellular carriers in 2009, and cellular density in 2000 and 2009. Per capita income information is included for each country as of 1992—that is, before most cellular systems were launched. The 1992 per-capita-income figure is a purchasing-power-parity adjusted number (in current international dollars), which accounts for differences in the cost of living across countries.[32]

The table shows that many African countries implemented some sort of telecommunications reform during the 1990s.[33] By 1998, thirty-eight out of the forty-eight countries surveyed had passed reform legislation. A few pioneering countries even introduced reform legislation during the 1980s. There were often different stages of reform, and not every country that embarked on the reform process completed all stages. Governments often passed legislation separating telecommunications from postal services, and additional laws "corporatizing" the entity in charge of supplying fixedline telephony services. In addition, by 1998, twenty-six countries had created a separate regulator. This was an entity—separate from the telecommunications ministry of the country—that was in charge of regulating certain dimensions of telecommunications services, such as numbering plans, tariffs, technical standards, interconnection rates, and frequency allocation.

That the regulator was "separate" from the telecommunication ministry did not always mean that it made decisions free of government interference. The degree of independence of the telecommunications regulator varied considerably across countries. Regulators were more independent from governments in countries where regulatory agencies had their own sources of funds. Further, regulators were more independent in countries where the senior staff of the regulatory agency was not appointed by the government and its ministers.

In Ghana, for example, the reform legislation passed in 1996 created a regulator with substantial responsibilities, the National Communications Authority. The law, however, failed to safeguard the independence of the regulator: all members of its board of directors were appointed by the president of the country and could be removed by the president at any time. Further, the law also gave the Ministry of Telecommunications considerable authority over the regulator.[34] In the Ivory Coast, by contrast, where two regulators were created by legislation passed in the mid-1990s, at least one of them, the Telecommunications Council, operated independently from the government and its ministers.[35]

Although the table does not show it, only five countries—the Democratic Republic of the Congo, Ghana, Madagascar, Nigeria, and Tanzania—had introduced some degree of competition in fixedline telecommunications services (local or long-distance) by 1998. By 1998, however, eighteen countries had introduced competition in either analog or digital cellular, or both. More than 60 percent of the countries that launched cellular service in or before 1998 had some form of competition in cellular services by 1998.

Although the table is just a sample of twelve countries out of forty-eight, the information presented suggests that, by 2009, cellular density depended on several factors, including how early cellular systems had been introduced in each country, how wealthy each country was, how early telecommunications reform had been implemented, and how comprehensive such reform was. Figure 11.3 shows the relationship between per capita income in 1992 (in purchasing-power-parity terms) and

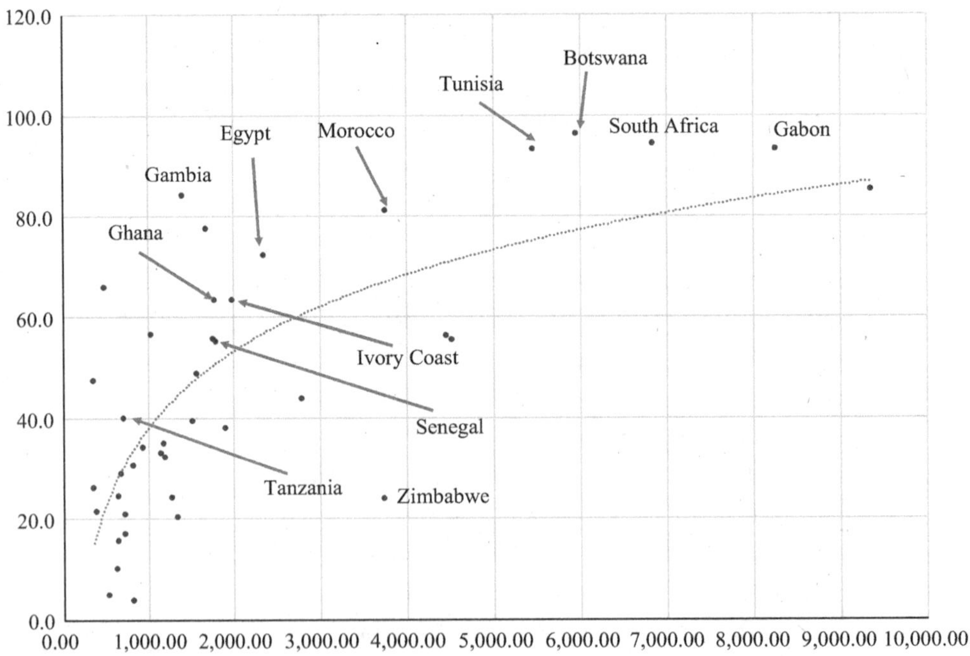

11.3 Per capita income in 1992 (in international dollars) and cellular density in 2009 in Africa. *Sources*: Penn World Tables; ITU database.

cellular density in 2009. The figure shows that two groups of countries had the highest densities in 2009. The first group included the North African countries that introduced cellular systems during the 1980s (Egypt, Morocco, and Tunisia), and the second included the wealthiest African countries, such as Gabon, South Africa, and Botswana.

But it was not just per capita income that mattered. Early and comprehensive telecommunications reform played an important role in accelerating the growth of cellular density. A World Bank study identified a group of ten countries that were early reformers: they created a telecommunications regulator, privatized their fixedline incumbents, and had a competitive cellular market for at least five years as of 2009.[36] They were Ghana, the Ivory Coast, Lesotho, Madagascar, Mauritania, Mauritius, Senegal, South Africa, Tanzania, and Uganda. Most of these countries passed telecommunications reform laws and created a separate regulator during the 1990s, and most of them had competitive cellular markets already in 1998. The World Bank study showed that, by 2009, these countries had a considerably higher cellular density—about 11 percentage points higher, on average—than would have been expected purely on the basis of their per capita income.

The World Bank study also found that African cellular markets grew faster, on average, after a second cellular carrier started providing service. Thus, countries that allowed multiple operators to enter simultaneously when cellular service was introduced performed better than those in which the entry of multiple carriers happened slowly over time. The Ivory Coast, for example, allowed competition from the very beginning in its cellular market and had cellular density of 2.93 by the year 2000. Ghana, which allowed cellular competitors to enter sequentially over time, had cellular density of 0.69 in the year 2000. This was so even though the first cellular network in the Ivory Coast was set up about three years later than in Ghana.[37]

NETWORK-SERVICES MARKETS IN AFRICA

As explained earlier, in thirteen of the twenty-seven African countries that launched cellular systems before 1997, the domestic PTT administration was the first entity to introduce cellular phones, and it operated the cellular network as a monopolist for several years. Algeria, Egypt, Morocco, and Tunisia—all of them in North Africa—are examples of countries in which cellular systems were launched between 1985 and 1990, and in which the domestic PTT administration of each country was the only cellular carrier in the market for a period of time after the initial launch of service. In three countries, the PTT administration formed a joint venture with one or more private organizations to launch the first cellular network: Nigeria (1992), Tanzania (1994), and Namibia (1995). In the remaining eleven countries that introduced cellular service before 1997, a private operator set up the first network. Telecel, for example, an African cellular pioneer, launched the first cellular networks in Zaire, Burundi, Guinea, and Madagascar, and Millicom, a multinational carrier headquartered in Luxembourg, introduced cellular service in Ghana and Mauritius.

Three types of private operators entered African cellular markets: some of them were born and headquartered in Africa, others were based in Europe, and yet others were headquartered in wealthy regions outside the OECD such as the Arabian Peninsula or in emerging markets such as India. Many of the multinational operators that entered cellular markets in sub-Saharan Africa from outside the region did so by acquiring one of the Africa-based cellular pioneers.[38]

THE FIRST WAVE OF PRIVATE CARRIERS: THE AFRICAN PIONEERS TELECEL, CELTEL, AND MTN

Network-services markets in Africa witnessed the rise of cellular carriers born and headquartered in Africa with operations in multiple African countries. These are often labeled "emerging-market multinational corporations." The pioneers were Telecel, Celtel, and MTN.[39]

In 1987 Miko Rwayitare and Joseph Gatt formed a joint venture in Zaire under the name Telecel International. Rwayitare was an engineer who had grown up in Zaire and had studied in Germany, and Gatt was an American pilot and businessman. After Gatt learned of the AMPS cellular technology in the United States, he and his partner, Rwayitare, decided to explore the possibility of setting up an AMPS network in Zaire, a country where fixedline phones were almost nonexistent.[40]

With limited funding, the founders managed to convince the government of Zaire to grant Telecel a cellular license, and the first Telecel AMPS network—and one of the first cellular networks on the African continent—was launched in 1988. In a few years, the company expanded to Burundi, Guinea, and Madagascar, where AMPS networks were inaugurated in 1994. Thus, Telecel was responsible for setting up the first cellular network in four of the eleven African countries where a private operator was responsible for introducing cellular phones before 1997. Where there was no cellular service, Telecel built networks from scratch; where there was cellular service, Telecel entered the local market by acquiring the assets of existing licensees. Its business model focused on offering postpaid cellular service to wealthy Africans mostly in urban environments.

In the late 1990s and early 2000s, the growth of 2G and prepaid cellular service allowed African carriers to expand the reach of service to rural areas and less affluent subscribers, which weakened Telecel's position in many African markets. Rwayitare and Gatt demerged their shares of the business in 1998. The Telecel networks were sold, and a good portion of the technical staff ended up working for Celtel, another Africa-based operator.

Celtel was the second African pioneer in cellular communications. In 1993 Mobile Systems International (MSI), a telecommunications consulting company, formed a joint venture with Vodafone to launch a cellular network in Uganda. Celtel was born years later as a spinoff from the venture. Mo Ibrahim, the MSI and Celtel founder, was a Sudanese who obtained a PhD in England and worked for Cellnet, the subsidiary of British Telecom that introduced cellular service in Britain. In 1989 Ibrahim and a group of engineers left Cellnet to start MSI, a company originally focused on developing software to design cellular networks. The software that MSI created, marketed as Planet, was used to plan many of the GSM networks set up around the world.

The postpaid cellular network MSI-Vodafone launched in Uganda had to deal with unpaid bills and debt-collection problems. MTN, another African operator, entered the Ugandan market in 1997 with a prepaid cellular model that allowed it to reach rural and less affluent subscribers and to displace MSI-Vodafone from its leadership position. In 1998 the company that would become Celtel was spun off from the joint venture under the name MSI-CI. (The name Celtel was not adopted until 2004.)

Having witnessed the success of prepaid cellular service in Uganda, MSI-CI (Celtel) expanded to other African countries using the prepaid business model. It grew by buying licenses in politically unstable countries that carriers from developed countries considered too risky. By 2004, Celtel was one of the largest cellular carriers by number of subscribers in sub-Saharan Africa. In 2005 Zain, a multinational carrier based in Kuwait, acquired the Celtel cellular networks.

The third pioneer, MTN, was founded in 1993 in South Africa, when the government decided to award two digital-cellular licenses. One of them went to Vodacom, a joint venture between Telkom, the PTT, and Vodafone. The other went to MTN, a partnership between CWIM, the Cable & Wireless international mobile subsidiary, and a few local companies. The founding team included Karel Pienaar, Robert Nisbet, and Buckley McGrath. Pienaar became chief technology officer of MTN, Nisbet chief financial officer, and McGrath chief of operations. Years later, Cable & Wireless sold its shares in the venture to concentrate on other markets.

Most of MTN's founding members came from Multichoice, a company that supplied satellite-television services to people of limited financial resources. MTN focused its business on delivering cellular services to those socioeconomic groups. When Vodacom and MTN launched their networks in South Africa in 1993, Vodacom rolled out a postpaid cellular platform in the wealthier regions of the country, while MTN targeted the low-income regions neglected by Vodacom. In 1996 MTN set up a prepaid cellular system. It also installed community payphones in schools and hospitals that connected to the MTN network, and it introduced an application that allowed credit sharing among cellular subscribers.

In 1998, after Cable & Wireless sold its stake in MTN, the local owners started expanding outside South Africa. By 2005 MTN had operations in eleven sub-Saharan countries. In that year, it attempted to acquire Celtel but was outmaneuvered by Zain, a Kuwait-based operator, which ended up acquiring the Celtel networks. MTN acquired the second-largest operator in Africa, Investcom, which had networks in ten countries. With the acquisition, MTN went from being an African operator to becoming an emerging-market leading carrier: after the transaction, the company had networks in South and East Africa, West and Central Africa, and North Africa and the Middle East. The largest MTN cellular markets were South Africa in South and East Africa, Nigeria and Ghana in West and Central Africa, and Syria, Yemen, and Sudan in North Africa and the Middle East.

THE EUROPEAN MEMBERS OF THE FIRST WAVE: VODAFONE, MILLICOM, AND FRANCE TÉLÉCOM

During the 1990s, Africa did not attract the kind of foreign investment in telecommunications infrastructure that flowed to other regions of the world. In that decade, many joint ventures and partnerships were formed involving North American and Western European companies, and many of those entities invested outside their home regions—but few of them invested in Africa. In 1996, while 32 percent of all telecommunications ventures invested in Eastern Europe and 26 percent did so in South America, only 2 percent invested in Africa. Some African countries were reluctant to allow international investment, and Africa was perceived as politically unstable.[41]

Three European operators had a modest presence in Africa in the early phase of cellular markets on that continent: Vodafone, Millicom International, and France Télécom. Vodafone was born in Britain in 1985 when the government licensed a private operator to compete with Cellnet, the cellular subsidiary of British Telecom. During the 1990s, Vodafone embarked on a process of international expansion, participating in partnerships, consortia, and joint ventures all over the world. In Africa, it formed a joint venture with Telkom, the PTT, to inaugurate a GSM network in South Africa in 1993, and another with MSI, the cellular consulting and software firm founded by Ibrahim, to set up a cellular network in Uganda. The Vodafone-Telkom venture, which operated under the name Vodacom, eventually launched cellular systems in other sub-Saharan countries, initially to the south of the equator (because of an agreement the venture had made with Vodafone).[42] Vodafone itself participated in cellular launches north of the equator.

Another European operator with a limited presence in sub-Saharan Africa in the early 1990s was the Luxembourg-based Millicom International, a company founded in 1990 when two cellular firms—Kinnevik from Sweden and Millicom Incorporated from the United States—merged their cellular assets in twelve countries.[43] Millicom International entered sub-Saharan Africa in 1992, when it set up the first cellular network in Ghana under the name Mobitel.[44]

The third European operator that invested in Africa in the early years of cellular markets on that continent was France Télécom, which underwent a privatization process starting in 1997.[45] Starting in the mid-1990s, France Télécom acquired portions of telecommunications entities in several African countries that had been French colonies until the 1960s.[46] In 1996 the company acquired a stake in the incumbent operator and its cellular subsidiary in the Ivory Coast, and the next year it did the same in Senegal. Further, France Télécom entered the Egyptian telecommunications market in 1998 by partnering with Orascom and Motorola to create the mobile operator MobiNil. By 2004 France Télécom had cellular networks in eleven countries, and by 2009 it had them in sixteen.[47]

NORTH AFRICA'S FORAYS INTO SUB-SAHARAN AFRICA IN THE EARLY 2000S: GLOBAL TELECOM HOLDING ACQUIRES TELECEL

In the early 2000s sub-Saharan Africa witnessed the irruption of a second wave of private operators, all of them headquartered in either North Africa or the Middle East. One of the most important was Orascom Telecom (later renamed Global Telecom Holding). Orascom was a conglomerate of companies founded in 1950 and headquartered in Egypt.[48] The conglomerate eventually comprised five core operating companies: Orascom Construction Industries, Orascom Projects and Touristic Development, Orascom Hotel Holding, Orascom Technology Systems, and Orascom Telecom.

Orascom Telecom was founded in Egypt in 1998. It obtained cellular licenses in North Africa and the Middle East, becoming an important mobile operator in the Arab world. Right before its initial public offering in the year 2000, it attempted to acquire Celtel, the African cellular pioneer. After the transaction fell through, Orascom Telecom acquired what it viewed as the next most attractive target—Telecel, the company founded in Zaire by Rwayitare and Gatt. Through this acquisition, Orascom became one of the top cellular players in sub-Saharan Africa.[49]

MIDDLE EASTERN EXCURSIONS INTO SUB-SAHARAN AFRICA IN THE MID-2000S: ZAIN ACQUIRES CELTEL, AND ETISALAT BUYS ATLANTIQUE TELECOM

In the mid-2000s sub-Saharan Africa witnessed an invasion of cellular carriers based in the Arabian Peninsula.[50] Zain, then called MTC, had pioneered the development of mobile communications in Kuwait, where it was based, and in the Persian Gulf region.[51] It was founded in 1983 as a joint venture between the government of Kuwait and private partners to supply cellular service to Kuwaiti citizens. It had a cellular monopoly in Kuwait until the year 2000, when the government issued a second license. After a competitor entered, MTC lost a large share of the Kuwaiti market. The government sold a large majority of its MTC shares to private owners in 2002.[52]

In 2005 MTC (Zain) entered sub-Saharan Africa by acquiring Celtel in a USD 3.4 billion transaction that was the largest Middle Eastern investment in Africa at the time. As explained earlier, Celtel had grown rapidly by purchasing cheap licenses in countries that large multinational carriers from OECD countries avoided.

When Zain purchased Celtel, it acquired cellular networks in multiple countries. The transaction drew the attention of large carriers from the Middle East and multinationals based in OECD countries. To prepare for the invasion of these powerful competitors, in September 2006 Zain launched the One Network project encompassing Kenya, Tanzania, Uganda, the Democratic Republic of Congo, the Republic of the Congo, and Gabon. The One Network service allowed customers to make calls and send text messages at local rates without roaming charges in any of the six countries. With One Network, Zain attempted to make it feasible for people with limited financial resources to have access to cellular service. Zain went from being a government-controlled monopoly in Kuwait with about six hundred thousand subscribers in 2002 to becoming the world's fastest-growing telecommunications company, with thirty-two million customers in twenty-two countries in the Middle East and Africa in 2008.

Etisalat was another African newcomer from the Middle East. The Etisalat telecommunications company was founded in 1976 by a British company and its local partners in the United Arab Emirates. In 1983 the company's ownership changed, with the United Arab Emirates government receiving 60 percent of the shares and the rest sold in a public auction. In 1991 the government passed legislation that made Etisalat a monopolist in the supply of fixedline and cellular services in the United Arab Emirates. In 1994 the company launched the first GSM network in the Middle East. A decade later, in 2004, Etisalat launched the first 3G network in the region.[53] In 2005 the company entered sub-Saharan Africa's cellular markets by acquiring a stake in Atlantique Telecom. At the time, Atlantique Telecom had majority shareholdings in cellular operators in six West African countries (Benin, Burkina Faso, Gabon, Niger, Togo, and the Ivory Coast).[54]

Table 11.3 summarizes the growth of the main cellular operators in Africa between 2003 and 2009. The carriers are ranked by the number of African subscribers they had at the end of 2008.

The table reveals several important facts. First, the operators at the top of the table (MTN, Zain, and Vodacom) were focused on Africa in that more than 70 percent of their total subscribers in late 2008 were there.

Table 11.3 Cellular operators in Africa, number of networks, by year, 2003–2009

	2003	2004	2005	2006	2007	2008	2009	African subscribers (m)	African subscribers as a percent of total
MTN	7	7	11	17	17	17	17	65.4	79.8
Zain	0	0	14	15	15	16	17	35.0	74.6
Vodacom	5	5	5	5	5	5	5	33.6	100.0
Vodafone	9	9	9	9	9	10	10	32.0	11.4
Orascom	8	7	5	5	5	6	7	23.4	37.6
France Télécom	11	11	11	11	14	15	16	18.1	18.4
Millicom	5	5	7	7	7	7	7	8.9	34.4
Etisalat	2	2	8	8	9	11	11	7.3	24.8
Portugal Telecom	5	5	5	5	5	5	5	3.8	15.0
Atlantique Telecom	6	6	6	0	0	0	0	0.0	0.0
Celtel	13	13	14	0	0	0	0	0.0	0.0
Investcom	5	4	5	6	0	0	0	0.0	0.0
Other	17	11	13	13	18	17	18	17.3	n/a

Source: P. Curwen and J. Whalley, *Mobile Telecommunications in a High-Speed World* (Farnham, UK: Gower, 2010), 201

Second, the operators at the top of the table were also the ones responsible for acquiring networks in sub-Saharan Africa in the 2000s, whereas those at the bottom were acquisition targets. The table reflects how MTN grew its networks from eleven in 2005 to seventeen in 2006 by acquiring Investcom, how Zain went from being absent in 2004 to owning fourteen networks in 2005 through its purchase of Celtel, one of the African pioneers, and how Etisalat went from two networks in 2004 to eight in 2005 via its acquisition of Atlantique Telecom. Put differently, the table captures the invasion of foreign operators in the mid-2000s and how they established a presence in sub-Saharan African markets by acquiring the cellular assets of several African pioneers.

INDIAN MULTINATIONALS IN SUB-SAHARAN AFRICA IN THE EARLY 2010S: BHARTI AIRTEL ACQUIRES ZAIN

Starting in 2010, Africa witnessed the incursion of a large cellular carrier from India: Bharti Airtel. The Bharti Group was founded in India in 1976. It started as a small manufacturer of bicycle parts and grew to become one of the largest business conglomerates in the country, with operations in financial services, retail trade, food, and telecommunications. Bharti Airtel was founded in 1995 under the name Bharti Tele-Ventures. In that year, Bharti Tele-Ventures launched New Delhi's first GSM network under the Airtel brand.[55] The company went public in 2002, and by 2003 it had the largest General Packet Radio Services network in India. (General Packet Radio Services was an evolution of the 2G GSM standard.) In 2006 Airtel became the first Indian carrier to launch 3G services when it established a 3G network in the Seychelles, an archipelago of more than a hundred islands in the Indian Ocean off the coast of East Africa. By 2010 Bharti Airtel was the leading cellular-service provider in India, with a market share of about 21 percent.[56]

In 2008 Bharti Airtel attempted to acquire the South African cellular pioneer MTN, but the transaction fell through. In its second attempt to enter Africa, Bharti acquired the Zain cellular networks in fifteen countries in 2010 for USD 10.7 billion. It was Bharti's largest acquisition at the time, and the largest ever cross-border transaction from one emerging market to another.[57] Based in Kuwait, Zain had entered Africa by acquiring the Celtel networks in 2005. Although Zain had been the market leader in ten of the fifteen countries where it had operations, its African cellular networks had not been consistently profitable.

Bharti's interest in Africa was driven by a combination of factors: a rapidly growing population that was becoming increasingly urbanized and a relatively low cellular density. Further, because personal-computer penetration was low and demand for Internet services was high, Bharti anticipated that demand for 3G cellular services, including mobile data services, would be consistently high in African markets.[58] It also believed it could make the former Zain networks operate profitably by cutting costs and changing the corporate culture.

PRIVATE OPERATORS IN AFRICA: TAKING STOCK

Private carriers entered sub-Saharan cellular markets in waves. The first wave comprised three African pioneers (Telecel, Celtel, and MTN) and three European multinationals with a modest presence in Africa (Vodafone, Millicom, and France Télécom). Later waves involved carriers from North Africa such as Orascom, which bought the Telecel networks; operators from the Arabian Peninsula such as Zain and Etisalat, which acquired Celtel and Atlantique Telecom, respectively; and later the large Indian carrier Bharti Airtel, which bought the Zain cellular assets in Africa.

By 2015 the five largest operators in Africa were MTN, Bharti Airtel, Vodacom, France Télécom (rebranded Orange), and Etisalat. MTN was the only African pioneer that had survived the invasion of foreign multinationals. The other leading firms reflected either the early European presence in African markets (Vodacom and France Télécom/Orange) or that of the more recently arrived multinational carriers from the Arabian Peninsula (Etisalat) and India (Bharti).[59]

CELLULAR INFRASTRUCTURE IN AFRICA

Between 1998 and 2008, about USD 50 billion was invested in telecommunications infrastructure in sub-Saharan Africa.[60] Cellular networks accounted for most of this investment. Of the USD 50 billion, about USD 37 billion went to greenfield mobile investments, which were those in which investors obtained new cellular licenses from African governments or acquired existing cellular licenses from their owners and then expanded the existing networks.[61]

The sponsors of investments in cellular infrastructure in Africa were the shareholders that had controlling stakes in cellular networks. As explained earlier, several of them, including Telecel, Celtel, and MTN, were from sub-Saharan Africa; others, such as Orascom (Global Telecom Holding), were headquartered in North Africa; yet others, including Zain and Etisalat, came from the Arabian Peninsula; and in 2010, a cellular carrier from India, Bharti Airtel, made an incursion in Africa. Between 1998 and 2008, 50 percent of the sponsors were from sub-Saharan Africa, 25 percent from North Africa and the Middle East, and the remaining 25 percent from the rest of the world (especially Western Europe, represented by Vodafone, Millicom, and France Télécom). The three most important transactions were the acquisition of Celtel by MTC (renamed Zain) in 2005, MTN's purchase of Investcom in 2006, and Bharti Airtel's acquisition of Zain in 2010.

Cellular-infrastructure financing came from private and government sources often headquartered in countries or regions that were different from those of the sponsors themselves. Between 1998 and 2008, most of the investment in African cellular networks was financed by private (as opposed to government) sources. Private sources of financing included bank loans, equities, and corporate bonds. In all three, organizations headquartered in South Africa played an important role—major South African banks that arranged syndicated loans for telecommunications infrastructure, the Johannesburg Securities Exchange, and the Bond Exchange of South Africa.[62]

Government financing of African cellular infrastructure has been the subject of abundant academic and news coverage because of the increasingly important role played by China, its telecommunications equipment manufacturers, and its state-owned financial institutions. Economists at the World Bank compiled a list of telecommunications-infrastructure projects in Africa that were financed by Chinese financial entities between 2001 and 2007.[63] They identified thirty-three such projects covering fixed-line networks, mobile systems, and backbone infrastructure in twenty-one African countries. Three Chinese infrastructure suppliers were involved in these projects, and the total financed over the relevant period was about USD 2.7 billion.

In those projects, Huawei, a private company; ZTE, a state-owned company; and Alcatel-Shanghai Bell, a joint venture involving France and China with state and private participation, supplied cellular and fiber-optic networks to both public and private telecommunications operators in Africa. What is special about these transactions is that they were mostly financed with funds originating in Eximbank and the China Development Bank, two financial institutions owned by the Chinese state. These investments, and others that occurred after 2007, have generated rampant speculation on the ultimate geopolitical and economic goals of the Chinese state on the African continent.[64]

CELLULAR PHONES IN AFRICA: TAKING STOCK

Cellular phones arrived late to the African continent. Most African countries set up cellular systems for the first time during the 1990s. Only a few countries in North Africa and two countries in sub-Saharan Africa—South Africa and Zaire—managed to launch cellular networks during the 1980s.

Cellular carriers in Africa had to adapt their business models to the socioeconomic features of the continent. Early on, cellular operators relied on postpaid cellular platforms: people used their phones first and paid later. This model, however, created problems: only wealthy Africans had access to mobile phones, and, despite this fact, cellular bills often went unpaid (or were paid much later than they were due).

The key business-model innovation was prepaid cellular services, introduced in Africa in the late 1990s. Minutes of cellular use were bought and sold just like fast-moving consumer goods: Africans bought low-denomination scratch cards several times during the month. This enabled people of limited financial resources to gain access to cellular phones.

Cellular phones changed the everyday lives of Africans. Before their arrival, the costs of collecting and transmitting information were high. Cellular phones changed this radically.

Mobile money was a cellular-based innovation that had a profound impact on people's lives. Systems such as M-PESA transformed the way people in Africa approached money. With M-PESA, Africans for the first time had a tool that allowed them to save, to pay bills and other debts, and to send money to relatives in a way that was safer and more efficient than using cash. Mobile phones also contributed to fueling political unrest in countries such as Tunisia and Egypt, especially during periods of economic decline. The Arab Spring has been labeled "the first smartphone revolution."[65]

Sub-Saharan Africa witnessed the rise of indigenous operators—including Telecel, Celtel, and MTN—that pioneered cellular services on the continent. Over time, carriers from North Africa, the Arabian Peninsula,

11.4 People receiving SMS health advice in South Africa. Over 475,000 South Africans receive SMS health advice on their mobile phones, resulting from the US President's Emergency Plan for AIDS Relief and the United States Agency for International Development. Photo from 2015. *Source*: Wikimedia.

and India made forays in sub-Saharan Africa and became important cellular players on the continent. In recent years, the top cellular carriers in Africa have included the indigenous pioneer MTN, some of the European carriers that made early investments in Africa (such as Vodafone and France Télécom/Orange), and some of the newly arrived operators (such as Bharti Airtel). In addition, the presence of Chinese manufacturers of telecommunications equipment, such as Huawei and ZTE, has given rise to speculation about the objectives of the Chinese state in Africa, especially because state-owned financial institutions have financed many investments in telecommunications infrastructure on the continent.

12

COMPETING IN 4G CELLULAR MARKETS

12.1 *Mobile Lovers*, by Banksy. The street-art image appeared on the door of a youth club in Bristol, England, in April 2014. It was removed for safekeeping, displayed at the Bristol Museum and Art Gallery, and later sold to a private buyer. The proceeds were used to support the youth club. Courtesy Sally Ilett.

The share of smartphones in all cellular devices rose at a fast pace in the transition from the 2000s to the 2010s. The growth of smartphones accelerated a trend that had appeared for the first time in the 3G era: cellular competition involved not just network operators such as AT&T and Verizon in the United States, or Vodafone and Telefónica in Europe, or device makers such as Nokia and Ericsson, but also creators and sponsors of smartphone operating systems such as Apple and Google. Smartphone operating systems quickly became platforms in which users and app developers participated. In a classic case of indirect network effects, creators and sponsors of mobile operating systems attracted consumers in part by attracting app developers and attracted app developers in part by attracting consumers.

A complex mobile ecosystem developed that included device makers, infrastructure manufacturers, network carriers, creators and sponsors of mobile operating systems, app developers, creators of mobile ads, and mobile-semiconductor designers and manufacturers. These groups collaborated with one another to support the mobile ecosystem. At the same time, members of these groups competed with one another in specific markets. For example, as a cellular-phone manufacturer, Samsung collaborated with Google, the sponsor of the smartphone operating system, and with Verizon, the network operator, while competing with other phone makers such as Nokia, Huawei, and Apple. In turn, designers and sponsors of mobile operating systems, such as Apple and Google, competed with one another by offering attractive bundles of features to consumers and by attracting app developers, while at the same time cooperating with app developers, network operators, and, in the case of Google, device makers.

While Apple and Google were radically transforming the smartphone operating-system market, several Chinese firms and a Taiwanese company were becoming important contenders in other cellular markets. Huawei and ZTE displaced many of the traditional leading firms in cellular infrastructure, Huawei and Xiaomi rose to the top of the cellular-phone market, and MediaTek and Spreadtrum started challenging Qualcomm in the baseband-processor market. All these companies were based in mainland China except for MediaTek, a Taiwanese firm.

COMPETITION AMONG SMARTPHONE OPERATING SYSTEMS

Until the late 2000s, competition among smartphone operating systems involved Symbian, Research in Motion (RIM), Palm, and Microsoft. In 2007 and 2008 two new players came to the forefront: Apple (with the iPhone and its iPhone OS) and Google (with Android).

BUSINESS MODELS

By the early 2010s, the worldwide market for smartphone operating systems had several important players. Two of them were rising (Android and iOS), while the rest (Symbian, BlackBerry OS, Palm OS, and Windows Mobile) were in decline. The operating systems' creators and sponsors chose a variety of business models to promote their products.[1]

In the first business model, the mobile operating system was closed source, and its owner was a corporation that also functioned as the handset maker. Put differently, the corporation that owned the proprietary operating system sold an integrated bundle of software and hardware. This was the approach adopted by RIM for BlackBerry OS, Palm for Palm OS, and Apple for iOS.

The second model—adopted by Microsoft for Windows Mobile—was one in which the mobile operating system was closed source, and its owner was a corporation that licensed it to handset makers for a price. The operating system's owner did not supply the hardware and was thus forced to interact with another set of partners (or customers), the handset makers.

The third model—implemented by Symbian—was one in which the mobile operating system was closed source, and its owner was a consortium or joint venture of handset makers that licensed it for a price. This model was somewhere in between the first and the second in that the owners of the proprietary operating system also supplied the hardware, but at the same time the system was licensed to handset makers other than its owners.

The fourth model—adopted by Google for Android—was one in which the mobile operating system was open source and was licensed to handset makers for free. Its evolution was dictated by an alliance of companies that included its main sponsor—Google—and a variety of partners, including handset makers, application developers, network operators, and even chip makers.

The fifth, and final, model was the one adopted by the Symbian Foundation. In this case, the operating system was open source and its development was fostered and supervised by a nonprofit organization whose founding members included handset makers, chip makers, and at least one network operator.

The source of revenues distinguished business models from one another. Apple and RIM generated revenues by selling a bundle of hardware and software. Microsoft monetized its operating system through licensing, and so did Symbian. Google received revenues neither from a hardware-software bundle nor from licensing the operating system but rather from ads shown on the platform.

Each business model had its own trade-offs. On one extreme was the Apple model: a closed-source mobile operating system whose owner was also the hardware maker. This approach provided the owner the maximum degree of control—control over the evolution of the operating system and over all aspects of the user experience. On the opposite extreme was the Google model: an open-source mobile operating system whose sponsor licensed it for free to the handset makers. This approach limited the control the sponsor had over the evolution of the operating system and over the user experience.

The Google model had one fundamental weakness: it encouraged fragmentation. New versions of the mobile operating system were released often, sometimes just a few months apart from one another, and handset makers tended to modify the interface to achieve some degree of product differentiation. These two facts, combined, led to a world in which many flavors of Android coexisted, which sometimes led to consumer (and developer) confusion.[2] Google counteracted the tendency toward fragmentation by keeping tight control over the evolution of the operating system itself.[3]

The Android model led to widespread adoption. The handset makers that adopted it as their operating system invested in promoting it. In addition, Android's fragmentation was overcome, to some extent, by the emergence of a clearly dominant handset partner: by mid-2012, Samsung reportedly accounted for about 40 percent of all Android devices in the market.

THE TAKEOFF

Every smartphone operating system had to solve the chicken-and-egg problem—it had to find a strategy to attract consumers, app developers, and (sometimes) handset manufacturers to grow at the starting point. Apple had one feature in common with RIM: both implemented a business model in which the owner of the operating system was also the handset maker. Thus, neither faced the need to attract hardware partners to the mobile operating system at the time of launch.

Apple and RIM managed to take off by selling an integrated bundle of hardware and software that, at the starting point, was particularly effective at attracting consumers. BlackBerry supplied one feature—seamless access to corporate email—that attracted users, especially business users. The original iPhone combined a superior (touch-based) interface with a mature Internet experience on a mobile device.[4] Both RIM and Apple launched their devices with a focus on getting consumers on board.

When Apple introduced the iPhone in mid-2007, it did not try to recruit as many network operators as possible to carry the phone: it relied exclusively on one network carrier—Cingular, wholly owned by AT&T at the time.[5] (Apple obtained significant concessions from AT&T in exchange for the exclusivity it granted AT&T, including a ten-dollar monthly payment from AT&T for each AT&T customer who carried an iPhone.)[6] Further, no third-party applications were present on the iPhone until the introduction of the iPhone 3G in 2008.[7] Apple's lift-off strategy was successful. Apple shipped almost six million devices worldwide even before opening the App Store in mid-2008. Shipments declined in the second quarter of 2008, right before the introduction of the iPhone 3G, but skyrocketed in the third quarter of 2008, when Apple shipped almost five million units worldwide.

Google's takeoff strategy was different. Rather than attracting consumers directly, the company focused on attracting handset makers with an operating system licensed to them for free, and on attracting third-party developers with an open-source operating system for which it was relatively easy to develop applications.

ATTRACTING USERS WITH FEATURES AND APPS

Although at the starting point Apple and Google resorted to different strategies to gain acceptance among consumers for their operating systems and devices, they both had to rely on constant innovation to retain users and gain new ones. Because Apple did not have hardware partners, the company was forced to innovate constantly in both software and hardware. Google, by contrast, left the hardware innovation mostly to the device manufacturers and focused on improving the Android operating system over time.

Starting in 2007, Apple introduced a new version of the iPhone every year. Each new version of the device had new software features, including a new version of the operating system. Apple, however, did not modify the design of the device every year, but rather every other year. The original iPhone was unique, but successive generations included catch-up features as well, which were responses to innovative features introduced previously by Google and its hardware partners. Table 12.1 presents the iPhone versions between 2007 and 2012 with their new software and hardware features.[8]

Creators and sponsors of smartphone operating systems competed not only by incorporating new features every year but also by offering an increasingly diverse stock of apps to consumers. In late 2011, for example, the selection of apps available was the third most important factor driving a smartphone purchase in the US market.[9] In the first half of 2012, 62 percent of smartphone owners in the United States used apps on their smartphones.[10] Among smartphone owners who used apps, 85 percent used social networking apps, 65 percent used apps to watch videos and movies, and 84 percent used maps, navigation, and search apps.

Apple introduced an app store—the App Store—in July 2008, featuring applications for both the iPhone and the iPod Touch music player. By early September 2008, two months after launch, over three thousand apps were available on the App Store, and users had already completed more than one hundred million app downloads.[11]

Google's app store, Android Market, was announced in August 2008 but was only made available to users in October of that year. Support for paid apps on Android Market was introduced in February 2009. By early 2009, third-party apps had become "the future of the smartphone business," as one analyst put it, and all the major smartphone manufacturers and mobile operating system owners—including BlackBerry and Microsoft—were either considering or already involved in launching an app store

Table 12.1 Innovation in hardware and operating-system software: The iPhone between 2007 and 2012

Device	Price	OS	Design	New features	Catch-up features
iPhone (June 2007)	$499 or $599	OS X	Largest and highest-resolution screen on the market	Touch-screen interface, outstanding Internet browser, full-blown iPod player	
iPhone 3G (July 2008)	$199 or $299	OS 2.0	No major changes	Online store for third-party apps	Access to 3G network
iPhone 3GS (July 2009)	$199 or $299	OS 3.0	No major changes	More speed, more memory, more battery life, video recording, better camera	Copy and paste, system-wide search
iPhone 4 (June 2010)	$199 or $299	iOS 4	Thinner handset, sharper screen	Faster processor, better battery life, better rear camera, second camera	Limited multitasking, folder for apps, unified inbox
iPhone 4S (October 2011)	$199, $299, or $399	iOS 5	No major changes	New camera, 4G download speeds, Siri, available on multiple networks (carriers)	
iPhone 5 (September 2012)	$199, $299, or $399	iOS 6	Lighter and thinner, larger screen with six rows of icons		One of the last smartphones to access LTE networks

Sources: W. Mossberg's reviews (see details in endnotes)

to compete with Apple's.[12] BlackBerry inaugurated its app store in April 2009, Nokia in May 2009, Samsung in September 2009, and Microsoft in October 2010.

Table 12.2 tracks the number of apps available on, and the number of app downloads completed from, Apple's App Store and Google's Android Market between 2008 and early 2013. By September 2012, about 700,000 apps were available on the App Store and users had made roughly thirty-five billion app downloads.[13] But Android Market (renamed Google Play in 2012) was catching up: it had 675,000 apps available and twenty-five billion app downloads.[14]

With their unique features and apps, Apple and Google radically transformed the global smartphone operating system market. Figure 12.2 shows the shares of the main players in this market in units sold between 2008 and 2017.

The figure shows the rapid rise of the iPhone and the Android phones, and the collapse of the incumbents. At first, the decline of companies such as BlackBerry was relative rather than absolute—their sales were rising while their shares were declining. At some point, however, their decline became both relative and absolute—BlackBerry's sales, for example, fell considerably between 2011 and 2012. By 2017 Android was the leading mobile operating system in the world, and Apple's iOS was its only competitor.

CELLULAR INFRASTRUCTURE: FROM CHINA TO THE WORLD

While Apple and Google were radically transforming the smartphone market, several important changes were happening in the cellular infrastructure, devices, and semiconductor markets. Many of those changes were associated with the rise of Chinese and Taiwanese competitors in businesses traditionally dominated by American and European companies.

Chinese and Taiwanese firms in the cellular infrastructure, device, and semiconductor markets frequently started by supplying products to customers in the Chinese domestic market, one of the largest cellular markets in the world. Once they established a footing in China, they began making incursions into other areas of the world, and especially into developing countries.

Table 12.2 Apps available on, and app downloads from, the App Store and Google Play, 2008–2012

	App Store		**Google Play**	
	Apps available	**Downloads to date (M)**	**Apps available**	**Downloads to date (M)**
July 2008	800	10		
January 2009	15,000	500		
March 2009	25,000	800	2,300	
June 2009	50,000	1,000		
July 2009	65,000	1,500		
November 2009	100,000	2,000		
December 2009			16,000	
March 2010	150,000	3,000	30,000	
April 2010	200,000	4,500	38,000	
August 2010			80,000	1,000
September 2010	250,000	6,500		
October 2010	300,000	7,000	100,000	
January 2011	350,000	10,000		
June 2011	425,000	14,000		
July 2011	425,000	15,000	250,000	6,000
October 2011	500,000	18,000	319,000	
December 2011			380,297	10,000
February 2012	500,000	24,000	450,000	
March 2012	550,000	25,000		
June 2012	650,000	30,000	600,000	20,000
September 2012	700,000	35,000	675,000	25,000

Sources: Multiple articles in the general and trade press.

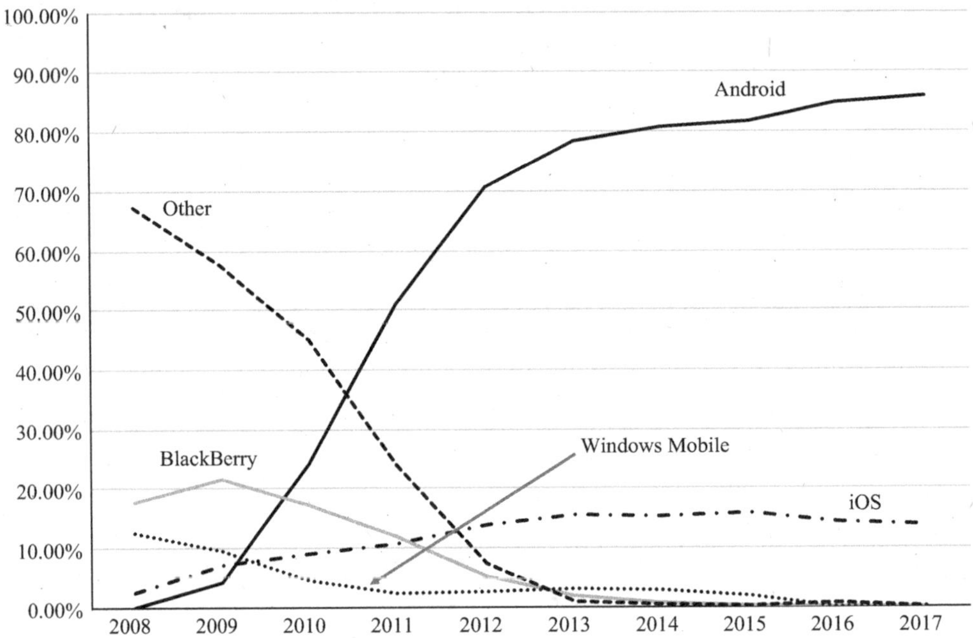

12.2 Global shares of smartphone operating systems, in units sold, 2008–2017. *Sources*: Gartner press releases.

Chinese companies have played a leading role not only in product and service markets but also in research and development (R&D). According to some studies, companies such as Huawei and ZTE are among the main owners of standard-essential patents pertaining to the 5G standard.[15] The possibility that China may be among the world leaders in 5G-related intellectual property has exacerbated geopolitical tensions between China and the United States.[16]

ENTREPRENEURSHIP WAVES IN CHINA

In the last four decades, China has witnessed four waves of entrepreneurship, and several of these waves have given rise to companies that have played an important role in the cellular industry.[17] The 1980s witnessed the first wave. Many of the first-wave entrepreneurs had no formal education beyond high school and little, if any, experience in business when they started. Two important cellular companies were launched in the

12.3 A child taking a selfie with Prime Minister Modi in Naya Raipur, India, in 2016. *Selfie* became the *Oxford Dictionary*'s word of the year in 2013. The editors reported that its frequency in the English language had increased by 17,000 percent during the year. *Source*: Wikimedia.

1980s: Legend (later rebaptized Lenovo) in 1984 and Huawei in 1988. The second wave came a few years later when, after Deng Xiaoping completed his southern tour of the country in 1992 to relaunch his economic reforms, individuals left jobs in government and academia to start new businesses. The third wave, between the late 1990s and the mid-2000s, was associated with China's entry into the World Trade Organization in 2001 and the process leading to it: companies founded in this period include Tencent in 1998, Alibaba in 1999, and Baidu in 2000. The three of them would eventually rank among the largest Internet companies in the world. The fourth wave started in the late 2000s. Companies belonging to this wave are using the mobile Internet to reach the massive Chinese consumer market. An important member of the fourth wave is Xiaomi, founded in 2011, which at the time of writing ranks among the largest global smartphone vendors.

HUAWEI

In the second half of the 2000s, as the market for smartphone operating systems was being transformed by the arrival of the iPhone and the Android devices, the cellular-infrastructure industry underwent its own transformation: an industry that had been dominated for decades by the likes of Ericsson, Siemens, and Alcatel saw the rise of Chinese competitors Huawei and ZTE.[18] Telecommunications equipment was one of the first industries in which Chinese firms expanded beyond the boundaries of their own domestic market. Such expansion started in the late 1990s, and by the early 2010s firms such as Huawei and ZTE accounted for more than 30 percent of the global market. The first Chinese company to capture portions of foreign markets was Huawei, with ZTE lagging in timing, technology, perceived quality, and market share.

Established in Shenzhen in 1988, in its early years Huawei sold fire alarms, supplied consulting services to machinery producers, and acted as a distributor of the HAX digital telecommunications switch for landlines produced by a Hong Kong manufacturer. Enticed by the profitability of foreign switch makers selling in the Chinese market, it decided to explore switching markets. It started by developing a small switching system through reverse engineering, but it soon found that many other domestic manufacturers were doing the same.

To differentiate itself from the rest, Huawei tackled a more challenging product—a large-scale switch system. At the time, multinational telecommunications firms were setting up joint ventures and wholly owned subsidiaries in China to take advantage of the rising demand for telecommunications equipment in the Chinese market, but none was willing to transfer the large-switch technology to Chinese firms. Huawei was thus forced to conduct in-house R&D to develop the product. It managed to recruit researchers from domestic universities and research institutes, and by taking risky loans at high interest rates, it funded the development of its own large-scale switch, the C&C08, which was ready by 1993. This product enabled Huawei to become the leading telecommunications-equipment firm in the domestic market. In the early days, Huawei's customers were in small cities and rural markets with less need of cutting-edge technology.

Relying mostly on in-house R&D, Huawei reverse engineered more advanced technologies, including data-networking equipment. In 1995 the company established research institutes in Shanghai and Beijing to focus on mobile and data communications. By 1998, when cellular communications took center stage in China, Huawei already had the capability to supply Global System for Mobile Communications (GSM) cellular equipment. By the year 2000, when the demand for voice over Internet Protocol and broadband access to the Internet started rising in the country, Huawei was able to supply the relevant data-networking equipment.

ZTE

ZTE was founded in 1985 in Shenzhen as a semiconductor company under the Ministry of Aerospace, and it initially produced watches, electric keyboards, and telephones. Encouraged by the success of foreign switch vendors in the Chinese market, the company focused all its financial resources on developing switch technology by imitation. Like Huawei, ZTE started by producing a small-scale telephone switch for China's rural markets. In the early 1990s it managed to produce more advanced telephone switches and eventually developed the ZXJ10, one of the key domestic digital-switching systems. Sales of the ZXJ10 fueled ZTE's growth through the mid-1990s.

ZTE used the revenues obtained from switch sales to diversify into other segments of the telecommunications-equipment business, including transmission and video-conferencing equipment. Between the mid- and late 1990s, the company set up nine R&D labs, including one in Shanghai for GSM systems and terminals, and three in the United States and one in Seoul for code division multiple access (CDMA) technologies. By the time the switching-equipment market became saturated in the late 1990s, ZTE had already developed substantial capabilities in other segments of the telecommunications-equipment industry.

Like Huawei, ZTE went from imitation to innovation. By mid-2007 it had been awarded 871 mobile-phone-related patents. At the time, it had seven R&D units in China and eight in the rest of the world. As of the early 2010s, both Huawei and ZTE were conducting research on various cellular technologies, including 3G technologies such as cdma2000, wideband CDMA (WCDMA), and time division–synchronous code division multiple access (TD-SCDMA). They differed in the way they funded their innovation: whereas Huawei used large portions of its own revenue to support R&D investments, ZTE raised financial capital through the stock market. In 1997 ZTE became the first domestic telecommunications-equipment firm to be listed in China's stock markets.

INTERNATIONAL EXPANSION OF HUAWEI AND ZTE

It would have been difficult for Huawei and ZTE to start their incursions into foreign markets by displacing the established players in Europe and North America. Thus, they began their international expansion by focusing on other emerging markets. In 1998 Huawei built a mobile network in Kenya and ZTE won a large project in Pakistan. Huawei's international sales grew rapidly during the 2000s—between 2003 and 2007, they rose from 27 percent to 72 percent of total company sales. By 2006 Huawei had contracts in forty African countries that, combined, exceeded USD 2 billion. By 2010 it had overtaken Nokia Siemens Networks to become the second-largest telecommunications-equipment vendor in the world behind Ericsson.

ZTE started its international expansion in the late 1990s. It began by targeting emerging markets in Asia and Africa, and by the mid-2000s it had become one of the main telecommunications-equipment suppliers in Russia and India. By 2006 it was also one of the top vendors in the global CDMA-equipment markets. In the late 2000s it penetrated mature markets in Western Europe and North America.

Huawei and ZTE competed aggressively on price. ZTE, for example, undercut competitors by as much as 50 percent when bidding for projects in the Chinese market. Their global growth strategies shared several ingredients. Both companies started by offering products of reasonable quality for the low-end Chinese market, often reverse engineered from those of leading international competitors. They then expanded into other emerging markets with products adapted to the local needs and budgets, which were often similar to China's. Only afterward did they make forays into mature markets, mainly by supplying products to low-cost operators in those regions. While expanding in foreign markets, they improved their products and the quality of their services through R&D, joint ventures, and acquisitions. In the final stage they attempted to compete directly with leading telecommunications-equipment vendors in all markets.

In the second half of the 2000s, the international growth of Huawei and ZTE changed the competitive dynamics of the telecommunications-equipment industry, including the cellular-infrastructure sector.[19] The rise of the Chinese competitors, combined with the global financial crisis of 2008 and technology shifts in the cellular world, brought about substantial changes in the industry. In 2004, the year when Huawei won its first major contract in Europe, there were seven large telecommunications-equipment vendors in the world: Ericsson, Nokia, Siemens, Alcatel, Lucent, Nortel, and Motorola. By the early 2010s only Ericsson had retained its original corporate structure. The price pressure exerted by the Chinese competitors forced companies to cut costs, which they did in part by combining their businesses. Alcatel and Lucent Technologies merged in November 2006, and soon thereafter Nokia and Siemens combined their network-equipment businesses into Nokia Siemens Networks.[20] Two other important vendors exited the industry altogether: Nortel filed for bankruptcy in 2009, and Motorola sold its network business in 2011.

Aggressive pricing by Huawei and ZTE led to substantial price declines starting in the mid-2000s. Between 2005 and 2011, the price of telecommunications equipment fell by about 10 percent per year. Through steady R&D investments, Huawei and ZTE were eventually able to develop products that were competitive at the high end of the market, which in turn spread the price pressure to those products as well.

By the mid-2010s Huawei and ZTE had become diversified telecommunications conglomerates that had transformed the world telecommunications-equipment market, and especially the cellular-infrastructure market. By 2015 Huawei had three business segments, organized on the basis of who the final customers were: the carrier segment accounted for 59 percent of the company's revenues and included cellular and fixed networks; the enterprise segment generated 7 percent of revenues and served industries such as finance, transportation, and energy; and the consumer segment accounted for 33 percent of revenues and included smartphones and smart wearables more generally. More than 50 percent of Huawei's revenues came from countries other than China.[21]

By 2018 Huawei was the leader in the global cellular-infrastructure market comprising 2G, 3G, and 4G equipment. The company had a 31 percent share of the market and was followed by Ericsson (27 percent), Nokia (22 percent), and ZTE (11 percent).[22]

THE RISE OF CHINESE COMPETITORS IN CELLULAR PHONES

Until 2008 the Chinese cellular-phone market was dominated by *shanzhai* handsets—cheap devices manufactured by domestic (and often non-branded) producers.[23] (Shanzhai originally means mountain fortress, but later came to mean counterfeit or imitation.) The growth of shanzhai handsets in China was the result of indigenous innovation stimulated by local demand, especially from those segments of the market that were most sensitive to price.

The market changed in the late 2000s, when users started migrating toward 3G systems. Chinese brands such as Xiaomi, Lenovo, Coolpad, Huawei, and ZTE outperformed not only the domestic shanzhai vendors but also international rivals such as Apple, Samsung, and Nokia.

THE 2G ERA IN CHINA AND THE GROWTH OF SHANZHAI MOBILE-PHONE VENDORS

Shanzhai devices were low-cost phones that outperformed local and international brands by taking advantage of MediaTek's turnkey semiconductor offerings, which incorporated operating-system and application software on a single chip. (MediaTek is a fabless semiconductor company founded in Taiwan in 1997 and discussed later in this chapter.) Relying on MediaTek's chips, shanzhai manufacturers were able to focus on innovation in mobile-phone design and on supplying a large variety of models. Some of the shanzhai phones had multiple SIM slots to accommodate different standards, a feature that was particularly useful in China, where each of the major cellular carriers operated on a different 3G standard. A few of the shanzhai manufacturers, such as Tianyu, were able to develop a strong (feature-phone) brand before the arrival of smartphones in the late 2000s.

Shanzhai cellular-phone makers clustered in or around the Shenzhen metropolitan area.[24] Shenzhen, a city located in the Pearl River Delta close to Hong Kong, experienced rapid economic growth starting around 1978, when the "reform and opening policy" was introduced.

Over time, electronics manufacturing migrated to Shenzhen, and by the late 2000s this region had become the largest manufacturing center for cellular phones in the world. In 2008, 180 million devices were exported from Shenzhen with an average price of around USD 68. At the time, an estimated ten thousand firms were involved in the production, distribution, and sale of mobile phones, components, or related services in the Shenzhen area.

After the Chinese government liberalized license regulations on the manufacturing and sale of handsets in 2007, sales of shanzhai cellular phones grew at a fast pace and captured a share of about 40 percent of the Chinese market, with even larger shares in the low end of the market. The golden age of the shanzhai manufacturers lasted until around 2010, when users started transitioning to 3G services and smartphones.

THE 3G ERA AND THE RISE OF CHINESE SMARTPHONE BRANDS

China launched 3G cellular services in 2009, when the domestic standard TD-SCDMA became ready for deployment. In the transition to 3G, China became a country with three standards: China Mobile adopted TD-SCDMA; China Telecom, cdma2000; and China Unicom, WCDMA.[25] By the end of 2012, about 233 million users (out of a total of 1.11 billion) had 3G cellular phones in China. Smartphone penetration in China was made possible by the advent of low-cost smartphones for the mass market, especially those running on the Android operating system. As of 2012, 25 percent of smartphones sold in the Chinese market were entry-level devices priced between CNY 1,000 and CNY 2,000 (roughly between

USD 150 and USD 300). In 2013 about 86 percent of smartphones produced by Chinese firms were Android based.

Although the key vendors in the global smartphone market—Samsung and Apple, especially—were present in the Chinese market in the early 2010s, they did not play as prevalent a role as they did in other markets. By 2012 Samsung and Apple were dominant brands in the world smartphone market, with about 40 percent and 25 percent of the market, respectively. By contrast, Samsung had just 14 percent of the Chinese market and Apple only 8 percent. In China, domestic firms such as Lenovo, Coolpad, ZTE, and Huawei had no less than 10 percent of the market each. Other companies, most of them domestic vendors, had, in combination, another 35 percent of the market.[26] Among these vendors were many that had started as shanzhai, nonbranded manufacturers but had managed to develop a degree of brand recognition, including K-Touch, Gionee, Meizu, Tianyu, Oppo, and Bubugao. In addition, some of the Chinese Internet giants, such as Baidu and Alibaba, entered the smartphone market with customized devices.

Among the Chinese smartphone manufacturers with a strong brand in the early 2010s were Lenovo, Huawei, ZTE, and Xiaomi. Lenovo had been an important supplier of 2G cellular phones and personal computers. Huawei and ZTE had a long history of supplying low-cost cellular equipment in China and other emerging markets. Xiaomi was a telecommunications newcomer. Founded in 2010 by Lei Jun, an entrepreneur well known in China, Xiaomi offered Android-based smartphones that had the look and feel of the iPhone. By 2014 Xiaomi was among the top smartphone vendors in the Chinese and global markets.

XIAOMI

Xiaomi was founded in 2010 by entrepreneur Lei Jun, who has often been described in the business press as the Chinese Steve Jobs.[27] The company quickly developed a reputation in the Chinese market as the top maker of low-cost smartphones. In the mid-2010s it started expanding in foreign markets, including Hong Kong, Singapore, Taiwan, the Philippines, and India.

Lei graduated from Wuhan University in the late 1980s with a degree in computer science. He moved to Shanghai and joined Kingsoft, the developer of what at the time was the top word-processing program in China, eventually becoming the company's CEO in 1998. While leading Kingsoft to a listing in the Hong Kong Stock Exchange, Lei founded several online startups, including an online video-sharing platform that listed on Nasdaq in the early 2010s. This background helped Lei find backers for Xiaomi, which he founded with a former Google engineer and five other partners.

Xiaomi followed the Apple model of outsourcing device production to contract manufacturers such as Taiwan's Foxconn. Whereas Apple sold smartphones at a premium, however, Xiaomi sold them barely above cost and generated most of its revenues from sales of software, services, advertising, and accessories. Xiaomi's founder has described the company's business model as closer to Amazon's than to Apple's.

Xiaomi adopted a novel marketing strategy: it sold most of its phones online, either through its own website or through Sina Weibo, one of the largest social-media platforms in China. Initially, to keep inventory and costs down, it sold handsets in weekly batches ranging from a few thousand to a few hundred thousand units.[28] Every week, it updated the user interface, labeled MIUI, of its Android phones based on users' suggestions. The company's success became evident in mid-2013, when it announced its Red Rice smartphone, designed to compete with Apple's iPhone 5C. Priced at $130, less than a quarter of the 5C's price, Xiaomi's Red Rice device was preordered by seven million users.

Figure 12.4 shows the evolution of the global smartphone market between 2012 and 2018. The figure shows that, by the end of the period, Xiaomi had become one of the top smartphone vendors in the world, trailing only Samsung, Apple, and Huawei.

THE RISE OF CHINESE AND TAIWANESE COMPETITORS IN BASEBAND CHIPS

The role of semiconductors in mobile handsets increased considerably with the transition from feature phones to smartphones. Smartphones have four crucial semiconductor components: baseband, application processor, connectivity, and memory. Baseband processors were, as of the early 2010s, the second most expensive item (after displays) in the

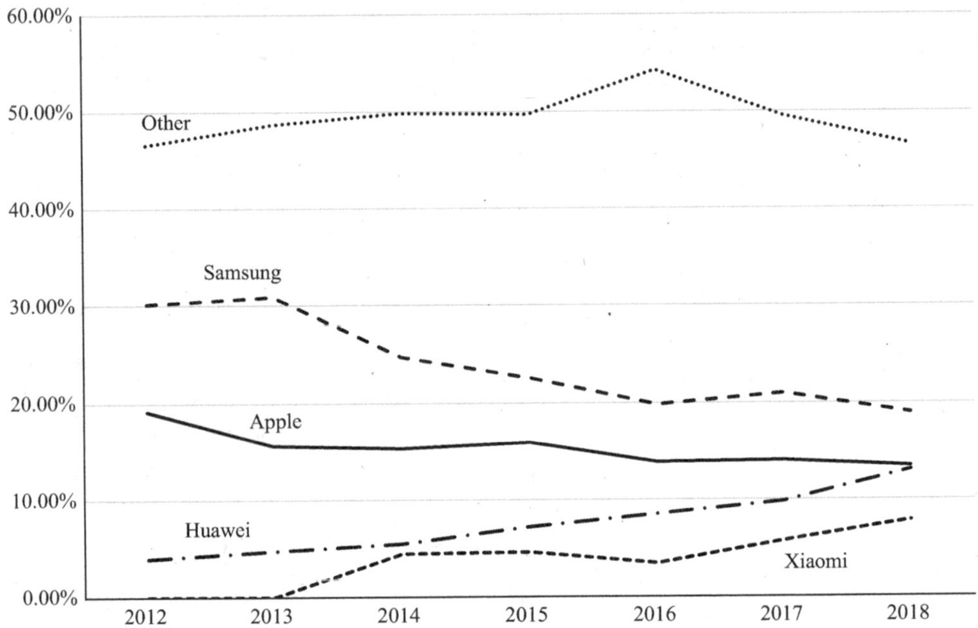

12.4 World smartphone sales, shares in units sold, 2012–2018. *Sources*: Gartner press releases.

iPhone's bill of materials.[29] Baseband processors manage the radio-control functions, including signal modulation, encoding, and radio-frequency shifting. They also manage the transmission of signals.

As the industry transitioned from 3G to 4G, the complexity of baseband processors increased because newer phones needed to remain backward compatible with old standards. An LTE mobile phone needs to function not only on LTE frequency bands but also on those allocated to 2G and 3G standards. Here we examine the evolution of the global baseband-processor market in the transition from 3G to 4G, with a special focus on the rise of MediaTek and Spreadtrum.

MEDIATEK

MediaTek was established in 1997 as a spinoff from the large Taiwanese semiconductor company United Microelectronics Corporation, which was Taiwan's first indigenous semiconductor company and itself a spinoff from the famous Industrial Technology Research Institute. Although originally United Microelectronics Corporation set out to be both a semiconductor designer and manufacturer, it later decided to spin off its integrated-circuit design activities into several units, and MediaTek was the unit that focused on multimedia devices.[30]

In its early years, MediaTek made controller chips for CD and DVD drives. In 2004 the company entered the mobile-phone business, making chipsets for the rapidly expanding feature-phone market in China. It later started selling mobile semiconductors in Indonesia, India, Brazil, Russia, and other developing markets.

MediaTek grew to become one of the top mobile-semiconductor design houses in the world by selling "reference designs" for feature phones—combinations of hardware and software that included the chips themselves, a simple operating system, a camera, and a display. The hardware portion of the MediaTek reference design often included several mobile semiconductor chips: one for power management, another for Bluetooth and Wi-Fi, a memory chip, and yet another that combined the application processor and baseband functions. Feature-phone manufacturers then took the hardware-software bundle supplied by MediaTek and customized it to cater to the needs of specific customers and markets. In the first decade of the twenty-first century, shanzhai manufacturers supplied the Chinese market with feature phones built on MediaTek reference designs.

In 2010 MediaTek sold 500 million semiconductors for feature phones, and the next year it moved from feature phones to smartphones.[31] It sold 10 million semiconductors for smartphones in 2011 and 110 million in 2012. Many of its smartphone customers in the early 2010s were Chinese manufacturers such as Oppo and Xiaomi.

SPREADTRUM

Spreadtrum was founded in 2001 by a group of engineers and entrepreneurs who returned to China after spending some time working in Silicon Valley.[32] In its early years, Spreadtrum developed mobile semiconductors for the GSM standard and its General Packet Radio Services evolution. In 2002 the company entered the Chinese baseband market with a GSM / General Packet Radio Services chip.

The founders then decided to explore the 3G market. Originally, they planned to design mobile semiconductors for the WCDMA standard, which was in the process of achieving widespread acceptance all over the world. In 2003, however, they chose to focus on TD-SCDMA, the Chinese native standard, and by 2004 they had a TD-SCDMA chip ready. Although Spreadtrum lost money for several years, it had a successful Nasdaq initial public offering in 2007.

Spreadtrum's financial situation began to improve only after the Chinese government formally adopted TD-SCDMA as one of its 3G mobile standards. Because Spreadtrum was a pioneer in the TD-SCDMA market, its sales grew at a fast pace with the adoption of the standard, from USD 100 million in 2009 to more than USD 1 billion in 2013. By 2012 Spreadtrum was the leader in the market for TD-SCDMA baseband chips with a 40 percent share, followed by Marvel (20 percent), and MediaTek (20 percent).[33] By the mid-2010s Spreadtrum had become the third-largest baseband-processor vendor in the world, after Qualcomm and MediaTek.

MEDIATEK AND SPREADTRUM VERSUS QUALCOMM

Although MediaTek and Spreadtrum grew at a remarkable pace in the first decade of the twenty-first century, they were not able to catch up with Qualcomm, the leader in the global market for baseband processors. Founded in San Diego in 1985, Qualcomm developed the CDMA technology that became one of the 2G standards in the United States and elsewhere during the 1990s. In its early years, Qualcomm was involved not only in the baseband semiconductor business but also in the production of mobile phones and base stations. In the late 1990s, however, the company sold off its handset and base-station business and focused on the design and manufacturing of mobile semiconductors, especially baseband chips.

Over the years, Qualcomm developed an important portfolio of patents covering different aspects of the baseband-processor business. Based on declarations made by intellectual-property owners to standard-setting organizations, Qualcomm had more than 90 percent of the CDMA standard-essential patents, about 27 percent of the WCDMA standard-essential patents, and about 16 percent of the LTE standard-essential patents.[34] Although its share of the total portfolio of standard-essential patents declined as the industry transitioned from 2G to 3G and later to 4G, Qualcomm remained the leading intellectual-property owner in baseband semiconductors.

Between the late 2000s and the mid-2010s, as the world market for baseband processors became more concentrated through exits and acquisitions, Qualcomm managed to consolidate its leading role. Figure 12.5 shows the market shares of the main players between 2008 and 2015.

The figure shows that Qualcomm grew its share in the global baseband-chip market from about 37 percent in 2008 to almost 60 percent in 2015. Its main competitors in this period were MediaTek, whose share rose from about 12 percent to about 19 percent, and Spreadtrum, which went from about 1 percent to about 7 percent. Other important vendors included Samsung, which designed and manufactured baseband processors mainly for use in its own cellular phones, and Intel, which grew by acquiring the German company Infineon in 2011 but was not able to maintain its share.[35]

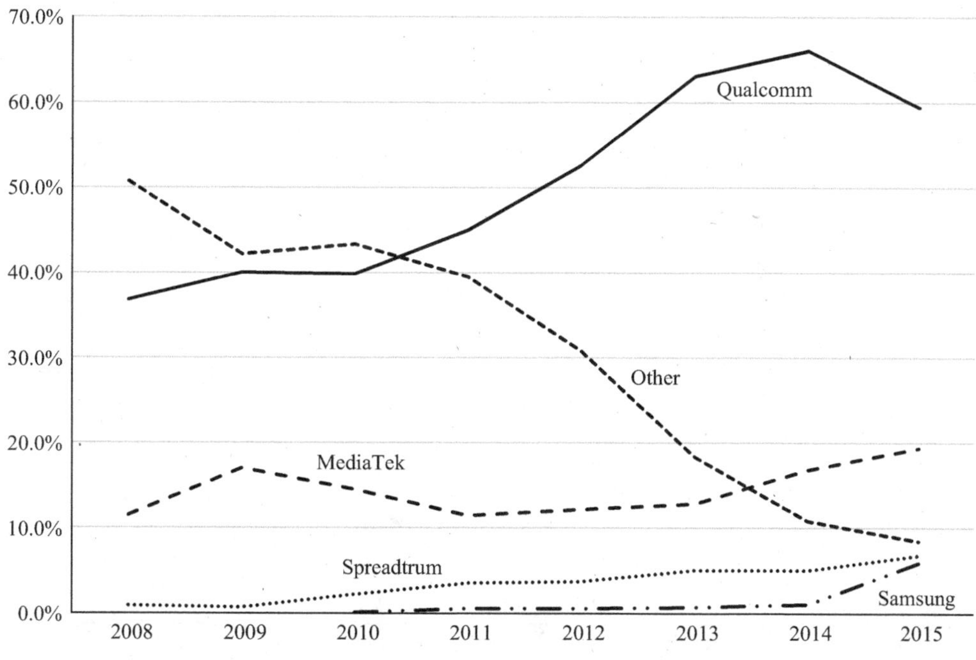

12.5 World baseband chip market, USD revenue shares, 2008–2015. *Source*: Korea Fair Trade Commission, In re: Alleged Abuse of Market Dominance of Qualcomm Incorporated, Decision No. 2017-0-25, January 20, 2017, http://www.theamericanconsumer.org/wp-content/uploads/2017/03/2017-01-20_KFTC-Decision_2017-0-25.pdf.

Figures 12.6 and 12.7 show that MediaTek and Spreadtrum managed to challenge Qualcomm's leadership more consistently in 3G (WCDMA) than in 4G (LTE) baseband processors.

Figure 12.6 shows that, in the WCDMA baseband market, Qualcomm's share peaked at about 55 percent in 2011 and declined thereafter to around 30 percent in 2015. The company's principal challengers were MediaTek and Spreadtrum. By 2015 MediaTek had a slightly larger share than Qualcomm in this market. Figure 12.7 shows that, in the LTE baseband market, Qualcomm's share peaked in 2013 at about 95 percent and declined thereafter. By 2015, however, Qualcomm still had a 70 percent share of this market.

MediaTek and Spreadtrum appear to have implemented a strategy of not challenging Qualcomm in baseband processors at the technological frontier—the LTE standard as of the mid-2010s—but rather catching up

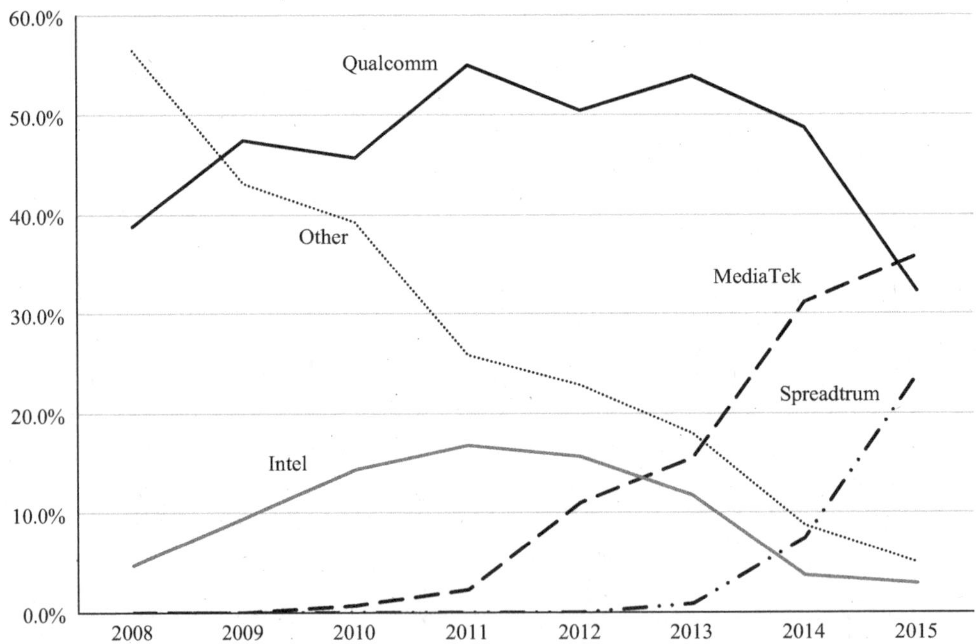

12.6 World WCDMA baseband chip market, USD revenue shares, 2008–2015. *Source*: Korea Fair Trade Commission, In re: Alleged Abuse of Market Dominance of Qualcomm Incorporated, Decision No. 2017-0-25, January 20, 2017, http://www.theamericanconsumer.org/wp-content/uploads/2017/03/2017-01-20_KFTC-Decision_2017-0-25.pdf.

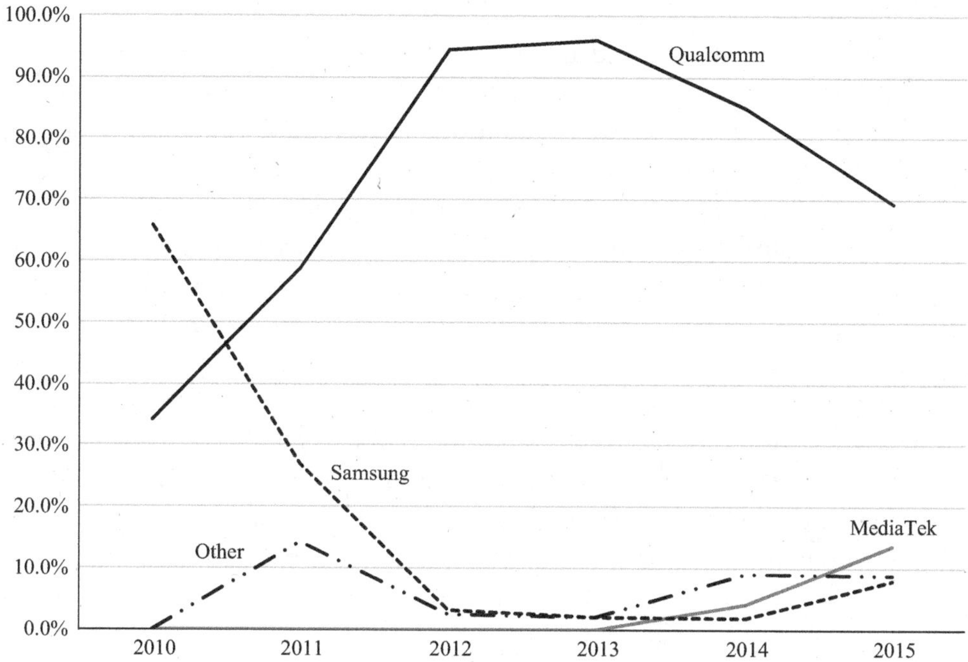

12.7 World LTE baseband chip market, USD revenues shares, 2010–2015. *Source*: Korea Fair Trade Commission, In re: Alleged Abuse of Market Dominance of Qualcomm Incorporated, Decision No. 2017-0-25, January 20, 2017, http://www.theamericanconsumer.org/wp-content/uploads/2017/03/2017-01-20_KFTC-Decision_2017-0-25.pdf.

with Qualcomm by gaining share in legacy standards. As of 2012 MediaTek and Spreadtrum had relatively large shares in 2G and 3G markets. For example, MediaTek had 37 percent of the 2G GSM market and Spreadtrum had 16 percent; further, Spreadtrum had 40 percent of the 3G TD-SCDMA market and MediaTek had 20 percent.[36] By 2015 MediaTek had caught up with Qualcomm in the 3G WCDMA market and Spreadtrum was on its way to doing so, but both lagged behind Qualcomm significantly in the 4G LTE market.

COMPETING IN 4G CELLULAR MARKETS: TAKING STOCK

It seems natural to think that the transformations that occurred in the smartphone operating system market in the late 2000s were the most consequential in the history of the mobile-phone industry. In 2007 Apple introduced a device, the iPhone, that revolutionized the world of mobile phones and laid the foundation for the world of mobile apps that we are accustomed to today. Soon thereafter, the first Android smartphone was launched, and Android soon became the most widely used smartphone operating system in the world.

At about the same time, however, radical changes were also taking place in other cellular markets, including infrastructure, handsets, and baseband processors. The common denominator in all these changes was that companies based in mainland China (and in neighboring Taiwan in one specific case) were becoming serious competitors for the European and American companies that had traditionally dominated those markets. The future of the mobile-phone industry may be shaped by the race between some of these Chinese and Taiwanese newcomers and some of the old incumbents to develop the intellectual property for the new cellular standard, and the products and services compatible with it.

CONCLUSIONS

Over the last four decades, the cellular industry has transformed the way we communicate with one another, travel, listen to music, check the weather forecast, and do our banking, among many other things. In our historical analysis of the industry, we have tracked the following themes: the role of standards, competition among standards and among firms within standards, the many ways national governments have shaped the industry in different countries, and the changing ways people have used mobile phones in their daily lives.

THE ROLE OF CELLULAR STANDARDS

Cellular standards are the rules of interaction between different components of the cellular system. The cellular industry would have been infeasible without standards. Although other high-tech industries such as the computer industry also have had standards (and still do), the key difference is that computer standards have tended to be market based. By contrast, many of the cellular standards discussed in this book were developed by committees. Although these committees usually included representatives from the cellular industry, their decisions were independent of any specific company.

The cellular industry evolved from a world with many standards in the 1980s to one with a single standard, Long-Term Evolution (LTE), in the 2010s. This evolution was not uniform across continents. Europe went from many standards in the 1980s to one, Global System for Mobile Communications (GSM), in the 1990s. By contrast, the United States went from a single standard, Advanced Mobile Phone Service (AMPS), in the 1980s to many in the 1990s. Over time, however, the world converged to a single cellular standard.

Cellular standards were of two types: open and closed. Open standards were developed through the interaction of many organizations, usually in committees, and their technical specifications were made available to all comers. By their very nature, open standards—including AMPS and Nordic Mobile Telephone (NMT) in the first generation and GSM in the second generation of cellular systems—facilitated competition among firms within standards and thus led to lower prices for handsets and infrastructure. Lower prices, in turn, facilitated adoption. Closed standards, by contrast, were developed by a company that usually monopolized the intellectual property at the core of the standard and, for a period of time at least, the manufacturing of handsets and infrastructure for the standard. In this case, prices remained high and adoption proceeded more slowly. This was the case in Germany, France, and Italy in the 1980s, during the first generation of cellular systems.

Open standards were adopted widely for another reason: network effects. Lower prices, especially for handsets, made it easier for people to become cellular subscribers, and adoption led to further adoption, not only in the countries where open standards were created but also abroad. AMPS, NMT, and GSM were heavily adopted outside their countries (or regions) of origin—the United States, Scandinavia, and Western Europe, respectively.

COMPETITION AMONG STANDARDS AND AMONG FIRMS WITHIN STANDARDS

In the history of the cellular industry, standards have competed with one another to gain adoption. During the 1990s, for example, in the transition from the first (analog) to the second (digital) generation of cellular phones, the US market witnessed an intense contest between promoters of alternative digital standards, D-AMPS and code division multiple access (CDMA). Similar battles were happening in other parts of the world. In South Korea, for example, two government ministries confronted each other during the 1990s with contrasting views on the ideal cellular standard for the country. The Ministry of Communications promoted CDMA, while the Ministry of Trade, Industry and Energy preferred a standard based on time division multiple access (TDMA).

In the transition to the third generation of cellular phones, two organizations were formed—the Third Generation Partnership Project (3GPP) and the Third Generation Partnership Project 2 (3GPP2)—that promoted alternative standards. The former championed what eventually became wideband CDMA (WCDMA), and the latter supported cdma2000. In the transition to the fourth generation, a confrontation between WiMAX (Worldwide Interoperability for Microwave Access) and LTE ended with the victory of the latter.

At the same time, firms have competed with one another within the confines of specific standards. Ericsson, Nokia, and Motorola, for example, were among the leading contenders in markets for cellular terminals for various 1G and 2G standards. Ericsson and Nokia were also strong competitors in markets for cellular infrastructure—both were among the leading firms in, for example, the market for GSM switching systems and the market for GSM base stations in the 1990s. Further, Motorola and Ericsson competed with Japanese equipment suppliers in the infrastructure market for the Pacific Digital Cellular (PDC) standard in Japan during the 1990s. More recently, MediaTek and Spreadtrum have started competing with Qualcomm in markets for baseband chips for the WCDMA and LTE standards.

Two different aspects of competition among firms within standards made it possible for large swaths of the population to gain access to cellular services. Competition within standards in the market for terminals (later handsets) usually lowered device prices, and competition within standards in the market for network services frequently lowered service prices. In a few countries—the United States and Britain among them—competition in network services started when the first cellular networks were launched (or soon thereafter).[1] Many other countries had to wait until the 1990s to realize the benefits of competition in cellular network services. In most Western European countries, for example, the transition from 1G to 2G cellular systems was accompanied by the transition from monopoly to competition in the network-services market. Most Western European governments allowed international consortia to enter domestic markets and compete with the cellular incumbent. As a result, service prices fell, and cellular adoption increased rapidly.[2]

During the 1990s, it was not just Western Europe that went from monopoly to competition in network services. Countries as diverse as New Zealand, Brazil, Mexico, South Korea, China, Singapore, Taiwan, and Israel also experienced, for the first time, the impact of competition among cellular carriers. There were nuances to how domestic markets evolved. In Israel, the new entrant lowered subscription prices dramatically and cellular density skyrocketed. In China, cellular competition was "bureaucratic competition"—that is, it involved government ministries confronting one another. In New Zealand, unlike what happened in many other countries, the government introduced competition among operators but did not create a regulator to act as an arbiter of the competitive process. In all countries, however, declining service prices facilitated the transition from a world in which cellular phones were only for the wealthy and well connected to a world in which they were accessible to most people.

THE ROLE OF NATIONAL GOVERNMENTS

National governments have shaped the evolution of the cellular industry all over the world and have done so in multiple ways. Governments allocated bands of spectrum to cellular communications—as opposed to other military and civilian uses. Two aspects of this process had an impact on the pace of cellular growth: the timing of the allocation and how generous it was.

In the early days of the cellular industry, the incumbent cellular carrier was usually a subsidiary of the incumbent fixedline operator—the national postal, telegraph, and telephone administration. Further, when network-services competition was introduced, it was the national government of each country that determined how many licenses would be awarded to new entrants. Thus, governments contributed to shaping the structure of network-services markets in each country.

National governments usually promoted the adoption of a specific standard, and in some cases the choice of standard was driven by industrial policy considerations: the goal was to foster the development of a domestic cellular industry. For example, during the first generation of cellular systems, Siemens developed the standard in Germany, Italtel did in Italy, and Matra did in France—each of them was a national champion in its country. In Japan, the government usually relied on a select group of conglomerates to supply handsets and infrastructure for cellular systems. In South Korea, the government chose a specific standard, CDMA, in the 1990s because it wanted to give domestic companies such as Samsung the opportunity to become solid competitors in the global infrastructure and handset markets.

In many countries, national governments created regulatory agencies to act as arbiters of the competitive process in network-services markets. This was particularly important for interconnection disputes, which in the early years often involved relationships between cellular carriers and the fixedline operator. The presumption was that, because the fixed-line incumbent often participated in the cellular market through its subsidiary, it had incentives to treat its own cellular subsidiary better than it treated the private cellular competitors. In countries such as India, an independent telecommunications regulator played an important role in defending the private cellular carriers from the attacks of the traditional telecommunications bureaucracy and in securing the long-term survival of the private carriers.

CELLULAR PHONES IN DAILY LIFE

The roles that cellular phones have played in people's lives have changed dramatically from the 1980s. During the 1980s cellular phones were used for voice communications. In the 1990s cellular subscribers started relying on short messaging services (texting) in addition to voice communications. The 2000s witnessed the rise of mobile email and the mobile Internet, the former associated with the BlackBerry name and the latter with the i-mode service in Japan. After 2007, with the arrival of mature smartphones such as the iPhone and the Android phones, the Internet accessed through apps started playing a major role among mobile users.

In 2015 the Pew Research Center conducted an "experience sampling" survey of smartphone owners in the United States: they were contacted twice a day over a one-week period and asked what they had used their mobile phones for in the hour before the survey. Texting was the most frequently used feature. Then came, in descending order, Internet use, voice and video calls, social networking, watching videos, and listening to music.[3] If we had told cellular users in the early 1980s that these are the kinds of things people would be doing on their cellular phones thirty-five years later, they would have thought we were out of our minds.

Even though cellular phones have had a substantial impact on people's lives in developed countries, their impact has probably been even deeper in many developing countries in Africa and elsewhere. African countries barely had a telecommunications infrastructure before cellular phones arrived. Thus, the mobile phone was the one and only communications device that most Africans knew and used. In Africa, cellular phones lowered the cost of collecting and transmitting information in a radical way. In addition, mobile phones gave many Africans a new way of understanding money. Using mobile money to pay bills, make remittances to relatives, and save had a profound impact on the lives of many people in Africa and elsewhere who never had access to a bank and a bank account. Cellular phones have come a long way from the time, in the early 1980s, when only wealthy people used them in their cars.

THE CELLULAR FUTURE

At the time of writing, carriers in different parts of the world are announcing that handsets for the next generation of cellular phones, 5G, are becoming available. The three novel features of the fifth generation of cellular phones are projected to be 100 Mbps minimum data rates for consumers, pervasive machine-to-machine connections (also known as the Internet of Things or IoT), and very low latency.[4]

5G is not just about technology—it is also about intellectual property rights over the technology and about geopolitics. Despite being the largest cellular market in the world, China has owned relatively few patents declared essential to any of the prior cellular standards. 5G could change that: preliminary analyses suggest that Chinese entities may own about 10 percent of patents that are essential to the 5G standard.[5]

The US government has reacted to what it perceives as the Chinese threat in frontier telecommunications technology. In May 2019, arguing that Huawei had violated sanctions imposed on Iran, the US government introduced measures to prevent Huawei from receiving US components that are at the core of the systems the Chinese company sells. Severe as they were, the restrictions had loopholes: suppliers could keep on selling components to Huawei as long as they were made outside the United States. A year later, in May 2020, the US government escalated its attack on Huawei: it introduced new measures to prevent companies around the world from using software or hardware originating in the United States to make components based on Huawei's designs.[6] The stated goal of these policies has been to "purge" cellular networks around the world of Huawei's components.

The confrontations between China and the United States have been forcing other countries to take sides, sometimes reluctantly. Australia banned Huawei's 5G equipment in 2018. In June 2020 carriers in Canada and Singapore declared that 5G networks in those countries would be built with equipment provided by Nokia and Ericsson, Huawei's main rivals. And countries in Western Europe have also started taking steps to limit Huawei's presence in their 5G cellular networks.[7]

How the 5G race will unfold is not for us to predict. Historians are practitioners of hindsight, not foresight. Just as cellular users in the early 1980s could not have imagined the many uses we give our cellular phones today, cellular users of today—the authors of this book included—cannot know what people will be doing with their cellular phones thirty-five or forty years from now. Nor do we have any certainty about which companies will be the leading suppliers of handsets, operating systems, infrastructure, and semiconductors for future generations of cellular phones.

The history of the cellular industry explored in this book suggests, however, that a few things are likely, if not certain, about the cellular future. First, future generations of cellular phones will create opportunities for new firms to rise and displace established incumbents. Second, technologies that we cannot imagine today will change the appearance and functionality of mobile phones. And finally, people will use cellular phones for purposes that we cannot even start to envision today.

NOTES

INTRODUCTION

1. On cellular standards, see especially J. Funk, *Global Competition between and within Standards: The Case of Mobile Phones* (Houndmills, UK: Palgrave, 2002); and R. Bekkers, *Mobile Telecommunications Standards: GSM, UMTS, TETRA, and ERMES* (Boston: Artech House, 2001). On the history of the standard-setting movement since the late nineteenth century, see especially J. Yates and C. Murphy, *Engineering Rules: Global Standard Setting since 1880* (Baltimore: Johns Hopkins University Press, 2019).

2. See J. Laffont, P. Rey, and J. Tirole, "Network Competition: I. Overview and Nondiscriminatory Pricing," *RAND Journal of Economics* 29, no. 1 (Spring 1998): 1–37; and J. Laffont, P. Rey, and J. Tirole, "Network Competition: II. Price Discrimination," *RAND Journal of Economics* 29, no. 1 (Spring 1998): 38–56.

3. See M. Campbell-Kelly and D. Garcia-Swartz, *From Mainframes to Smartphones: A History of the International Computer Industry* (Cambridge, MA: Harvard University Press, 2015), 57–102.

4. G. Garrard, *Cellular Communications: Worldwide Market Development* (Boston: Artech House, 1998); G. Klemens, *The Cellphone: History of the Gadget That Changed the World* (Jefferson, NC: McFarland, 2010); J. Agar, *Constant Touch: A History of the Global Mobile Phone* (London: ICON Books, 2013).

5. M. Cooper, *Cutting the Cord* (New York: RosettaBooks, 2020).

6. J. Meurling and R. Jeans, *The Mobile Phone Book* (London: Ericsson Radio Systems, 1994); K. Poupée, *La téléphonie mobile* (Paris: Presses universitaires de France, 2003).

7. Funk, *Global Competition*; H. Gruber, *The Economics of Mobile Telecommunications* (Cambridge: Cambridge University Press, 2005); P. Curwen and J. Whalley, *The Internationalisation of Mobile Telecommunications: Strategic Challenges in a Global*

Market (Cheltenham, UK: Edward Elgar, 2008); P. Curwen and J. Whalley, *Mobile Telecommunications in a High-Speed World: Industry Structure, Strategic Behavior and Socio-economic Impact* (Farnham, UK: Gower, 2010).

8. J. Hausman, "Mobile Telephone," in *Handbook of Telecommunications Economics*, vol. 1, *Structure, Regulation, and Competition*, ed. M. Cave, S. Majumdar, and I. Vogelsang (Amsterdam: Elsevier, 2002), 563–604; J. Church and N. Gandal, "Platform Competition in Telecommunications," in *Handbook of Telecommunications Economics*, vol. 2, *Technical Evolution and the Internet*, ed. S. Majumdar, I. Vogelsang, and M. Cave (Amsterdam: Elsevier, 2005), 119–53; J. Gans, S. King, and J. Wright, "Wireless Communications," in Majumdar, Vogelsang, and Cave, *Handbook of Telecommunications Economics*, 241–285.

9. J.-J. Laffont and J. Tirole, *Competition in Telecommunications* (Cambridge, MA: MIT Press, 1999); C. Cambini, P. Ravazzi, and T. Valletti, *Il mercato delle telecomunicazioni* (Bologna: Il Mulino, 2003).

CHAPTER 1

1. A. Mehrotra, *Cellular Radio: Analog and Digital Systems* (Boston: Artech House, 1994), 146–150. In his recent book on the origins of cellular phones, Martin Cooper argues that there were proto-cellular systems many years before 1978. He mentions a system set up in the late 1960s for the New York City–Washington, DC, Metroliner train that included a "handoff" feature from one system to another. He also mentions a system that Motorola set up for the Chicago Police Department in the 1960s and that included all the cellular features with the exception of the call "handoff." See M. Cooper, *Cutting the Cord* (New York: RosettaBooks, 2020), esp. 49–50, 57–64. These systems may well have been precursors of cellular networks, but they were not nearly as fully developed as the Chicago system of 1978.

2. As of mid-1988, forty-two countries had cellular systems, according to US Department of Commerce, *A Competitive Assessment of the U.S. Cellular Radiotelephone Industry* (Washington, DC: US Department of Commerce, 1988), 37–38, 76. There were forty-five countries with at least one cellular system by the end of 1988, according to J. Funk, *Global Competition between and within Standards: The Case of Mobile Phones* (Houndmills, UK: Palgrave, 2002), 43–45. The 2.5-million subscriber number comes from US Department of Commerce, *Competitive Assessment*, xi. According to the International Telecommunications Union (ITU) database, there were more than four million cellular subscriptions by the end of 1988.

3. The story of how an NMT 450 system came to be launched in Saudi Arabia very early in the history of cellular networks can be found in J. Meurling and R. Jeans, *The Mobile Phone Book* (London: Ericsson Radio Systems, 1994), 55–57.

4. On the history and economics of spectrum allocation in the United States, see especially T. Hazlett, *The Political Spectrum* (New Haven, CT: Yale University Press, 2017).

5. H. Aitken, *Syntony and Spark: The Origins of Radio* (New York: John Wiley and Sons, 1976), 179–297. This book also has excellent chapters on Heinrich Hertz and Oliver Lodge. On electromagnetic experimental and theoretical developments in the

nineteenth century, see B. Hunt, *Pursuing Power and Light: Technology and Physics from James Watts to Albert Einstein* (Baltimore: Johns Hopkins University Press, 2010); and "Electrical Theory and Practice in the Nineteenth Century," in *The Cambridge History of Science*, vol. 5, *The Modern Physical and Mathematical Sciences*, ed. M. Nye (Cambridge: Cambridge University Press, 2003), 311–327. For the mathematics of Maxwell's equations, see D. Fleisch, *A Student's Guide to Maxwell's Equations* (Cambridge: Cambridge University Press, 2008).

6. Aitken, *Syntony and Spark*, 202.

7. Marconi's patent, No. 763,772, "Apparatus for Wireless Telegraphy," was filed on November 10, 1900. This patent generated disputes with the Serbian-born American engineer Nikola Tesla, whose two important radio-related patents were granted in 1900: No. 645,576, "System of Transmission of Electrical Energy," and No. 649,621, "Apparatus for Transmission of Electrical Energy." The patent office denied Marconi's patent in 1903 on the basis of Tesla's preexisting patents. However, in 1904, and without any explanation, the patent office reversed its decision and granted Marconi's patent. The disputes continued for decades. See G. Klemens, *The Cellphone* (Jefferson, NC: McFarland, 2010), esp. 34–37.

8. When the *Titanic* hit an iceberg on April 14, 1912, the crew used a Marconi wireless telegraph system to send distress signals to other ships. The system, which was capable of transmitting signals as far as 350 miles, was used on the ship for two purposes: to send "marconigrams" (wireless telegrams) on behalf of the first-class passengers, and to send and receive weather reports and ice warnings. See J. Ouellette, "US Court Grants Permission to Recover Marconi Telegraph from *Titanic* Wreckage," Ars Technica, May 28, 2020, https://arstechnica.com/science/2020/05/us-court-grants-permission-to-recover-marconi-telegraph-from-titanic-wreckage/.

9. H. Aitken, *The Continuous Wave: Technology and American Radio, 1900–1932* (Princeton, NJ: Princeton University Press, 1985), esp. 1–27.

10. For details on these devices, see Aitken, 28–161; and F. Nebeker, *Dawn of the Electronic Age* (Hoboken, NJ: John Wiley and Sons, 2009), 23–24, 27, 37.

11. W. Young, "Advanced Mobile Phone Service: Introduction, Background, and Objectives," *Bell System Technical Journal* 58, no. 1 (January 1979): 1–14; H. Kargman, "Land Mobile Communications: The Historical Roots," in *Communications for a Mobile Society*, ed. R. Bowers et al. (Beverly Hills, CA: Sage, 1978), 19–34; G. Calhoun, *Digital Cellular Radio* (Norwood, MA: Artech House, 1988), 23–37; E. F. O'Neill, ed., *A History of Engineering and Science in the Bell System: Transmission Technology (1925–1975)* (Murray Hill, NJ: AT&T Bell Laboratories, 1985), 401–418.

12. D. Noble, "The History of Land-Mobile Radio Communications," *Proceeding of the IRE* 50, no. 5 (May 1962): 1405–1414.

13. For details on technical terminology, see especially H. Newton, *Newton's Telecom Dictionary*, 26th ed., with S. Schoen (New York: Flatiron, 2011).

14. Calhoun, *Digital Cellular Radio*, 27–28. See also R. Macario, *Cellular Radio: Principles and Design* (Houndmills, UK: Macmillan, 1993), 18–20.

15. Noble, "History," 1408–1409.

16. Young, "Advanced Mobile Phone Service," 2.

17. Young, 3.

18. Calhoun, *Digital Cellular Radio,* 32–35; Young, "Advanced Mobile Phone Service," 3–5.

19. Kargman, "Land Mobile Communications," 29–31.

20. Young, "Advanced Mobile Phone Service," 6. The number of fixed-telephone subscriptions in the United States comes from the ITU database.

21. G. Garrard, *Cellular Communications: Worldwide Market Development* (Boston: Artech House, 1998), 15–16.

22. Garrard, 17.

23. B. Mölleryd, *The Building of a World Industry: The Impact of Entrepreneurship on Swedish Mobile Telephony* (Stockholm: Teldok, 1997).

24. Garrard, *Cellular Communications,* 20.

25. T. Pempel, "Land Mobile Communications in Japan: Technical Developments and Issues of International Trade," in Bowers et al., *Communications for a Mobile Society,* 317–343.

26. V. Mac Donald, "The Cellular Concept," *Bell System Technical Journal* 58, no. 1 (January 1979): 15–41; Calhoun, *Digital Cellular Radio,* 39–44.

27. S. Gibson, *Cellular Mobile Radiotelephones* (Englewood Cliffs, NJ: Prentice-Hall, 1987), 7, argues that, before cellular systems, at most seven hundred customers could be accommodated in New York City, and those customers could complete one out of every two telephone calls the first time they tried. Other sources refer to about a dozen simultaneous conversations.

28. Young, "Advanced Mobile Phone Service," 7–8. See also A. Madrigal, "The 1947 Paper That First Described a Cell-Phone Network," *Atlantic,* September 16, 2011, https://www.theatlantic.com/technology/archive/2011/09/the-1947-paper-that-first-described-a-cell-phone-network/245222/.

29. On the clustering of cells in cellular systems, see, for example, Mehrotra, *Cellular Radio,* 31–46.

30. It can be shown that clusters contain C cells, where C is an integer given by the expression $C=i^2+j^2+ij$. In this expression, $i=0, 1, 2, \ldots$ and $j=0, 1, 2, \ldots$ The AMPS system uses a frequency-reuse cluster of 7, whereas GSM uses 3 or 4. See M. Schwartz, *Mobile Wireless Communications* (Cambridge: Cambridge University Press, 2005), 66–67.

31. The dictum comes from G. Box and N. Draper, *Empirical Model Building and Response Surfaces* (New York: John Wiley and Sons, 1987), 424.

32. Calhoun, *Digital Cellular Radio,* 92–111. On coverage problems in Britain, see Garrard, *Cellular Communications,* 188.

33. On the technicalities of cellular-system design, see Mehrotra, *Cellular Radio.* Other technical sources include Gibson, *Cellular Mobile Radiotelephones;* and V. Garg and J. Wilkes, *Wireless and Personal Communications Systems* (Upper Saddle River, NJ:

Prentice Hall, 1996). An excellent summary of the technical dimensions of cellular phones is in W. Webb, *Understanding Cellular Radio* (Boston: Artech House, 1998).

34. The traditional unit of telephone traffic is the Erlang, and measures of spectral efficiency include Erlangs/MHz, Erlangs/channel, and Erlangs/MHz/square mile. On the spectral efficiency of precellular mobile-phone systems compared with that of early cellular systems, see especially J. Frey and A. Lee, "Technologies for Land Mobile Communications: 900-MHz Systems," in Bowers et al., *Communications for a Mobile Society*, 60–63.

35. The fixedline density numbers come from the ITU database. In 1989, there were 67.20 fixed-telephone subscriptions per 100 people in Sweden, 51.97 in Finland, 49.06 in Norway, 55.52 in Denmark, and 52.20 in the United States. There were 43.05 in Japan.

36. Meurling and Jeans, *Mobile Phone Book*, esp. 24–26.

37. In addition, inexpensive frequency synthesizers, which allow mobile phones to switch across frequencies, were developed only starting in the 1970s. See, for example, Young, "Advanced Mobile Phone Service," 7.

CHAPTER 2

1. See J. Meurling and R. Jeans, *The Mobile Phone Book* (London: Ericsson Radio Systems, 1994), esp. 7–14, 99–100.

2. Here we follow G. Brock, *The Telecommunications Industry: The Dynamics of Market Structure* (Cambridge, MA: Harvard University Press, 1981).

3. Between 1902 and 1912, fixedline density in the United States grew from 2.3 percent to 8.8 percent, much faster than in Europe, where it rose from 0.3 to 0.7 percent. See AT&T, *Telephone Statistics of the World* (New York: AT&T, 1912), 8. In Europe, the Scandinavian countries were the ones with the highest telephone density; see AT&T, 15.

4. See United Nations, *Core ICT Indicators* (November 2005), https://www.itu.int/ITU-D/ict/partnership/material/CoreICTIndicators.pdf.

5. G. Calhoun, *Digital Cellular Radio* (Norwood, MA: Artech House, 1988), esp. 35–36.

6. E. Noam, *Telecommunications in Europe* (New York: Oxford University Press, 1992), 3–25.

7. Noam, 4.

8. S. Vogel, *Freer Markets, More Rules* (Ithaca, NY: Cornell University Press, 1996), 139–142. See also Y. Ito and A. Iwata, "Japan: Creating the Domestic and International Network," in *Telecommunications in the Pacific Basin: An Evolutionary Approach*, ed. E. Noam, S. Komatsuzaki, and D. Conn (New York: Oxford University Press, 1994), 440–457.

9. Calhoun, *Digital Cellular Radio*, 39–49.

10. The key references on the evolution of the US cellular system in the 1980s include G. Garrard, *Cellular Communications: Worldwide Market Development* (Boston:

Artech House, 1998), 30–47; and H. Gruber, *The Economics of Mobile Telecommunications* (Cambridge: Cambridge University Press, 2005), 123–135. For a business history focused on Craig McCaw, one of the most successful cellular entrepreneurs of the 1980s, see O. Corr, *Money from Thin Air* (New York: Crown Business, 2000). An excellent company history centered on Pacific Telesis is in L. Galambos and E. Abrahamson, *Anytime, Anywhere: Entrepreneurship and the Creation of a Wireless World* (Cambridge: Cambridge University Press, 2002).

11. For details, see A. Mehrotra, *Cellular Radio: Analog and Digital Systems* (Boston: Artech House, 1994), 146–150.

12. The Federal Communications Commission licensed cellular operators in 734 markets—306 MSAs and 428 rural statistical areas. See Federal Communications Commission, *Annual Report and Analysis of Competitive Market Conditions with Respect to Mobile Commercial Services* (August 18, 1995), 4, https://us-fcc.app.box.com/s/47g5d49y4d2jfarnhrcy2npj7hu17w5k.

13. The seven Regional Bell Operating Companies were Ameritech, Bell Atlantic, Bell South, Nynex, Pacific Telesis, Southwestern Bell, and US West.

14. S. Gibson, *Cellular Mobile Radiotelephones* (Englewood Cliffs, NJ: Prentice-Hall, 1987), 19–22.

15. S. Hultén and B. Mölleryd, "Mobile Telecommunications in Sweden," in *Mobile Telecommunications: Emerging European Markets*, ed. K. Schenk, J. Müller, and T. Schnöring (Boston: Artech House, 1995), 1–28; Meurling and Jeans, *Mobile Phone Book*, 27–33; Garrard, *Cellular Communications*, 48–61. See also T. Haug, "A Commentary on Standardization Practices: Lessons from the NMT and GSM Mobile Telephone Standards Histories," *Telecommunications Policy* 26, nos. 3–4 (April–May 2002): 101–107; and J. Lehenkari and R. Miettinen, "Standardisation in the Construction of a Large Technological System—the Case of the Nordic Mobile Telephone System," *Telecommunications Policy* 26, nos. 3–4 (April–May 2002): 109–127.

16. US International Trade Commission, *Global Competitiveness of U.S. Advanced-Technology Industries: Cellular Communications* (Washington, DC: US International Trade Commission, 1993), esp. 4–4; Garrard, *Cellular Communications*, 359–361.

17. Noam, *Telecommunications in Europe*, 103–132. On the deregulation process in Britain more generally, see M. Armstrong, S. Cowan, and J. Vickers, *Regulatory Reform: Economic Analysis and British Experience* (Cambridge, MA: MIT Press, 1994).

18. See especially T. Valletti and M. Cave, "Competition in UK Mobile Communications," *Telecommunications Policy* 22, no. 2 (March 1998): 109–131. See also Gruber, *Economics of Mobile Telecommunications*, 98–105; Garrard, *Cellular Communications*, 95–124; Meurling and Jeans, *Mobile Phone Book*, 89–97.

19. In France, a competing network was allowed to operate in the late 1980s.

20. M. Berlage and T. Schnöring, "The Introduction of Competition in German Mobile Communications Markets," in Schenk, Müller, and Schnöring, *Mobile Telecommunications*, 29–49; US International Trade Commission, *Global Competitiveness*, 4–6, 4–7; Gruber, *Economics of Mobile Telecommunications*, 105–108.; Garrard, *Cellular Communications*, 74–76.

21. Gruber, *Economics of Mobile Telecommunications*, 108–111.; US International Trade Commission, *Global Competitiveness*, 4–7; Garrard, *Cellular Communications*, 76–79.

22. Gruber, *Economics of Mobile Telecommunications*, 112–115; US International Trade Commission, *Global Competitiveness*, 4–7; Garrard, *Cellular Communications*, 79–83.

23. Another discussion of the French system is in J. Manguian, *Les radiocommunicatons* (Paris: Presses universitaires de France, 1993), chap. 4.

24. In Hong Kong, a TACS system was launched in 1984, even before one was formally introduced in Britain in 1985. This was possible because TACS was simply a modified version of an existing standard, the American AMPS. The fact that Cable & Wireless, a British telecommunications company, owned the cellular firm that introduced TACS in Hong Kong likely helped as well. See Garrard, *Cellular Communications*, 394.

25. K. Lyytinen and V. Fomin, "Achieving High Momentum in the Evolution of Wireless Infrastructures: The Battle over the 1G Solutions," *Telecommunications Policy* 26, no. 3–4 (April–May 2002), 149–170; J. Funk, *Global Competition between and within Standards: The Case of Mobile Phones* (Houndmills, UK: Palgrave, 2002), 36–92.

26. It happens to be the case that the United States, Sweden, and Britain were among the few countries that had duopolistic competition in cellular network services in the 1980s. This likely helped to foster cellular adoption during the 1980s, not so much by lowering subscription tariffs—at the time, nominal tariffs were stable in Britain and Sweden, and declined slowly in the United States—but rather by increasing network capacity.

CHAPTER 3

1. On the equipment manufacturers, see D. Steinbock, *Wireless Horizon* (New York: Amacom, 2003), chaps. 8–10; D. Steinbock, *The Nokia Revolution* (New York: Amacom, 2001); J. Meurling and R. Jeans, *The Mobile Phone Book* (London: Ericsson Radio Systems, 1994); and J. Morone, *Winning in High-Tech Markets* (Boston: Harvard Business School Press, 1993), chap. 3.

2. F. Nebeker, *Dawn of the Electronic Age* (Hoboken, NJ: John Wiley and Sons, 2009), esp. 298–302.

3. E. Noam, *Telecommunications in Europe* (New York: Oxford University Press, 1992), 206–209.

4. Meurling and Jeans, *Mobile Phone Book*, 45–49.

5. Steinbock, *Wireless Horizon*, 242.

6. M. Haikio, *Nokia: The Inside Story* (London: Prentice Hall, 2002), 54–61.

7. C. Palmberg, "Industrial Transformation through Public Technology Procurement? The Case of Nokia and the Finnish Telecommunications Industry," in *Public Technology Procurement and Innovation*, ed. C. Edquist et al. (New York: Springer Science, 2000), 167–196.

8. Steinbock, *Wireless Horizon*, 213–241; Morone, *Winning in High-Tech Markets*, 65–124.

9. A. Affrunty Sr., *A Personal Journal: 50 Years at Motorola* (Rolling Meadows, IL: Motorola University Press, 1994), esp. 1–41.

10. Affrunty, 42–46.

11. N. Linge and A. Sutton, *30 Years of Mobile Phones in the UK* (Gloucestershire, UK: Amberley, 2015), 8.

12. For a description of the technical details of the "first cellular call," see M. Cooper, *Cutting the Cord* (New York: RosettaBooks, 2020), esp. 105–130.

13. The distinction between committee-based and market-based competition comes from J. Funk, *Global Competition between and within Standards: The Case of Mobile Phones* (Houndmills, UK: Palgrave, 2002). We draw frequently from this book in our study of competition in infrastructure and handsets in the 1980s and 1990s.

14. US International Trade Commission, *Global Competitiveness of U.S. Advanced-Technology Industries: Cellular Communications* (Washington, DC: US International Trade Commission, 1993), esp. 5–12 through 5–23.

15. US International Trade Commission, 5–12, 5–13.

16. US International Trade Commission, esp. 5–25 through 5–37.

17. US Department of Commerce, *A Competitive Assessment of the U.S. Cellular Radiotelephone Industry* (Washington, DC: US Department of Commerce, 1988), 3.

18. US International Trade Commission, *Global Competitiveness*, 5–25 through 5–37.

19. US International Trade Commission, 5–30.

20. Morone, *Winning in High-Tech Markets*, 77.

21. US International Trade Commission, *Global Competitiveness*, 5–31.

22. T. Merriden, *Rollercoaster: The Turbulent Life and Times of Vodafone and Chris Gent* (Oxford: Capstone, 2003), esp. 11–35.

23. G. Garrard, *Cellular Communications: Worldwide Market Development* (Boston: Artech House, 1998), 112–124.

24. T. Valletti and M. Cave, "Competition in UK Mobile Communications," *Telecommunications Policy* 22, no. 2 (March 1998): 109–131.

25. We draw frequently from L. Galambos and E. Abrahamson, *Anytime, Anywhere: Entrepreneurship and the Creation of a Wireless World* (Cambridge: Cambridge University Press, 2002), esp. 34–40; and O. Corr, *Money from Thin Air: The Story of Craig McCaw, the Visionary Who Invented the Cell Phone Industry, and His Next Billion-Dollar Idea* (New York: Crown Business, 2000).

26. C. Lazzareschi, "Pacific Telesis Closes Deal for Cellular Firm," *Los Angeles Times*, March 1, 1986, https://www.latimes.com/archives/la-xpm-1986-03-01-fi-13297-story.html.

27. E. Tucker, "Mobile Phone Rivals Are Being Consolidated," *Washington Post*, August 11, 1986, https://www.washingtonpost.com/archive/business/1986/08/11/mobile-phone-rivals-are-being-consolidated/629b5593-4311-41cf-8f2e-ad3926c224a6/.

28. R. Gregson, "McCaw's Mark," *RCR Wireless News*, March 12, 2001, https://www.rcrwireless.com/20010312/carriers/mccaws-mark-2.

29. G. Fabrikant, "Lin Backs Takeover by McCaw," *New York Times*, December 5, 1989, https://www.nytimes.com/1989/12/05/business/lin-backs-takeover-by-mccaw.html. See also Corr, *Money from Thin Air*, 178–192.

30. G. Barth, "Cellular Phones: Is There Really Competition?" (incidental paper, Harvard University, Center for Information Policy Research, 1994), 21–22.

31. US General Accounting Office, *Concerns about Competition in the Cellular Telephone Service Industry* (Washington, DC: US General Accounting Office, July 1992).

32. P. Parker and L. Röller, "Collusive Conduct in Duopolies: Multimarket Contact and Cross-Ownership in the Mobile Telephone Industry," *RAND Journal of Economics* 28, no. 2 (Summer 1997): 304–322. See also M. Busse, "Multimarket Contact and Price Coordination in the Cellular Telephone Industry," *Journal of Economics and Management Strategy* 9, no. 3 (2000): 287–320.

33. US General Accounting Office, *Concerns about Competition*, 2.

34. US International Trade Commission, *Global Competitiveness*, 3–3.

35. For this analysis, we rely on data presented in H. Fullerton, "Duopoly Market Structure in a Communications Industry: The U.S. Cellular Telephone Experience" (PhD diss., University of Florida, 1996), appendix A.

36. On the evolution of subscription tariffs in the United States, see especially Parker and Röller, "Collusive Conduct in Duopolies"; and Busse, "Multimarket Contact." On Britain, see Valletti and Cave, "Competition in UK Mobile Communications." On Sweden, see S. Hultén and B. Mölleryd, "Mobile Telecommunications in Sweden," in *Mobile Telecommunications: Emerging European Markets*, ed. K. Schenk, J. Müller, and T. Schnöring (Boston: Artech House, 1995), 1–28. For a detailed analysis of subscription tariffs in Britain, Sweden, the United States, Germany, and Israel in the 1980s and 1990s, see N. Garcia-Vicente, D. Garcia-Swartz, and M. Campbell-Kelly, "What Do We Know about Duopolies? Insights from the History of Cellular Phones," *Telecommunications Policy*, published ahead of print, October 7, 2021, https://www.sciencedirect.com/science/article/abs/pii/S0308596121001634.

CHAPTER 4

1. D. Geradin and M. Kerf, *Controlling Market Power in Telecommunications* (Oxford: Oxford University Press, 2003), esp. 134–136.

2. J. Meurling and R. Jeans, *The Mobile Phone Book* (London: Ericsson Radio Systems, 1994), 99–100; G. Calhoun, *Digital Cellular Radio* (Norwood, MA: Artech House, 1988), 143–196.

3. The process is called speech coding; see W. Webb, *Understanding Cellular Radio* (Boston: Artech House, 1998), 102–106.

4. A. Mehrotra, *Cellular Radio: Analog and Digital Systems* (Boston: Artech House, 1994), 264–267.

5. H. Gruber, *The Economics of Mobile Telecommunications* (Cambridge: Cambridge University Press, 2005), 22–23; V. Garg and J. Wilkes, *Wireless and Personal Communications Systems* (Upper Saddle River, NJ: Prentice Hall, 1996), chaps. 2–3.

6. Webb, *Understanding Cellular Radio*, 118–132.

7. D. Mock, *The Qualcomm Equation* (New York: Amacom, 2005), 70–72.

8. There is a reasonably large literature on the GSM standard. We rely on Meurling and Jeans, *Mobile Phone Book*, 101–111; G. Cattaneo, "The Making of a Pan-European Network as a Path-Dependency Process: The Case of GSM versus IBC (Integrated Broadband Communication) Network," in *Global Telecommunications Strategies and Technological Changes*, ed. G. Pogorel (Amsterdam: North-Holland, 1994), 59–79; R. Bekkers, *Mobile Telecommunications Standards: GSM, UMTS, TETRA, and ERMES* (Boston: Artech House, 2001), 261–362; F. Hillebrand, *Die Pionierzeit des GSM-Standards und des D1-Netzes* (Wilburgstetten, Germany: Schlembach, 2013); F. Hillebrand, ed., *GSM and UMTS: The Creation of Global Mobile Communication* (New York: Wiley, 2002); G. Garrard, *Cellular Communications: Worldwide Market Development* (Boston: Artech House, 1998), 125–170; and T. Haug, "A Commentary on Standardization Practices: Lessons from the NMT and GSM Mobile Telephone Standards Histories," *Telecommunications Policy* 26, nos. 3–4 (April–May 2002): 101–107. See also A. Manninen, "Elaboration of NMT and GSM Standards: From Idea to Market" (PhD diss., Faculty of Humanities of the University of Jyväsklylä, Finland, 2002). Some of the most important GSM original documents are still available online: "History of GSM: Birth of the Mobile Revolution," accessed January 8, 2018, http://www.gsmhistory.com/rare-gsm-documents/.

9. Department of Trade and Industry, "Phones on the Move: Personal Communications in the 1990s" (Department of Trade and Industry, London, 1989). See T. Valletti and M. Cave, "Competition in UK Mobile Communications," *Telecommunications Policy* 22, no. 2 (March 1998): 109–131; and Bekkers, *Mobile Telecommunications Standards*, 296–299.

10. The details are in Mock, *Qualcomm Equation*, 21–134. Shorter treatments are in Garrard, *Cellular Communications*, 313–355; and Bekkers, *Mobile Telecommunications Standards*, 347–358.

11. Gruber, *Economics of Mobile Telecommunications*, 131–132.

12. Bekkers, *Mobile Telecommunications Standards*, 353–355.

13. We draw mainly from T. Salmon, "Spectrum Auctions by the United States Federal Communications Commission," in *Auctioning Public Assets*, ed. M. Janssen (Cambridge: Cambridge University Press, 2004), 147–176; and P. Cramton, "Spectrum Auctions," in *Handbook of Telecommunications Economics*, vol. 1, *Structure, Regulation, and Competition*, ed. M. Cave, S. Majumdar, and I. Vogelsang (Amsterdam: Elsevier, 2002), 605–639.

14. On auction formats, see, among others, P. Milgrom, "Auctions and Bidding: A Primer," *Journal of Economic Perspectives* 3, no. 3 (1989): 3–22.

15. Y. Yim, A. Kanafani, and J. Ygnace, "Expanding Usage of Cellular Phones: User Profile and Transportation Issues" (PATH Research Report, Institute of Transportation Studies, University of California at Berkeley, December 1991), 3–4.

16. Yim, Kanafani, and Ygnace, 3.

17. T. Parish, *Case Studies of Market Research for Three Transportation Communication Products* (Cambridge, MA: Arthur D. Little, March 1994), 35.

18. Yim, Kanafani, and Ygnace, "Expanding Usage of Cellular Phones," 2.

19. Garrard, *Cellular Communications*, 259.

20. Organisation de coopération et de développement économiques (OCDE), *Téléphonie mobile: Structures et tendances en matière de prix* (Paris: OCDE, 2000).

21. FCC, *6th Annual Commercial Mobile Services Competition Report* (Washington, DC: FCC, 2001), 27–34.

22. For background on the dissolution of the Soviet Union, see especially S. Kotkin, *Armageddon Averted: The Soviet Collapse, 1970–2000* (Oxford: Oxford University Press, 2008).

23. Garrard, *Cellular Communications*, 377–378.

24. OCDE, *Téléphonie mobile*, 69–75; A. Taylor and J. Vincent, "An SMS History," in *Mobile World: Past, Present and Future*, ed. L. Hamill and A. Lasen (New York: Springer, 2005), 75–91; P. Curwen, *The Future of Mobile Communications: Awaiting the Third Generation* (New York: Palgrave Macmillan, 2002), 16–18.

25. On the very first text message, see OCDE, *Téléphonie mobile*, 69. The source of the statement regarding the very first SMS appears to be S. Buckingham, "Success 4 SMS," Mobile Streams, February 2001, https://ftp.unpad.ac.id/orari/library/library-ref-eng/ref-eng-2/physical/wireless/cellullar/sms/sms.pdf.

26. On the history of the Internet, see M. Campbell-Kelly and D. Garcia-Swartz, "The History of the Internet: The Missing Narratives," *Journal of Information Technology* 28, no. 1 (2013): 18–33.

27. Taylor and Vincent, "SMS History," 82–84.

CHAPTER 5

1. Sweden, France, and Japan also had competitive markets in network services during the 1980s, but with restrictions.

2. T. Valletti and M. Cave, "Competition in UK Mobile Communications," *Telecommunications Policy* 22, no. 2 (March 1998): 109–131; H. Gruber, *The Economics of Mobile Telecommunications* (Cambridge: Cambridge University Press, 2005), 98–105; G. Garrard, *Cellular Communications: Worldwide Market Development* (Boston: Artech House, 1998), 171–217.

3 Department of Trade and Industry, "Phones on the Move: Personal Communications in the 1990s" (Department of Trade and Industry, London, 1989).

4. R. Bekkers, *Mobile Telecommunications Standards: GSM, UMTS, TETRA, and ERMES* (Boston: Artech House, 2001), 297.

5. Bekkers, 296–298.

6. For details on the evolution of prices in the British cellular market, see Valletti and Cave, "Competition in UK Mobile Communications," esp. 128–131.

7. The key references on the evolution of cellular in the United States are the Federal Communications Commission Commercial Mobile Radio Services Competition reports, the first one of which was published in 1995. Excellent summaries are in Gruber, *Economics of Mobile Telecommunications*, 123–135; and Garrard, *Cellular Communications*, 313–355.

8. M. Berlage and T. Schnöring, "The Introduction of Competition in German Mobile Telecommunications Markets," in *Mobile Telecommunications: Emerging European Markets*, ed. K. Schenk, J. Müller, and T. Schnöring (Boston: Artech House, 1995), 29–49; Gruber, *Economics of Mobile Telecommunications*, 105–108; Garrard, *Cellular Communications*, 252–265. See also M. Stoetzer and D. Tewes, "Competition in the German Cellular Market? Lessons of Duopoly," *Telecommunications Policy* 20, no. 4 (May 1996): 303–310.

9. Whether BMPT was a truly independent regulator is a matter of debate in the literature. See, for example, D. Gillick, "Telecommunications Policies and Regulatory Structures," *Telecommunications Policy* 16, no .9 (December 1992): 726–732. Gillick suggests that, because BMPT was a government ministry, it was not a truly independent regulator.

10. On interconnect and leased-line fees, see Garrard, *Cellular Communications*, 228–229.

11. See esp. Stoetzer and Tewes, "Competition?"

12. Gruber, *Economics of Mobile Telecommunications*, 108–111; Garrard, *Cellular Communications*, 287–293.

13. C. Gambini, P. Ravazzi, and T. Valletti, *Il mercato delle telecomunicazioni* (Bologna: Mulino, 2003), 123–126.

14. On the confrontations between the European Commission and several European countries regarding telecommunications liberalization issues, see P. Curwen, *Restructuring Telecommunications: A Study of Europe in a Global Context* (New York: St. Martin's, 1977), 64–68.

15. Gruber, *Economics of Mobile Telecommunications*, 116–117; Garrard, *Cellular Communications*, 278–280.

16. H. Sousa, "Telecommunications Reforms in Portugal and the European Union," *Telecommunications Policy* 20, no. 9 (November 1996): 653–667.

17. For an excellent history of China up to the early 2010s, see K. Mühlhahn, *Making China Modern: From the Great Qing to Xi Jinping* (Cambridge, MA: Belknap Press of Harvard University Press, 2019). For background on the Chinese economy, see esp. B. Naughton, *The Chinese Economy: Transitions and Growth* (Cambridge, MA: MIT Press, 2007). On the evolution of the Chinese telecommunications system, see L. Xiong-Jian and Z. You-Nong, "China," in *Telecommunications in the Pacific Basin: An Evolutionary Approach*, ed. E. Noam, S. Komatsuzaki, and D. Conn (New York: Oxford University Press, 1994), 73–85; K. Zita, "China: Steps toward Political and Financial Reform," in Noam, Komatsuzaki, and Conn, *Telecommunications in the Pacific Basin*, 86–105; M. Mueller and Z. Tan, *China in the Information Age: Telecommunications and the Dilemmas of Reform* (Westport, CT: Praeger, 1997); X. Yan and

D. Pitt, *Chinese Telecommunications Policy* (Boston: Artech House, 2002); D. Lu and C. Wong, *China's Telecommunications Market: Entering a New Competitive Age* (Cheltenham, UK: Edward Elgar, 2003); S. Guan, *China's Telecommunications Reform: From Monopoly towards Competition* (New York: Nova Science, 2003); and esp. E. Harwitt, *China's Telecommunications Revolution* (Oxford: Oxford University Press, 2008). On mobile phones in China, see esp. X. Wang, "Mobile Communications in China: An Analysis of the Diffusion Process" (PhD diss., Ohio State University, 1995); and J. Wang and C. Cheng, "The History of the Mobile Phone in China," in *Mobile Communication and Greater China*, ed. R. Chu et al. (London: Routledge, 2012), 64–79. For a recent summary of the evolution of Chinese telecommunications generally, see J. Xia, "China's Telecommunications Evolution, Institutions, and Policy on the Eve of 5G: A Two-Decade Retrospect and Prospect," *Telecommunications Policy* 41, no. 10 (November 2017): 931–947.

18. Harwitt, *China's Telecommunications Revolution*, 67–69.

19. For background on the Indian economy, see V. Joshi and I. Little, *India: Macroeconomics and Political Economy, 1964–1991* (Washington, DC: World Bank, 1994); and V. Joshi and I. Little, *India's Economic Reforms, 1991–2001* (Delhi: Oxford University Press, 1998). On the telecommunications system in India, see B. Mody, "Liberalization of Telecommunications in India in the Mid-1990s," in *Telecommunications in Western Asia and the Middle East*, ed. E. Noam (New York: Oxford University Press, 1997), 3–19; and T. Chowdary, "Telecommunications in India," in Noam, *Telecommunications in Western Asia*, 20–37. On the reforms of the 1990s, see M. Athreya, "India's Telecommunications Policy: A Paradigm Shift," *Telecommunications Policy* 20, no. 1 (January–February 1996): 11–22. Important sources on cellular phones in India are A. Desai, *India's Telecommunications Industry: History, Analysis, Diagnosis* (New Delhi: Sage, 2006); and A. Doron and R. Jeffrey, *The Great Indian Phone Book* (Cambridge, MA: Harvard University Press, 2013), esp. 39–62.

20. For a complete list of the companies that received licenses, see Desai, *India's Telecommunications Industry*, chap. 2, appendix 2, and appendix 3.

21. Desai, 84–89.

22. On the US regional markets during the 1980s, see P. Parker and L. Röller, "Collusive Conduct in Duopolies: Multimarket Contact and Cross-Ownership in the Mobile Telephone Industry," *RAND Journal of Economics* 28, no. 2 (Summer 1997): 304–322; and M. Busse, "Multimarket Contact and Price Coordination in the Cellular Telephone Industry," *Journal of Economics and Management Strategy* 9, no. 3 (2000): 287–320. On Britain, see Valletti and Cave, "Competition in UK Mobile Communications." On Sweden, see S. Hultén and B. Mölleryd, "Mobile Telecommunications in Sweden," in Schenk, Müller, and Schnöring, *Mobile Telecommunications*, 1–28. In Sweden, there was a duopoly from the early 1980s, but one of the duopolists (the PTT) was much larger than the other (Comvik) in number of subscribers. Comvik remained a small cellular carrier throughout the 1980s.

23. On price declines during the 1990s in the US regional markets, especially after 1995, see J. Hausman, "Cellular Telephone, New Products, and the CPI," *Journal of Business and Economic Statistics* 17, no. 2 (April 1999): 188–194. On price changes in

the British market in the 1990s, see Valletti and Cave, "Competition in UK Mobile Communications," esp. 128–131. On price declines in the German market in the 1990s (in the transition from monopoly to duopoly first and to oligopoly with more than two carriers later), see Stoetzer and Tewes, "Competition?" On price changes in the Swedish market in the 1990s, see Hultén and Mölleryd, "Mobile Telecommunications in Sweden."

CHAPTER 6

1. The distinction between committee-based and market-based competition in cellular phones comes from J. Funk, *Global Competition between and within Standards: The Case of Mobile Phones* (Houndmills, UK: Palgrave, 2002).

2. In our analysis of the history of AirTouch, we draw from L. Galambos and E. Abrahamson, *Anytime, Anywhere: Entrepreneurship and the Creation of a Wireless World* (Cambridge: Cambridge University Press, 2002).

3. G. Garrard, *Cellular Communications: Worldwide Market Development* (Boston: Artech House, 1998), 377–387.

4. Garrard, 368–377.

5. C. Lazzareschi, "AT&T Near Deal to Buy Third of McCaw Cellular," *Los Angeles Times*, November 5, 1992.

6. "Company News; PacTel Will Change Name to AirTouch after Spinoff," *New York Times*, February 17, 1994.

7. See "A Brief History of 2-Way Radio Communications & Push-to-Talk," *Peak PTT Blog*, April 17, 2021, https://blog.peakptt.com/a-brief-history-of-2-way-radio-communications-push-to-talk/. See also M. O'Brien, "Looking Back while Going Forward: How the Early Days of Nextel Reflect on Today," RCRWireless, February 17, 2009, https://www.rcrwireless.com/20090217/wireless/looking-back-while-going-forward-how-the-early-days-of-nextel-reflect-on-today. For details on the evolution of Nextel, see the FCC's Commercial Mobile Radio Services Competition reports for the 1995–2001 period.

8. L. Helm, "Evolving Wireless World: AirTouch, US West to Merge Domestic Cellular Operations," *Los Angeles Times*, July 26, 1994.

9. E. Andrews, "Company News; AT&T Completes Deal to Buy McCaw Cellular," *New York Times*, September 20, 1994.

10. FCC, *2nd Annual Commercial Mobile Services Competition Report* (Washington, DC: FCC, 1997), 11–13. See also "National Alliance Places First Call on PCS System," PR Newswire, October 31, 1995.

11. FCC, *2nd Annual*, table 4.

12. "PCS PrimeCo Picks Wireless Technology," *New York Times*, June 7, 1995.

13. D. Mock, *The Qualcomm Equation* (New York: Amacom, 2005), 128–130.

14. For details on the history of Vodafone, see T. Merriden, *Rollercoaster: The Turbulent Life and Times of Vodafone and Chris Gent* (Oxford: Capstone, 2003); and C. Ibbott, *Global Networks* (New York: Palgrave Macmillan, 2007).

15. "AT&T Launches Flat-Rate Wireless Service," *Los Angeles Times*, May 8, 1998.

16. L. Laird, "US Authorities Clear Vodafone Merger with AirTouch," *Guardian*, June 24, 1999.

17. L. Cauley, "Bell Atlantic and Vodafone Plan IPO of Verizon Wireless Joint Venture," *Wall Street Journal*, April 5, 2000.

18. G. Naik and A. Raghavan, "Vodafone, Mannesmann Set Takeover at $180.95 Billion after Long Struggle," *Wall Street Journal*, February 4, 2000. For a retrospective from a German perspective, see F. Hubik, "Wie der 'Haifisch' das 'Hirn' besiegte," *Handelsblatt*, April 2, 2015.

19. Here we draw from I. Schmideg, "Mobile Telecommunications in Hungary," in *Mobile Telecommunications: Emerging European Markets*, ed. K. Schenk, J. Müller, and T. Schnöring (Boston: Artech House, 1995), 89–118; and Garrard, *Cellular Communications*, 378–382.

20. Garrard, *Cellular Communications*, 378.

21. Garrard, 382–386; M. Kets de Vries et al., *The New Russian Business Leaders* (Northampton, MA: Edward Elgar, 2004), 176–197.

22. This section draws from Garrard, *Cellular Communications*, 231–237.

23. Organisation de coopération et de développement économiques, *Téléphonie mobile: Structures et tendances en matière des prix* (Paris: Organisation de coopération et de développement économiques, 2000), 87. The countries included in figure 6.3 are the following: the United States, Japan, South Korea, the Netherlands, Canada, Denmark, Germany, Britain, Australia, Austria, Sweden, Italy, Belgium, Finland, France, Greece, Ireland, Mexico, Poland, Portugal, Spain, Switzerland, Hungary, Turkey, the Czech Republic, Iceland, Luxembourg, New Zealand, and Norway.

24. Funk, *Global Competition*, 93–205.

25. R. Bekkers, *Mobile Telecommunications Standards: GSM, UMTS, TETRA, and ERMES* (Boston: Artech House, 2001), 313–347.

CHAPTER 7

1. R. Bekkers, *Mobile Telecommunications Standards: GSM, UMTS, TETRA, and ERMES* (Boston: Artech House, 2001), 583–584.

2. This paragraph is based on calculations made by the authors on the basis of information presented in P. Curwen and J. Whalley, *Mobile Telecommunications in a High-Speed World: Industry Structure, Strategic Behavior and Socio-economic Impact* (Farnham, UK: Gower, 2010), 68–69. A distinction needs to be made between networks that were launched on a specific standard and networks that were operating on that standard at any given point in time. A network that was originally launched on a standard but was later converted to one of the standard's "evolutions" was no longer operating on the original standard from the time of the conversion—it was operating on the standard's evolution.

3. PricewaterhouseCoopers, *Technology Forecast: 2002–2004*, vol. 2, *Emerging Patterns of Internet Computing* (Menlo Park, CA: PricewaterhouseCoopers Global Technology Centre, 2002), 293–354.

4. C. Giachetti, *Competitive Dynamics in the Mobile Phone Industry* (New York: Palgrave Macmillan, 2013), 63.

5. Risk and Policy Analysts, *Study on the Impact of the MoU on Harmonisation of Chargers for Mobile Telephones and to Assess Possible Future Options* (Loddon, UK: Risk and Policy Analysts, prepared for DG Enterprise and Industry, August 22, 2014), 12, https://rpaltd.co.uk/uploads/report_files/j829-mobile-chargers.pdf.

6. "Gartner Says Worldwide Smartphone Sales Reached Its Lowest Growth Rate with 3.7 Per Cent Increase in Fourth Quarter of 2008," PresseBox, March 11, 2009, https://www.pressebox.com/pressrelease/gartner-uk-ltd/Gartner-Says-Worldwide-Smartphone-Sales-Reached-Its-Lowest-Growth-Rate-With-3-7-Per-Cent-Increase-in-Fourth-Quarter-of-2008/boxid/246446.

7. "ComScore Reports December 2009 U.S. Mobile Subscriber Market Share," Comscore, February 8, 2010, https://www.comscore.com/Insights/Press-Releases/2010/2/comScore-Reports-December-2009-U.S.-Mobile-Subscriber-Market-Share.

8. Bekkers, *Mobile Telecommunications Standards*, 453–497. See also F. Hillebrand, ed., *GSM and UMTS: The Creation of Global Mobile Communication* (New York: Wiley, 2002); and J. Funk, *Global Competition between and within Standards: The Case of Mobile Phones* (Houndmills, UK: Palgrave, 2002), 206–245. The key technical features of the 3G standards are succinctly presented in K. Poupée, *La téléphonie mobile* (Paris: Presses universitaires de France, 2003), 75–88.

9. See, for example, P. Dupuis, "The UMTS Standardization Work in ETSI: Section 1: The Initial Work (up to Spring 1996)," and F. Hillebrand, "The UMTS Standardization Work in ETSI: Section 2: The Creation of the UMTS Foundation in ETSI from April 1996 to February 1999," both in Hillebrand, *GSM and UMTS*, 179–220.

10. For more technical details, see A. Ghosh et al., *Fundamentals of LTE* (Upper Saddle River, NJ: Prentice Hall, 2011), 13–14.

11. See esp. X. Yan, *3G Mobile Policy: The Case of China and Hong Kong*, ITU Telecommunication Case Studies (Office of the Secretary General of the International Telecommunications Union, June 2001), https://www.itu.int/osg/spu/ni/3G/casestudies/china/China_3g_Final.PDF.

12. "China's 3G Technology Gamble: Who Has the Last Laugh?," Knowledge@Wharton, University of Pennsylvania, Wharton School of Business, July 6, 2011, https://knowledge.wharton.upenn.edu/article/chinas-3g-technology-gamble-who-has-the-last-laugh/.

13. "China's 3G Network Deployment Update: CDMA and WCDMA on Their Way while TS-SCDMA Slower to Materialize," IHS Markit, September 21, 2009, https://technology.ihs.com/393444/chinas-3g-network-deployment-update.

14. J. Heggestuen, "There's a Race on for 3G Subscribers in China," Business Insider, September 25, 2013, https://www.businessinsider.com.au/the-3g-market-in-china-2013-9.

15. See J. Korhonen, *Introduction to 3G Mobile Communications*, 2nd ed. (Boston: Artech House, 2003), 5–8. For more details, see Poupée, *La téléphonie mobile*, 42–57; and Ghosh et al., *Fundamentals of LTE*, 1–43.

16. Risk and Policy Analysts, *Study on the Impact*, 16.

17. In our analysis of the UMTS auctions, we rely mostly on P. Klemperer, "How (Not) to Run Auctions: The European 3G Telecom Auctions," in *Spectrum Auctions and Competition in Telecommunications*, G. Illing and U. Klüh (Cambridge, MA: MIT Press, 2003), 97–118.

18. On auction types, see R. McAfee and J. McMillan, "Auctions and Bidding," *Journal of Economic Literature* 25 (June 1987): esp. 701–703.

19. For details on the simultaneous ascending design and the outcome of the British 3G auction, see especially K. Binmore and P. Klemperer, "The Biggest Auction Ever: The Sale of the British 3G Telecom Licenses," *Economic Journal* 112 (March 2002): 74–96.

20. Klemperer, "How (Not) to Run Auctions." More details on the Dutch auction are in E. van Damme, "The European UMTS Auctions," *European Economic Review* 46, no. 4–5 (May 2002): 846–858.

21. P. Curwen, *The Future of Mobile Telecommunications: Awaiting the Third Generation* (New York: Palgrave Macmillan, 2002), 91–127.

22. S. Tweedie, "The World's First Smartphone, Simon, Was Created 15 Years before the iPhone," Business Insider, June 14, 2015, https://www.businessinsider.com/worlds-first-smartphone-simon-launched-before-iphone-2015-6. See also C. O'Malley, "Simonizing the PDA," *Byte*, December 1994, http://web.archive.org/web/19990221174856/byte.com/art/9412/sec11/art3.htm.

23. Here we draw often from M. Campbell-Kelly et al., "Economics and Business Perspectives on Smartphones as Multi-sided Platforms," *Telecommunications Policy* 39, no. 8 (September 2015): 717–734.

24. The fundamental reference on Palm is A. Butter and D. Pogue, *Piloting Palm: The Inside Story of Palm, Handspring, and the Birth of the Billion-Dollar Handheld Industry* (New York: Wiley, 2002). See also D. Jackson, "Palm-to-Palm Combat," *Time*, June 24, 2001.

25. W. Greenwald, "R.I.P. Palm: A History of the Brand That Changed Smartphones and PDAs Forever," *PCMag*, February 11, 2011, https://www.pcmag.com/article2/0,2817,2379948,00.asp. See also Tom Spring, "Wireless Palm VII ships," CNN, May 24, 1999, http://www.cnn.com/TECH/computing/9905/24/palm7.idg/.

26. B. Brown and M. Brown, "Handspring Treo 180," *PCMag*, January 4, 2002, https://www.pcmag.com/article2/0,2817,2442,00.asp.

27. See C. Tilley, "The History of Microsoft Windows CE," HPC:Factor, February 18, 2001, https://www.hpcfactor.com/support/windowsce/.

28. On the history of Symbian, see A. Orlowski, "Symbian, the Secret History: Dark Star," Register, November 23, 2013, https://www.theregister.co.uk/2010/11/23/symbian_history_part_one_dark_star/; J. West and D. Wood, "Evolving an Open Ecosystem: The Rise and Fall of the Symbian Platform," in *Advances in Strategic Management*, vol. 30, *Collaboration and Competition in Business Ecosystems*, ed. R. Adner et al. (Bingley, UK: Emerald Group, 2013), 27–67; and esp. D. Wood, *Smartphones and Beyond: Lessons from the Remarkable Rise and Fall of Symbian* (Amazon Digital Services, 2014).

29. M. Campbell-Kelly and D. Garcia-Swartz, *From Mainframes to Smartphones: A History of the International Computer Industry* (Cambridge, MA: Harvard University Press, 2015), 105–123.

30. There are a number of references on the evolution of RIM. We draw heavily from J. McNish and S. Silcoff, *Losing the Signal: The Untold Story behind the Extraordinary Rise and Spectacular Fall of BlackBerry* (New York: Flatiron Books, 2015). For background on the evolution of electronic mail in the 1980s and the 1990s, see M. Campbell-Kelly and D. Garcia-Swartz, "The History of the Internet: The Missing Narratives," *Journal of Information Technology* 28, no. 1 (2013): 18–33; and M. Campbell-Kelly and D. Garcia-Swartz, "From Products to Services: The Software Industry in the Internet Era," *Business History Review* 81, no. 4 (2007): 735–764.

31. "Photos: BlackBerrys through the Ages," ZDNet, March 21, 2018, https://www.zdnet.com/pictures/photos-blackberrys-through-the-ages/.

32. Funk, *Global Competition*, 218.

33. The core references on i-mode are T. Natsuno, *I-mode Strategy* (Chichester, UK: Wiley, 2003); and T. Natsuno, *The i-mode Wireless Ecosystem* (Chichester, UK: Wiley, 2003). See also D. Steinbock, *The Mobile Revolution: The Making of Mobile Services Worldwide* (London: Kogan Page, 2005), 66–75; and I. Kunii, "Amazing DoCoMo," with S. Baker, *Business Week*, January 17, 2000. See also Poupée, *La téléphonie mobile*, 113–120.

34. For details on pricing and subscriber growth, see esp. Natsuno, *I-mode Strategy*, 1–29.

35. On this issue, see, for example, E. Markova, *Liberalization and Regulation of the Telecommunications Sector in Transition Countries: The Case of Russia* (Heidelberg: Physica-Verlag, 2009), 53.

CHAPTER 8

1. On telecommunications in New Zealand, see P. McCabe, "New Zealand: The Unique Experiment in Deregulation," in *Telecommunications in the Pacific Basin: An Evolutionary Approach*, ed. E. Noam, S. Komatsuzaki, and D. Conn (New York: Oxford University Press, 1994), 393–419; H. Dordick, "New Zealand: Testing the Limits of Nonregulation," in Noam, Komatsuzaki, and Conn, *Telecommunications in the Pacific Basin*, 420–439; and esp. D. Geradin and M. Kerf, *Controlling Market Power in Telecommunications* (Oxford: Oxford University Press, 2003), 119–162. On cellular phones in New Zealand, see G. Garrard, *Cellular Communications: Worldwide Market Development* (Boston: Artech House, 1998), 368.

2. Merrill Lynch, *Global Wireless Matrix 2Q04* (Merrill Lynch, September 29, 2004), 100.

3. P. Curwen, *The Future of Mobile Communications: Awaiting the Third Generation* (New York: Palgrave Macmillan, 2002), 92.

4. H. Gruber, *The Economics of Mobile Telecommunications* (Cambridge: Cambridge University Press, 2005), 135–139; Garrard, *Cellular Communications*, 359–363.

5. For details on the merger, see "Merger of DDI, KDD, and IDO," December 16, 1999, available at http://www.kddi.com/english/corporate/news_release/archive/kdd/press-e99/e99-072n.html.

6. Curwen, *Future of Mobile Communications*, 92.

7. Merrill Lynch, *Global Wireless Matrix 2Q04*, 90.

8. K. Choo and M. Kang, "South Korea: Structure and Changes," in Noam, Komatsuzaki, and Conn, *Telecommunications in the Pacific Basin*, 286–302. A crucial source on mobile phones in Korea is W. Jho, *Building Telecom Markets: Evolution of Governance in the Korean Mobile Telecommunications Market* (New York: Springer, 2014).

9. D. Mock, *The Qualcomm Equation* (New York, Amacom, 2005), 82–85.

10. Merrill Lynch, *Global Wireless Matrix 2Q04*, 92.

11. Curwen, *Future of Mobile Communications*, 92.

12. On Malaysia, see F. Onn, "Malaysia," in Noam, Komatsuzaki, and Conn, *Telecommunications in the Pacific Basin*, 132–154; Garrard, *Cellular Communications*, 397–398; and Organisation for Economic Co-operation and Development (OECD), *Mobile Cellular Communications: Pricing Strategies and Competition* (Paris: OECD, 1996), box 6. On Thailand, see S. Charmonman, "Thailand," in Noam, Komatsuzaki, and Conn, *Telecommunications in the Pacific Basin*, 201–216; Garrard, *Cellular Communications*, 400–401; and OECD, *Mobile Cellular Communications*, box 6. On Hong Kong, see J. Day, "Hong Kong," in Noam, Komatsuzaki, and Conn, *Telecommunications in the Pacific Basin*, 242–264; Garrard, *Cellular Communications*, 394–396; and OECD, *Mobile Cellular Communications*, box 6. On Singapore, see E. You, "Singapore," in Noam, Komatsuzaki, and Conn, *Telecommunications in the Pacific Basin*, 265–285; and Garrard, *Cellular Communications*, 399. On Taiwan, see T. Fan-Tung and M. Chi-Kuo, "Taiwan," in Noam, Komatsuzaki, and Conn, *Telecommunications in the Pacific Basin*, 315–335; Garrard, *Cellular Communications*, 400; and esp. S. Chen, "Telecommunications Liberalization: A Taiwanese Perspective," in *Deregulation and Interdependence in the Asia-Pacific Region*, ed. T. Ito and A. Krueger (Chicago: University of Chicago Press, 2000), 327–350.

13. Curwen, *Future of Mobile Communications*, 92–93.

14. For an excellent study of growth and development in East Asia, see D. Perkins, *East Asian Development: Foundations and Strategies* (Cambridge, MA: Harvard University Press, 2013). A table comparing GDP per capita growth in purchasing-power-parity terms at 2005 prices between 1960 and 2010 can be found in Perkins, 2. For a table with growth broken down by decade, see Perkins, 6.

15. M. Campbell-Kelly and D. Garcia-Swartz, *From Mainframes to Smartphones: A History of the International Computer Industry* (Cambridge, MA: Harvard University Press, 2015), 144–148.

16. Perkins, *East Asian Development*, 92–93.

17. On corporatization, see esp. E. Noam, *Telecommunications in Europe* (New York: Oxford University Press, 1992), 61–64.

18. D. Rosenne, B. Erez, and S. Hai, "Telecommunications in Israel: Advanced Technology, Evolving Competition," in *Telecommunications in Western Asia and the Middle East*, ed. E. Noam (New York: Oxford University Press, 1997), 90–125; N. Gandal, "New Horizons: Telecommunications Policy in Israel in the 21st Century," *Israel Economic Review* 2 (2003): 101–111; A. Schejter, "Israeli Cellular Telecommunications Policy," *Telecommunications Policy* 30 (2006): 14–28; Garrard, *Cellular Communications*, 407–408.

19. S. Teitelbaum, "Cellular Obsessions," *Wired*, January 1, 1997, https://www.wired.com/1997/01/ffisraeli/.

20. Merrill Lynch, *Global Wireless Matrix 2Q04*, 86.

21. Curwen, *Future of Mobile Communications*, 92.

22. J. Funk, *Global Competition between and within Standards: The Case of Mobile Phones* (Houndmills, UK: Palgrave, 2002), 43–44.

23. A. Botelho et al., "Brazil," in *Telecommunications in Latin America*, ed. E. Noam (New York: Oxford University Press, 1998), 227–250; M. Wohlers, "The Brazilian Way of Telecommunications Reform," in Noam, *Telecommunications in Latin America*, 251–259. On cellular phones in Brazil, see Garrard, *Cellular Communications*, 372–373; C. Piragibe, "Competition and Globalization: Brazilian Telecommunications Policy at the Crossroads," unpublished paper, 2001; and ANATEL, *Review 2001: Including the Excluded* (Brasília, 2002). A detailed source on the process of cellular liberalization in Brazil is M. Maciel, J. Whalley, and R. van der Meer, "Foreign Investment and Consolidation in the Brazilian Mobile Telecommunications Market," *info* 8, no. 3 (2006): 60–77.

24. Funk suggests that the first CDMA network in Brazil was established in 1996; see his *Global Competition*, 45. Maciel, Whalley, and van der Meer imply that the transition to digital cellular accelerated in or around 1998; see "Foreign Investment and Consolidation," 64.

25. ANATEL, *Review 2001*, 38–53.

26. GSMA, *Brazil Mobile Observatory 2012* (London: GSMA, September 2012), 18, https://www.gsma.com/spectrum/wp-content/uploads/2012/10/gsma_brazil_obs_web_09_12-1.pdf.

27. K. Griffith, "Mexico," in Noam, *Telecommunications in Latin America*, 166–191.

28. R. Schwartz, *Wireless Communications in Developing Countries* (Boston: Artech House, 1996), 39–40; Garrard, *Cellular Communications*, 369–371.

29. Merrill Lynch, *Global Wireless Matrix 2Q04*, 96.

30. OECD, *Review of Telecommunication Policy and Regulation in Mexico* (Paris: OECD, 2012), esp. 59–60.

31. GSMA, *Country Overview: Mexico* (London: GSMA, 2016), 11, https://www.gsma.com/latinamerica/wp-content/uploads/2016/06/report-mexico2016-EN.pdf.

32. According to the Penn World Tables, Mexico had higher per capita GDP than Brazil in purchasing-power-parity terms between 1980 and 2010 (although Brazil

likely experienced faster growth in per capita income than Mexico over the period). Further, population density in Mexico has been between two and three times higher than in Brazil for decades.

33. Recent studies have pointed out that cellular density grew more slowly in Mexico than in other Latin American countries also because of its socioeconomic conditions, especially wide income disparities and high poverty levels. In addition, the regulatory framework (scarce spectrum and high taxes on carriers) conspired against cellular growth. See GSMA, *Country Overview: Mexico,* 20.

CHAPTER 9

1. Here we draw from P. Curwen and J. Whalley, *Mobile Telecommunications in a High-Speed World: Industry Structure, Strategic Behavior and Socio-economic Impact* (Farnham, UK: Gower, 2010), 163–183.

2. Recent studies point out that Mexico's market structure is somewhat unusual in that the very few carriers in operation have highly unequal shares. As of 2016, Telcel (América Móvil) had 67 percent of the market, Movistar (Telefónica) had 24 percent, and AT&T, which acquired Iusacell and Nextel in 2015, had 9 percent. See GSMA, *Country Overview: Mexico* (London: GSMA, 2016), 24, https://www.gsma.com/latinamerica/wp-content/uploads/2016/06/report-mexico2016-EN.pdf.

3. For details, see esp. P. Curwen and J. Whalley, *The Internationalisation of Mobile Telecommunications: Strategic Challenges in a Global Market* (Cheltenham, UK: Edward Elgar, 2008), 165–188.

4. The idea that cellular network-services markets are "natural oligopolies" is suggested in H. Gruber, *The Economics of Mobile Telecommunications* (Cambridge: Cambridge University Press, 2005), 7.

5. Gruber, 50–51.

6. Gruber, 282–285.

7. See "Vodafone's Search for the G Spot," *Economist,* August 22, 2019, https://www.economist.com/business/2019/08/22/vodafones-search-for-the-g-spot.

8. An excellent source on the evolution of the Brazilian cellular market is M. Maciel, J. Whalley, and R. van der Meer, "Foreign Investment and Consolidation in the Brazilian Mobile Telecommunications Market," *info* 8, no. 3 (2006): 60–77.

9. Here we draw from M. Kets de Vries et al., *The New Russian Business Leaders* (Northampton, MA: Edward Elgar, 2004), chap. 6. Other important sources on cellular phones in Russia include N. Dobrovolskaya and A. Salueña, *Development of Russian Mobile Communications* (Lappeenranta, Finland: Lappeenranta University of Technology, Northern Dimension Research Centre, 2004); and E. Markova, *Liberalization and Regulation of the Telecommunications Sector in Transition Countries: The Case of Russia* (Heidelberg: Physica-Verlag, 2009), esp. 52–57.

10. Mobile TeleSystems, *MTS Annual Report 2005* (Moscow: Mobile TeleSystems, 2006), 3.

11. Merrill Lynch, *Global Wireless Matrix 2Q04* (Merrill Lynch, September 29, 2004), 112.

12. Gruber, *Economics of Mobile Telecommunications*, 178–179.

13. The idea that technological discontinuities in the evolution of cellular phones created opportunities for new device makers is developed in C. Giachetti, *Competitive Dynamics in the Mobile Phone Industry* (New York: Palgrave Macmillan, 2013).

14. See J. Gertner, "The Fall and Rise of Iridium," *Wall Street Journal*, June 3, 2016, https://www.wsj.com/articles/the-fall-and-rise-of-iridium-1464980784; and esp. J. Bloom, *Eccentric Orbits: The Iridium Story* (New York: Grove, 2017).

15. A key reference on this subject is Nomura Equity Research, *U.S. Semiconductor Primer* (Nomura Equity Research, December 11, 2013), http://pdf.zacks.com/pdf/NO/H4560776.PDF. See also Scott Wilson, *Rising Tide: Exploring Pathways to Growth in the Mobile Semiconductor Industry* (Deloitte University Press, 2013).

16. Nomura Equity Research, *U.S. Semiconductor Primer*, 133–147.

17. See Korea Fair Trade Commission, In re Alleged Abuse of Market Dominance of Qualcomm Incorporated, Decision No. 2017-0-25, January 20, 2017, http://www.theamericanconsumer.org/wp-content/uploads/2017/03/2017-01-20_KFTC-Decision_2017-0-25.pdf.

18. Korea Fair Trade Commission, 19.

19. B. Petrovik, "Texas Instruments Shareholders Ask for a Pullback from the Mobile Chip Business," Android Authority, August 13, 2012, https://www.androidauthority.com/texas-instruments-mobile-chip-pullback-107479/.

20. Korea Fair Trade Commission, In re Alleged Abuse, 26.

21. See, among many others, J. Brodkin, "Broadcom Will Sell or 'Wind Down' Its Cellular Baseband Business," Ars Technica, June 2, 2014, https://arstechnica.com/information-technology/2014/06/broadcom-will-sell-or-wind-down-its-cellular-baseband-business/.

CHAPTER 10

1. O. Thomas, "Apple: Hello, iPhone," CNN Money, January 9, 2007, https://money.cnn.com/2007/01/09/technology/apple_jobs/.

2. See A. Ghosh et al., *Fundamentals of LTE* (Upper Saddle River, NJ: Prentice Hall, 2011), 1–43.

3. Ghosh et al., 24–25.

4. P. Curwen and J. Whalley, *Fourth Generation Mobile Communication: The Path to Superfast Connectivity* (Heidelberg: Springer, 2013), 1–25. See also E. Dashevky, "FAQ: How Is LTE-Advanced Different from Regular LTE?," *PCWorld*, January 21, 2014, https://www.pcworld.com/article/2083981/faq-how-is-lte-advanced-different-from-regular-lte.html.

5. See, for example, E. Lee and T. Chau, *Telecom Services: The Geopolitics of 5G and IoT*, Jefferies Franchise Note (Jefferies, September 14, 2017), 8–9, https://www.jefferies

.com/CMSFiles/Jefferies.com/files/Insights/TelecomServ.pdf. See also Curwen and Whalley, *Fourth Generation Mobile Communication*, 4; and N. Hide, "Why Your Next 4G Phone Should Have FDD-LTE and TDD-LTE," *CNET*, August 2, 2013, https://www.cnet.com/news/why-your-next-4g-phone-should-have-fdd-lte-and-tdd-lte/.

6. Ghosh et al., *Fundamentals of LTE*, 28–29.

7. "Intel Ships Its Next-Generation WiMAX Chip with Support for Mobile Networks," Intel, October 11, 2006, https://newsroom.intel.com/news-releases/intel-ships-its-next-generation-wimax-chip-with-support-for-mobile-networks/.

8. B. Crothers, "Intel WiMax to Go Live, Will Devices Follow?," *CNET*, September 25, 2008, https://www.cnet.com/news/intel-wimax-to-go-live-will-devices-follow/.

9. "Clearwire Goes Official with Mobile WiMAX, Will Use 'Let's Be Clear' Branding," RCR Wireless, December 1, 2008, https://www.rcrwireless.com/20081201/wireless/clearwire-goes-official-with-mobile-wimax-will-use-lets-be-clear-branding.

10. "Qualcomm Halts UMB Project, Sees No Major Cuts," Reuters, November 13, 2008, https://www.reuters.com/article/qualcomm-umb/qualcomm-halts-umb-project-sees-no-major-job-cuts-idUSN1335969420081113.

11. S. Musil, "WiMax Backers Plan Patent License Pool," *CNET*, June 8, 2008, https://www.cnet.com/news/wimax-backers-plan-patent-license-pool/.

12. J. Philip, "Qualcomm Will Not Participate in Future Spectrum Auctions in India: Dr. Paul Jacobs," *Economic Times*, October 16, 2012, https://economictimes.indiatimes.com/industry/telecom/qualcomm-will-not-participate-in-future-spectrum-auctions-in-india-dr-paul-jacobs/articleshow/16841410.cms.

13. S. Kinney, "Today Is the Last Day of Sprint WiMAX Service," RCR Wireless, March 31, 2016, https://www.rcrwireless.com/20160331/network-infrastructure/today-last-day-sprint-wimax-service-tag17.

14. See Curwen and Whalley, *Fourth Generation Mobile Communication*, esp. 28–36.

15. We draw on F. Vogelstein, "The Untold Story: How the iPhone Blew Up the Wireless Industry," *Wired*, January 9, 2008, https://www.wired.com/2008/01/ff-iphone/; F. Vogelstein, "How the Android Ecosystem Threatens the iPhone," *Wired*, April 14, 2011, https://www.wired.com/2011/04/mf-android/; F. Vogelstein, *Dogfight: How Apple and Google Went to War and Started a Revolution* (New York: Sarah Crichton Books, 2013); and B. Merchant, *The One Device: The Secret History of the iPhone* (New York: Little, Brown, 2017).

16. See, for example, F. Vogelstein, "And Then Steve Said, 'Let There Be an iPhone,'" *New York Times Magazine*, October 4, 2013, https://www.nytimes.com/2013/10/06/magazine/and-then-steve-said-let-there-be-an-iphone.html.

17. See, for example, J. Morris, "Palm Treo Review," *CNET*, November 17, 2003, https://www.cnet.com/reviews/palm-treo-review/.

18. J. Lendino, "Happy 4th Birthday, Android," *PCMag UK*, September 20, 2012, https://uk.pcmag.com/smartphones/60263/news/happy-4th-birthday-android.

19. Markoff, "I, Robot."

20. D. Roth, "Google's Open Source Android OS Will Free the Wireless Web," *Wired*, June 23, 2008, https://www.wired.com/2008/06/ff-android/.

21. E. Schonfeld, "Breaking: Google Announces Android and Open Handset Alliance," *TechCrunch*, November 5, 2007, https://techcrunch.com/2007/11/05/breaking-google-announces-android-and-open-handset-alliance/.

22. K. German, "A Brief History of Android Phones," CNET, August 2, 2011, https://www.cnet.com/tech/mobile/a-brief-history-of-android-phones/.

23. J. Skillings, "Nokia to Buy Symbian Outright, Launches App Effort," *CNET*, June 24, 2008, https://www.cnet.com/news/nokia-to-buy-symbian-outright-launches-app-effort/.

24. S. Shankland, "Nokia, Microsoft Becoming Windows Phone Bedfellows," *CNET*, February 11, 2011, https://www.cnet.com/news/nokia-microsoft-becoming-windows-phone-bedfellows/.

25. R. Cheng, "Nokia on the Edge: Inside an Icon's Fight for Survival," *CNET*, December 18, 2012, https://www.cnet.com/news/nokia-on-the-edge-inside-an-icons-fight-for-survival/.

26. D. Reisinger, "RIM Sheds Old Name, Becomes BlackBerry," *CNET*, January 30, 2013, https://www.cnet.com/news/rim-sheds-old-name-becomes-blackberry/.

27. The key reference on the evolution of the Chinese economy is B. Naughton, *The Chinese Economy: Transitions and Growth* (Cambridge, MA: MIT Press, 2007); on the process of economic reform that started in 1978, see esp. Naughton, 85–98.

28. Naughton, 406–410.

29. In recent years, the idea of Africa as a tech frontier, especially for mobile entrepreneurs, has appeared often in the business and popular press. See, among many others, M. Rigosoglio, "Africa Is the Next Frontier for Entrepreneurs," Insights by Stanford Business, January 1, 2011, https://www.gsb.stanford.edu/insights/africa-next-frontier-entrepreneurs; J. Schenker, "Africa: Tech's New Frontier," *Innovator*, May 24, 2018, https://innovator.news/africa-techs-new-frontier-332b9767a1e2; and J. Kahn, "Africa Is Tech's Next Great Frontier, GitHub A.I. Expert Says," *Fortune*, July 17, 2019, https://fortune.com/2019/07/17/africa-is-techs-next-great-frontier-github-ai/.

30. "The African Century: Africa Is Changing So Rapidly, It Is Becoming Hard to Ignore," *Economist*, March 26, 2020, https://www.economist.com/special-report/2020/03/26/africa-is-changing-so-rapidly-it-is-becoming-hard-to-ignore.

CHAPTER 11

1. See G. Garrard, *Cellular Communications: Worldwide Market Development* (Boston: Artech House, 1998), 402–406; and J. Funk, *Global Competition between and within Standards: The Case of Mobile Phones* (Houndmills, UK: Palgrave, 2002), 43–45. Historical per capita gross national product figures for Africa can be found in B. Dudley, "Decolonisation and the Problems of Independence," in *The Cambridge History of Africa*, ed. M. Crowder (Cambridge: Cambridge University Press, 1984), 8:90–91. Some sources suggest that Mauritius, an island nation in the Indian Ocean,

launched its first cellular system in 1989. See N. Gough et al., *Africa: The Impact of Mobile Phones*, Vodafone Policy Paper Series, No. 2 (March 2005), https://sarpn.org/documents/d0001181/P1309-Vodafone_March2005.pdf.

2. Garrard, *Cellular Communications*, 404.

3. For an excellent synthesis of the impact of mobile phones in Africa, see J. Aker and I. Mbiti, "Mobile Phones and Economic Development in Africa," *Journal of Economic Perspectives* 24, no. 3 (Summer 2010): 207–232.

4. E. Noam, introduction to *Telecommunications in Africa*, ed. E. Noam (Oxford: Oxford University Press, 1999), 3–12.

5. For a history of African countries since independence, see M. Meredith, *The State of Africa* (London: Simon and Schuster, 2011).

6. An interesting description of Africa before and after the arrival of mobile phones is in R. Southwood, *Less Walk, More Talk: How Celtel and the Mobile Phone Changed Africa* (Chichester, UK: Wiley, 2008), esp. 38–49.

7. The African figures come from Noam, introduction, 4. The numbers for the United States, Japan, and Sweden were computed by the authors from the International Telecommunications Union (ITU) database.

8. Southwood, *Less Walk, More Talk*, xiv–xv, 39.

9. R. Alemu, *The Liberalisation of the Telecommunications Sector in Sub-Saharan African and Fostering Competition in Telecommunications Services Markets* (Berlin: Springer, 2018), 25–26.

10. Alemu, 28.

11. "Democracy in Africa: Generation Game," *Economist*, March 7, 2020, 44–46.

12. See esp. Southwood, *Less Walk, More Talk*, 38–49.

13. Southwood, 47.

14. Southwood, 66–77.

15. Southwood, 40–41. A thorough discussion of the advantages and disadvantages of prepaid cellular systems in Africa and elsewhere is in R. Layton, "The Prepaid Mobile Market in Africa," in *The African Mobile Story*, ed. K. Skouby and I. Williams (Aalborg, Denmark: River, 2014), 51–77.

16. Southwood, *Less Walk, More Talk*, 43–44.

17. Southwood, 43.

18. L. Silver and C. Johnson, "Majorities in Sub-Saharan Africa Own Mobile Phones, but Smartphone Adoption Is Modest," Pew Research Center, October 9, 2018, https://www.pewresearch.org/global/2018/10/09/majorities-in-sub-saharan-africa-own-mobile-phones-but-smartphone-adoption-is-modest/.

19. For the history of M-PESA, see esp. T. Omwansa and N. Sullivan, *Money, Real Quick: Kenya's Disruptive Mobile-Money Innovation* (Kent, UK: Balloon View, 2012). See also K. Piper, "What Kenya Can Teach Its Neighbors—and the US—about Improving the Lives of the 'Unbanked,'" Vox, September 11, 2020, https://www.vox.com/future-perfect/21420357/kenya-mobile-banking-unbanked-cellphone-money.

20. Omwansa and Sullivan, *Money, Real Quick*, 24–25.

21. The details about how M-PESA was developed and tested are in Omwansa and Sullivan, *Money, Real Quick*, chap. 1.

22. "M-PESA Frequently Asked Questions," Vodafone, accessed December 1, 2020, https://www.vodafone.com/what-we-do/services/m-pesa/m-pesa-faqs.

23. See, for example, T. Suri and W. Jack, "The Long-Run Poverty and Gender Impacts of Mobile Money," *Science*, December 9, 2016, 1288–1292, https://www.jefftk.com/suri2016.pdf.

24. See, for example, H. Ahmed and B. Cowan, "Mobile Money and Healthcare Use: Evidence from East Africa" (NBER Working Paper 25669, National Bureau of Economic Research, Cambridge, MA, March 2019), https://www.nber.org/system/files/working_papers/w25669/w25669.pdf.

25. For a short explanation of the Arab Spring, see E. Blakemore, "What Was the Arab Spring and How Did It Spread?," *National Geographic*, March 19, 2019, https://www.nationalgeographic.com/culture/topics/reference/arab-spring-cause/. For detailed historical and sociological interpretations, see A. Bayat, *Revolution without Revolutionaries: Making Sense of the Arab Spring* (Stanford, CA: Stanford University Press, 2017); and M. Castells, *Networks of Outrage and Hope: Social Movements in the Internet Age* (Cambridge, UK: Polity, 2012).

26. Bayat, *Revolution without Revolutionaries*, 8–10.

27. See, for example, Blakemore, "What Was the Arab Spring?"

28. See, for example, Meredith, *State of Africa*, esp. 699–700.

29. See P. Howard et al., "Opening Closed Regimes: What Was the Role of Social Media During the Arab Spring?" (Working Paper 2011.1, Project on Information Technology and Political Islam, Seattle, WA, 2011), https://papers.ssrn.com/sol3/papers.cfm?abstract_id=2595096.

30. M. Manacorda and A. Tesei, "Liberation Technology: Mobile Phones and Political Mobilization in Africa" (CESifo Working Paper 5904, Center for Economic Studies and Ifo Institute, Munich, June 2016), https://papers.ssrn.com/sol3/papers.cfm?abstract_id=2795957.

31. The information as of 1998 comes from ITU, *General Trends in Telecommunications Reform 1998*, vol. 2, *Africa* (Geneva: ITU, 1998); and ITU, *General Trends in Telecommunications Reform 1998*, vol. 4, *Arab States* (Geneva: ITU, 1998). The information as of 2009 comes from M. Williams, R. Mayer, and M. Minges, *Africa's ICT Infrastructure* (Washington, DC: World Bank, 2011).

32. For details on the purchasing-power-parity methodology, see I. Kravis et al., *A System of International Comparisons of Gross Product and Purchasing Power* (Baltimore: Johns Hopkins University Press, 1975); and I. Kravis, A. Heston, and R. Summers, *International Comparisons of Real Product and Purchasing Power* (Baltimore: Johns Hopkins University Press, 1978).

33. Studies of telecommunications reform in specific African countries include L. Haggarty, M. Shelley, and S. Wallsten, "Telecommunications Reform in Ghana"

(World Bank Policy Research Working Paper 2983, Washington, DC, 2003); J. Laffont and T. N'Guessan, "Telecommunications Reform in Cote d'Ivoire" (World Bank Policy Research Working Paper 2895, Washington, DC, 2002); G. Clarke, F. Gebreab, and H. Mgombelo, "Telecommunications Reform in Malawi" (World Bank Policy Research Working Paper 3036, Washington, DC, 2003); J. Azam, M. Guia, and T. N'Guessan, "Telecommunications Sector Reform in Senegal" (World Bank Policy Research Working Paper 2894, Washington, DC, 2002); and M. Shirley et al., "Telecommunications Reform in Uganda" (World Bank Policy Research Working Paper 2864, Washington, DC, 2002).

34. Haggarty, Shirley, and Wallsten, "Telecommunications Reform in Ghana," 22–24.

35. J. Laffont and T. N'Guessan, "Telecommunications Reform in Côte d'Ivoire," 19–21.

36. Williams, Mayer, and Minges, *Africa's ICT Infrastructure*, 108–109.

37. Ghana caught up with the Ivory Coast during the 2000s. By the early 2010s, both countries had similar cellular densities.

38. Many of these acquisitions are discussed in M. Jahanbakht, "The Role of Entrepreneurs in the Development of the African Mobile Telecommunications Industry" (PhD diss., Carnegie Mellon University, August 2015).

39. See M. Dike and E. Rose, "Mobile Telecoms in Sub-Saharan Africa: A Multimarket Competition Perspective," unpublished paper, October 2015; M. Dike and E. Rose, "Cross-Border Expansion and Competitive Interactions of Indigenous Mobile Network Operators in Sub-Saharan Africa," *Thunderbird International Business Review*, published ahead of print, February 27, 2018, https://doi.org/10.1002/tie.21971. and P. Curwen and J. Whalley, *Mobile Telecommunications in a High-Speed World: Industry Structure, Strategic Behavior and Socio-economic Impact* (Farnham, UK: Gower, 2010), chap. 10.

40. The history of Telecel, Celtel, and MTN is explored in Jahanbakht, "Role of Entrepreneurship." The most comprehensive history of Celtel is in Southwood, *Less Walk, More Talk*. Brief discussions of several of the African cellular pioneers can be found in D. Fick, *Entrepreneurship in Africa: A Study of Successes* (Westport, CT: Quorum Books, 2002). For biographies of two of the main African cellular entrepreneurs, Mo Ibrahim and Miko Rwayitare, see M. Makura, *Africa's Greatest Entrepreneurs* (Cape Town: Penguin Random House South Africa, 2008).

41. Noam, introduction, 6–7.

42. Southwood, *Less Walk, More Talk*, 9.

43. Dike and Rose, "Mobile Telecoms," 16–17.

44. Haggarty, Shelley, and Wallsten, "Telecommunications Reform in Ghana," 19–20.

45. The history of France Télécom, later known as Orange, is in US Securities and Exchange Commission, Exhibit 15.1, in *Orange 2017 Registration Document*, accessed November 21, 2020, https://www.sec.gov/Archives/edgar/data/1038143/000163462118000014/exhibit_15.1.htm. This document summarizes the history of the company's forays in various regions of the world, including Africa.

46. For the presence of France Télécom in sub-Saharan African markets in the second half of the 1990s, see US Securities and Exchange Commission. See also Laffont and N'Guessan, "Telecommunications Reform in Cote d'Ivoire"; Azam, Guia, and N'Guessan, "Telecommunications Reform in Senegal"; and Noam, introduction, 6.

47. Curwen and Whalley, *Mobile Telecommunications*, 201.

48. Fick, *Entrepreneurship in Africa*, 248–250.

49. Southwood, *Less Walk, More Talk*, 7–8. Two other important carriers launched networks in sub-Saharan Africa in the early 2000s: Globacom and Africell. Globacom was founded in Lagos, Nigeria, in 2003 and it expanded to other sub-Saharan countries starting in 2008. Africell, a holding company of telecommunications operations headquartered in London and run from Lebanon, entered sub-Saharan Arica when it set up a network in the Gambia in 2001. See Dike and Rose, "Mobile Telecoms," 16, 19.

50. Maroc Telecom, the incumbent telecommunications operator in Morocco, also entered sub-Saharan cellular markets in this period. See Dike and Rose, "Mobile Telecoms," 15–16.

51. T. Khanna and K. Palepu, *Winning in Emerging Markets* (Boston: Harvard Business School Press, 2010), 171–174.

52. Southwood, *Less Walk, More Talk*, 174.

53. See "Overview of the Etisalat Telecommunication Company," Dubai-freezone .ae, accessed November 14, 2020, https://en.dubai-freezone.ae/articles-about-bussines-in-uae/overview-of-the-etisalat-telecommunication-company.html.

54. "UAE's Etisalat Buys 50% of W Africa's Atlantique Telecom," Balancing Act, May 6, 2005, https://www.balancingact-africa.com/news/telecoms-en/7374/uaes-etisalat-buys-50-of-w-africas-atlantique-telecom.

55. For details on the evolution of Bharti Airtel and its acquisition of the Zain networks in Africa, see I. Bose and N. Celly, "The Indian Tiger Prowls in Africa: Bharti Airtel's Acquisition of Zain Africa" (University of Hong Kong, Asia Case Research Centre, 2011).

56. Bose and Celly, 23.

57. Bose and Celly, 1.

58. Bose and Celly, 9.

59. Edgardo Sepulveda, "Creating Mobile Telecom Markets in Africa," World Bank, International Finance Corporation, September 2016, https://openknowledge.worldbank.org/handle/10986/30345.

60. See esp. Williams, Mayer, and Minges, *Africa's ICT Infrastructure*, chap. 4.

61. Williams, Mayer, and Minges, 127.

62. Williams, Mayer, and Minges, 128–136.

63. Williams, Mayer, and Minges, 240–244.

64. For a summary of China's presence in Africa, especially in Africa's telecommunications sector, see E. Oreglia, *Africa's Many Chinas* (Intel Labs, 2012), http://www

.ercolino.eu/docs/Oreglia_Proj_AfricasManyChinas.pdf. Chinese interest in Africa was not restricted to telecommunications infrastructure. Between 2000 and 2010, trade between Africa and China rose tenfold to almost USD 115 billion. Chinese direct investment in Africa increased from less than USD 0.5 billion in 2003 to more than USD 9 billion in 2010. By 2010 China had become the most powerful foreign player in Africa. See Meredith, *State of Africa*, 696–697.

65. See, for example, F. Zeller, "Arab Spring: The First Smartphone Revolution," Yahoo! Finance, November 29, 2020, https://au.finance.yahoo.com/news/arab-spring-first-smartphone-revolution-021935287.html.

CHAPTER 12

1. See M. Campbell-Kelly et al., "Economics and Business Perspectives on Smartphones as Multi-sided Platforms," *Telecommunications Policy* 39, no. 8 (September 2015): 717–734.

2. E. Mack, "Here's What Android Fragmentation Really Looks Like," *CNET*, May 15, 2012, https://www.cnet.com/news/heres-what-android-fragmentation-really-looks-like/.

3. D. Garcia-Swartz and M. Campbell-Kelly, "Openness as a Business Strategy: Historical Perspectives on Openness in Computing and Mobile Phones," *Information Economics and Policy* 48, published ahead of print, July 11, 2019, https://doi.org/10.1016/j.infoecopol.2019.07.001.

4. The essential reference on the early growth of the iPhone is J. West and W. Mace, "Browsing as the Killer App: Explaining the Rapid Success of Apple's iPhone," *Telecommunications Policy* 34, nos. 5–6 (June–July 2010): 270–286.

5. M. Honan, "Apple Unveils iPhone," *Macworld*, January 9, 2007, https://www.macworld.com/article/1054769/smartphones/iphone.html.

6. F. Vogelstein, "The Untold Story: How the iPhone Blew Up the Wireless Industry," *Wired*, January 9, 2008, https://www.wired.com/2008/01/ff-iphone/.

7. T. Krazit, "Apple Unveils iPhone 2, Both the Phone and the Business," *CNET*, June 9, 2008, https://www.cnet.com/news/apple-unveils-iphone-2-both-the-phone-and-the-business/.

8. The sources for the table are the following: W. Mossberg and K. Boehret, "The iPhone is a Breakthrough Handheld Computer," AllThingsD, June 26, 2017, http://allthingsd.com/20070626/the-iphone-is-breakthrough-handheld-computer/; W. Mossberg and K. Boehret, "Testing out the iPhone," Wall Street Journal, June 27, 2007, https://www.wsj.com/articles/SB118289311361649057; W. Mossberg, "Newer, Faster, Cheaper iPhone 3G," The Wall Street Journal, July 9, 2008, https://www.wsj.com/articles/SB121555740704037313; W. Mossberg, "Apple iPhone 3GS is a Better Model--or Just Get OS 3.0," The Wall Street Journal, June 19, 2009, https://www.wsj.com/articles/SB124525573550923759; W. Mossberg, "Thinner, Faster, Smarter iPhone Raises the Stakes," The Wall Street Journal, June 20, 2010, https://www.wsj.com/articles/SB10001424052748704853404575322951290405346, W. Mossberg, "The iPhone Finds its Voice," AllThingsD, October 11, 2011, http://allthingsd.com

/20111011/the-iphone-finds-its-voice/; W. Mossberg, "The iPhone Takes to the Big Screen," The Wall Street Journal, September 18, 2012, https://www.wsj.com/articles/SB10000872396390444450004578004370248427736.

9. "2012 Mobile Future in Focus: Key Insights from 2011 and What They Mean for the Coming Year," Comscore, February 23, 2012, https://www.comscore.com/Insights/Presentations-and-Whitepapers/2012/2012-Mobile-Future-in-Focus.

10. Nielsen, *The Mobile Consumer: A Global Snapshot* (New York: Nielsen, February 2013), https://www.nielsen.com/wp-content/uploads/sites/3/2019/04/Mobile-Consumer-Report-2013-1.pdf.

11. P. Cohen, "App Store Downloads Top 100 Million, Games a Centerpiece," Macworld, September 9, 2008,https://www.macworld.com/article/192380/appstore-5.html.

12. T. Krazit, "The Once and Future App Store," *CNET*, April 3, 2009, https://www.cnet.com/news/the-once-and-future-app-store/.

13. On the factors that drove the growth of the iPhone platform, see D. Garcia-Swartz and F. Garcia-Vicente, "Network Effects on the iPhone Platform: An Empirical Examination," *Telecommunications Policy* 39, no. 10 (November 2015): 877–895; and D. Garcia-Swartz, M. Muhamedagic, and D. Saenz, "The Role of Prices and Network Effects in the Growth of the iPhone Platform," *Technological Forecasting and Social Change* 147 (October 2019): 110–122.

14. See C. Newton, "Google Play Store Hits 25 Billion Downloads, Launches Discounts," *CNET*, September 26, 2012, https://www.cnet.com/news/google-play-store-hits-25-billion-downloads-launches-discounts/; and P. Sloan, "Apple by the Numbers: 35B Apps Downloaded, 100M iPads Sold," *CNET*, October 23, 2012, https://www.cnet.com/news/apple-by-the-numbers-35b-apps-downloaded-100m-ipads-sold/.

15. See, for example, IPlytics, *Who Is Leading the 5G Patent Race? A Patent Landscape Analysis on Declared 5G Patents and 5G Standard Contributions* (IPlytics, November 2019), https://www.iplytics.com/wp-content/uploads/2019/01/Who-Leads-the-5G-Patent-Race_2019.pdf.

16. See E. Lee and T. Chau, *Telecom Services: The Geopolitics of 5G and IoT*, Jefferies Franchise Note (Jefferies, September 14, 2017), https://www.jefferies.com/CMSFiles/Jefferies.com/files/Insights/TelecomServ.pdf. Whether China had the majority of the standard-essential patents pertaining to its own native standard, time division–synchronous code division multiple access (TD-SCDMA), is a matter of dispute, since the standard relied heavily on code division multiple access (CDMA) and Qualcomm owned a large portion of the CDMA-related intellectual property. See the discussion in Lee and Chau, 8. Lee and Chau believe that China and Qualcomm reached a confidential agreement regarding TD-SCDMA standard-essential patent shares.

17. On the four waves of entrepreneurship in China, see E. Tse, *China's Disruptors* (New York: Portfolio/Penguin, 2015), esp. 56.

18. See P. Fan, "Catching Up through Developing Innovation Capability: Evidence from China's Telecom-Equipment Industry," *Technovation* 26 (2006): 359–368; and

P. Fan, "Innovation, Globalization, and Catch-Up of Latecomers: Cases of Chinese Telecom Firms," *Environment and Planning A* 43 (2011): 830–849. See also P. Fan and X. Gao, "Catching Up and Developing Innovation Capabilities in China's Telecommunication Equipment Industry," in *China as an Innovation Nation*, ed. Y. Zhou, W. Lazonick, and Y. Sun (Oxford: Oxford University Press, 2016), 215–239; and Y. Zhang and Y. Wu, "Huawei and Zhengfei Ren: Leadership in a Technology-Innovative Firm," in *Technological Entrepreneurship in China: How Does It Work?*, ed. C. Petti (Cheltenham, UK: Edward Elgar, 2012), 103–122.

19. See Stenvall Skoeld & Co., "As Chinese Companies Conquer Global Telecom Equipment, Other Industries Are Bound to Follow," Seeking Alpha, August 8, 2012, https://seekingalpha.com/article/789791-as-chinese-companies-conquer-global-telecom-equipment-other-industries-are-bound-to-follow.

20. "Nokia, Siemens to Merge Telecom Networks," DW, June 19, 2006, https://www.dw.com/en/nokia-siemens-to-merge-telecom-networks/a-2060590.

21. Huawei Investment and Holding, *Building a Better Connected World: 2015 Annual Report* (Shenzhen, China: Huawei Investment and Holding, 2016), 22, https://www-file.huawei.com/-/media/corporate/pdf/annual-report/annualreport2015_en.pdf.

22. S. Teral, "Global Mobile Infrastructure Market," *IEEE ComSoc Technology Blog*, April 3, 2019, https://techblog.comsoc.org/category/global-mobile-infrastructure-market/.

23. See S. Chen and P. Wen, "The Evolution of China's Mobile Phone Industry and Good-Enough Innovation," in Zhou, Lazonick, and Sun, *China as an Innovation Nation*, 261–282.

24. W. Shin, C. Chien, and J. Wang, "Shanzhai! MediaTek and the 'White Box' Handset Market" (Harvard Business School Case Study 9–610–081, Boston, December 22, 2010), 3.

25. S. Schwankert, "3G in China: One Country, Three Standards," *PCWorld*, September 11, 2008, https://www.pcworld.com/article/150985/article.html.

26. Chen and Wen, "Evolution," 269.

27. Tse, *China's Disruptors*, 67–70. See also D. Kline, "Behind the Fall and Rise of China's Xiaomi," *Wired*, December 22, 2017, https://www.wired.com/story/behind-the-fall-and-rise-of-china-xiaomi/.

28. M. Kahn, "China's Xiaomi Reports Booming Smartphone Sales," *PCWorld*, July 2, 2014, https://www.pcworld.com/article/2450220/chinas-xiaomi-reports-booming-smartphone-sales.html.

29. Nomura Equity Research, *U.S. Semiconductor Primer* (Nomura Equity Research, December 11, 2013), 134, http://pdf.zacks.com/pdf/NO/H4560776.PDF.

30. See Shin, Chien, and Wang, "Shanzhai!"; and M. Honan, "The Next Global Smartphone Revolution: Made in Taiwan," *Wired*, February 5, 2013, https://www.wired.com/2013/02/mediatek/.

31. MediaTek's transition from feature phones to smartphones is explored in "Serial Disrupter: MediaTek Has Burst into the Market for Smartphone Chips," *Economist*, August 24, 2013, https://www.economist.com/business/2013/08/24/serial-disrupter.

32. See Y. Sun, Z. Xue, and D. Du, "State, Multinational Corporations, Returnees, and Development of China's Integrated Circuit (IC) Design Industry," in Zhou, Lazonick, and Sun, *China as an Innovation Nation*, 240–260; and Y. Wan and S. Gao, "Spreadtrum: Aspiring to Be a World Class Chip Company," in Petti, *Technological Entrepreneurship in China*, 123–138.

33. Nomura Equity Research, *U.S. Semiconductor Primer*, 135.

34. See Korea Fair Trade Commission, In re Alleged Abuse of Market Dominance of Qualcomm Incorporated, Decision No. 2017-0-25, January 20, 2017, 19–20, http://www.theamericanconsumer.org/wp-content/uploads/2017/03/2017-01-20_KFTC-Decision_2017-0-25.pdf.

35. "Intel Completes Acquisition of Infineon's Wireless Solutions Business," Intel news release, January 31, 2011, https://newsroom.intel.com/news-releases/intel-completes-acquisition-of-infineons-wireless-solutions-business/#gs.9qc9di.

36. Nomura Equity Research, *U.S. Semiconductors Primer*, 135.

CONCLUSIONS

1. In the United States during the 1980s, there was frequently a short lag between the launch of the first cellular network in each regional market and the entry of the competitor.

2. In some contexts, competition between carriers did not lead to lower subscription tariffs. During the 1980s, for example, there was duopolistic competition in the United States, Britain, and Sweden. Nominal subscription prices remained stable in Britain and Sweden and declined slowly in the American regional markets. But duopolistic competition during the 1980s led to more awareness of cellular phones among consumers and also to increased capacity. Enhanced competition among carriers during the 1990s usually led to more substantial tariff declines.

3. See A. Smith, "U.S. Smartphone Use in 2015," Pew Research Center, April 1, 2015, https://www.pewresearch.org/internet/2015/04/01/us-smartphone-use-in-2015/.

4. See E. Lee and T. Chau, *Telecom Services: The Geopolitics of 5G and IoT*, Jefferies Franchise Note (Jefferies, September 14, 2017), 3, https://www.jefferies.com/CMSFiles/Jefferies.com/files/Insights/TelecomServ.pdf.

5. Lee and Chau, 5.

6. "The Geopolitics of 5G: America's War on Huawei Nears Its Endgame," *Economist*, July 16, 2020.

7. "Geopolitics of 5G."

INDEX

History of Computing

William Aspray and Thomas J. Misa, editors

Janet Abbate, *Recoding Gender: Women's Changing Participation in Computing*

John Agar, *The Government Machine: A Revolutionary History of the Computer*

William Aspray and Paul E. Ceruzzi, *The Internet and American Business*

William Aspray, *John von Neumann and the Origins of Modern Computing*

Charles J. Bashe, Lyle R. Johnson, John H. Palmer, and Emerson W. Pugh, *IBM's Early Computers*

Martin Campbell-Kelly, *From Airline Reservations to Sonic the Hedgehog: A History of the Software Industry*

Paul E. Ceruzzi, *A History of Modern Computing*

I. Bernard Cohen, *Howard Aiken: Portrait of a Computer Pioneer*

I. Bernard Cohen and Gregory W. Welch, editors, *Makin' Numbers: Howard Aiken and the Computer*

James Cortada, *IBM: The Rise and Fall and Reinvention of a Global Icon*

Nathan Ensmenger, *The Computer Boys Take Over: Computers, Programmers, and the Politics of Technical Expertise*

Thomas Haigh, Mark Priestley, and Crispin Rope, *ENIAC in Action: Making and Remaking the Modern Computer*

John Hendry, *Innovating for Failure: Government Policy and the Early British Computer Industry*

Mar Hicks, *Programmed Inequality: How Britain Discarded Women Technologists and Lost Its Edge in Computing*

Michael Lindgren, *Glory and Failure: The Difference Engines of Johann Müller, Charles Babbage, and Georg and Edvard Scheutz*

David E. Lundstrom, *A Few Good Men from Univac*

René Moreau, *The Computer Comes of Age: The People, the Hardware, and the Software*

Arthur L. Norberg, *Computers and Commerce: A Study of Technology and Management at Eckert-Mauchly Computer Company, Engineering Research Associates, and Remington Rand, 1946–1957*

Emerson W. Pugh, *Building IBM: Shaping an Industry and Its Technology*

Emerson W. Pugh, *Memories That Shaped an Industry*

Emerson W. Pugh, Lyle R. Johnson, and John H. Palmer, *IBM's Early Computers: A Technical History*

Kent C. Redmond and Thomas M. Smith, *From Whirlwind to MITRE: The R&D Story of the SAGE Air Defense Computer*

Alex Roland with Philip Shiman, *Strategic Computing: DARPA and the Quest for Machine Intelligence, 1983–1993*

Raúl Rojas and Ulf Hashagen, editors, *The First Computers—History and Architectures*

Corinna Schlombs, *Productivity Machines: German Appropriations of American Technology from Mass Production to Computer Automation*

Dinesh C. Sharma, *The Outsourcer: A Comprehensive History of India's IT Revolution*

Dorothy Stein, *Ada: A Life and a Legacy*

Christopher Tozzi, *For Fun and Profit: A History of the Free and Open Source Software Revolution*

John Vardalas, *The Computer Revolution in Canada: Building National Technological Competence, 1945–1980*

Maurice V. Wilkes, *Memoirs of a Computer Pioneer*

Jeffrey R. Yost, *Making IT Work: A History of the Computer Services Industry*

Thomas Haigh and Paul E. Ceruzzi, *A New History of Modern Computing*

Daniel D. Garcia-Swartz and Martin Campbell-Kelly, *Cellular: An Economic and Business History of the International Mobile-Phone Industry*